Spatial Ec

MONOGRAPHS IN POPULATION BIOLOGY

EDITED BY SIMON A. LEVIN AND HENRY S. HORN

Titles available in the series (by monograph number)

1. *The Theory of Island Biogeography*, by Robert H. MacArthur and Edward O. Wilson
6. *Stability and Complexity in Model Ecosystems*, by Robert M. May
10. *Geographic Variation, Speciation, and Clines*, by John A. Endler
14. *Some Adaptations of March-Nesting Blackbirds*, by Gordon H. Orians
16. *Cultural Transmission and Evolution: A Quantitative Approach*, by L. L. Cavalli-Sforza and M. W. Feldman
17. *Resource Competition and Community Structure*, by David Tilman
18. *The Theory of Sex Allocation*, by Eric L. Charnov
19. *Mate Choice in Plants: Tactics, Mechanisms, and Consequences*, by Mary F. Wilson and Nancy Burley
21. *Natural Selection in the Wild*, by John A. Endler
22. *Theoretical Studies on Sex Ratio Evolution*, by Samuel Karlin and Sabvin Lessard
23. *A Hierarchical Concept of Ecosystems*, by R. V. O'Neill, D. L. DeAngelis, J. B. Waide, and T. F. H. Allen
24. *Population Ecology of the Cooperatively Breeding Acorn Woodpecker*, by Walter D. Koening and Ronald L. Mumme
25. *Population Ecology of Individuals*, by Adam Lomnicki
26. *Plant Strategies and the Dynamics and Structure of Plant Communities*, by David Tilman
28. *The Ecologic Detective: Confronting Models with Data*, by Ray Hilborn and Marc Mangel
29. *Evolutionary Ecology across Three Trophic Levels: Goldenrods, Gallmakers, and Natural Enemies*, by Warren G. Abrahamson and Arthur E. Weis
30. *Spatial Ecology: The Role of Space in Population Dynamics and Interspecific Interactions*, edited by David Tilman and Peter Kareiva

Spatial Ecology

The Role of Space in Population Dynamics and Interspecific Interactions

EDITED BY
DAVID TILMAN
AND
PETER KAREIVA

PRINCETON UNIVERSITY PRESS
PRINCETON, NEW JERSEY
1997

Published by Princeton University Press, 41 William Street,
Princeton, New Jersey 08540
In the United Kingdom: Princeton University Press,
Chichester, West Sussex

Library of Congress Cataloging-in-Publication Data

Spatial ecology: the role of space in population dynamics and
interspecific interactions/edited by David Tilman and Peter
Kareiva.
p. cm. — (Monographs in population biology ; 30)
Includes bibliographical references (p.) and index.
ISBN 0-691-01653-4 (alk. paper). — ISBN 0-691-01652-6 (pbk. :
alk. paper)
1. Spatial ecology. I. Tilman, David, 1949- . II. Kareiva,
Peter M., 1951- . III. Series.
QH54.15.S62S62 1997
577.8′2—dc21 97-8460
CIP

This book has been composed in Baskerville

Princeton University Press books are printed on acid-free paper and
meet the guidelines for permanence and durability of the Committee
on Production Guidelines for Book Longevity of the Council on
Library Resources

Printed in the United States of America

10 9 8 7 6 5 4 3 2 1

10 9 8 7 6 5 4 3 2
(Pbk.)

http://pup.princeton.edu

Contents

PART III
Competition in a Spatial World

PART IV
The Final Analysis: Does Space Matter or Not? And How Will We Test Our Ideas?

Preface

Although the world is unavoidably spatial, and each organism is a discrete entity that exists and interacts only within its immediate neighborhood, these realities long have been ignored by most ecologists because they can greatly complicate field research and modeling. However, several lines of inquiry have highlighted the potentially critical roles of space and led to growing interest in spatial ecology. One of the earliest threads in this tapestry came from Gause's (1935) studies of predator-prey dynamics. Even though *Paramecium* and *Didinium* persist in innumerable seeps in nature, one or both species invariably went extinct in the laboratory after a period of unstable oscillations. Gause suggested that periodic reinvasion by one or both species, such as might happen in a subdivided natural habitat, was necessary to allow coexistence. Huffaker (1958) extended this in a study of a phytophagous mite and its predator. He also found that persistence was impossible in small homogeneous habitats but was prolonged in more complex habitats in which there were barriers that established distinct patches linked by slow dispersal. Spatial complexity was essential for the partial stabilization and persistence of this predator and prey interaction. Few papers have piqued such interest in spatial ecology as has Huffaker's classic study.

MacArthur and Wilson's (1967) theory of island biogeography added more threads to the spatial tapestry and established the abiding interest of conservation biologists in spatial processes. MacArthur and Wilson noted that human expansion results in the fragmentation of once continuous ecosystems, resulting in species extinctions. However, the number of species likely to be threatened by a given amount of habitat fragmentation should depend on the spatial patterning of the fragmentation, including such aspects as the sizes of remnant patches, their vicinity to each other, and the existence and qualities of corridors linking them. Although nonspatial models may give insights into such issues, each real world situation is unavoidably explicitly spatial, and land management aimed at habitat restoration or habitat preservation must be grounded in space.

Concerns about invasions by novel organisms (e.g., Elton 1958; Mooney and Drake 1986; Drake et al. 1989; Skellam 1951) and interest in the spread of novel genotypes also drew attention to such aspects of space as the role of geographic isolation, dispersal dynamics, and effective population sizes. Unprecedentedly rapid movements of modern humans among biogeographic realms has produced a crisis of exotic species invasions with dramatic ecological and economic consequences (OTA 1993). Many of these novel organisms have spread in a manner that Elton (1958) called "ecological explosions." The rate and the spatial patterning of these invasions has captured the interest of both resource managers and those interested in ecological patterns.

Conceptual issues, such as the paradox of diversity, are also woven into the tapestry of spatial ecology. Hutchinson (1961) noted that the open, seemingly well-mixed waters of lakes and oceans may contain a hundred or more species of phytoplanktonic algae, all of which presumably compete with each other for the same few limiting nutrients. Available theory, however, predicted that the number of coexisting species should not exceed the number of limiting resources. The same paradox applied to prairies, tropical rain forests, and other terrestrial plant communities in which hundreds of species coexist while competing for only a handful of limiting factors. Space seemed to have the potential of providing a solution to this paradox of diversity. Such spatial solutions have been proposed by Levin and Paine (1974); Horn and MacArthur (1972); Platt and Weis (1977); Tilman (1994); and others.

Finally, mathematical ecologists found that spatial models often predicted unexpected and intriguing spatial patterns and dynamics. Even in a model of a physically homogeneous habitat with extremely simple rules of organismal interactions and movement, spatial patterns unavoidably developed. Although the general expectation might be that individuals would become uniformly spread across such a homogeneous habitat, simple models in fact reveal a surprising, robust, and general prediction—clumping. A single species, growing in a density-dependent manner, for instance, occupies a homogeneous habitat in clumps and has other areas in which it does not occur. Such patterns were suggested by Turing (1952) as a possible explanation of the compartmentalization of animals that arises during embryogenesis. More complex extensions of this approach have revealed a tremendous variety of static and dynamic patterns resulting from a variety of local interactions coupled with dispersal in a uniform environment.

This book is designed to highlight the importance of space to all five of the aforementioned topics: stability, patterns of diversity, invasions, coexistence, and pattern generation. We aim to illustrate both the diversity of approaches used to study spatial ecology and the underlying similarities of these approaches. The book is not intended to be a thorough review of these issues, nor is it a cookbook designed to allow the reader to acquire the necessary skills for pursuing spatial ecology. Many important processes and approaches, especially detailed and explicit studies at the landscape level, are not included. Rather, this book offers a smorgasbord of concepts and approaches related to spatial ecology that, we hope, whets the appetite of the next generation of ecological researchers and entices them to incorporate spatial processes into their thinking.

In addition, this book contains two major crosscurrents. The strongest one concerns the trade-off between simplicity and realism in dealing with space. Within many chapters, and in comparison among chapters, space is dealt with on many levels of reality. Each level offers insights but imposes restrictions on what can be understood. Levin and Pacala (Chapter 12) and Steinberg and Kareiva (Chapter 14) revisit these issues, and synthesize them, at the end of the book. The real world has organisms that have specific *x*-, *y*-, and *z*-coordinates, fecundities, and probabilities of mortality that may change with time in ways that depend on their genotypes, the genotypes of other organisms in their neighborhoods, the coordinates of all other organisms in their neighborhoods, and the potentially changing physical attributes of the habitat. To understand these complexities, we have compartmentalized nature (and our minds) into the artificial subdisciplines of evolutionary, population, community, and ecosystem ecology. We also focus on particular interactions such as interspecific competition, predator-prey interactions, and host-parasitoid or host-disease interactions. Similarly, in studying the spatial aspects of these subdisiplines or biotic interactions, we have used various levels of realism in the description of space. Some approaches, such as Levins-like metapopulation models, treat space only implicitly by considering the proportion of potential sites occupied by individuals or populations. Others, such as cellular automata models, assume that organisms divide a habitat into a series of equal-sized, identical patches that occur in either a square or hexagonal grid. Still others, such as the analytical reaction-diffusion models, assume that space is a continuous variable, and that organisms "diffuse" through space. Finally, others, such as spatial simulators, include explicit spatial coordinates for each individual and for its region of influence.

A second crosscurrent also deals with the simplicity-realism trade-off, but with regard to the need to include evolutionary change in models of interspecific interactions. Both Antonovics et al. (Chapter 7) and Lehman and Tilman (Chapter 8) show that the addition of even simple genetic mechanisms can lead to major qualitative changes in the predictions of spatial models. Evolutionary models have long included spatial processes, but few attempts have been made to synthesize evolution, interspecific interactions, and space. These chapters suggest that such syntheses may offer many insights.

Although it is clear that space has been greatly overlooked in ecology, it is not evident how space should be incorporated into theory or fieldwork. Is it necessary to consider the explicit spatial location of all individuals? Or are cellular automata-like models reasonable approximations? Or might metapopulation-like pseudospatial models suffice? Are there added insights, and what are these insights, that come from increasingly more realistic considerations of space? Simple models, such as metapopulation models, are often analytically tractable because of their assumption of global dispersal. In contrast, the more realistic reaction-diffusion and other partial differential equation models are tractable only for cases that are often too simple to be of great ecological interest. Cellular automata models, and more spatially complex computer simulations, offer much greater ecological reality but suffer from the difficulty of generalization.

In total, this book demonstrates that spatial perspectives can provide significant insights into numerous ecological questions including the dynamics and conservation of a single species; the dynamics and outcome of competitive, predator-prey, host-parasite, and host-disease interactions; and the maintenance of biodiversity in fragmented landscapes. It is our hope that such insights will inform and inspire future ecological research.

This book would not have been possible without the support and encouragement of Bill Murdoch and of the National Center for Ecological Analysis and Synthesis in Santa Barbara, whom we gratefully acknowledge. Our NCEAS workshop on spatial ecology was deeply stimulating. Indeed, our discussions of the material that became chapters in this book reminded many of us of a doctoral oral examination. We thank Marilyn Snowball, Shari Staufenberg, and the rest of the NCEAS staff who made our gathering possible and enjoyable. Finally, we thank Nancy Larson for her immense assistance in preparing this manuscript. Her help was partially supported by the National Science Foundation and the Andrew Mellon Foundation.

We dedicate this book to our teachers and mentors. Don Farnum, a high school biology teacher in Benton Harbor, Michigan, shared his love of biology with David Tilman, and, at the University of Michigan, Steve Hubbell and the late Peter Kilham provided inspiration and nurture to a neophyte ecologist. Janis Antonovics showed Peter Kareiva the virtues of biology over political science, Richard MacMillen emphasized the value of doing fieldwork instead of just talking about it, and Simon Levin and Richard Root were an unlikely but perfect "couple" as a Ph.D. advisory team.

Contributors

Roy M. Anderson is from the Department of Zoology at Oxford University, Centre for the Epidemiology of Infectious Disease, South Parks Road, Oxford, OX1 3PS, United Kingdom.

Janis Antonovics is from the Department of Botany at Duke University, Durham, North Carolina 27708.

Howard V. Cornell is from the Department of Biological Sciences at the University of Delaware, 117 Wolf Hall, Newark, Delaware 19716.

Neil M. Ferguson is from the Department of Zoology at Oxford University, Centre for the Epidemiology of Infectious Disease, South Parks Road, Oxford, OX1 3PS, United Kingdom.

Ilkka Hanski is from the Department of Ecology and Systematics, Division of Population Biology, P.O. Box 17 (Arkadiankatu 7), FIN-00014, at the University of Helsinki, in Finland.

Michael P. Hassell is from the Department of Biology at Imperial College, Silwood Park, Ascot, Berks, SL5 7PY, United Kingdom.

Elizabeth Eli Holmes is from the Department of Biology at Colorado State University, Fort Collins, Colorado 80523.

Andrew M. Jarosz is from the Department of Botany and Plant Pathology at Michigan State University, East Lansing, Michigan 48824.

Peter Kareiva is from the Department of Zoology at the University of Washington, NJ-15, Box 351800, 24 Kincaid Hall, Seattle, Washington 98195-1800.

Ronald H. Karlson is from the Department of Biological Sciences at the University of Delaware, 117 Wolf Hall, Newark, Delaware 19716.

Clarence L. Lehman is from the Department of Ecology, Evolution, and Behavior at the University of Minnesota, 100 Ecology Building, 1987 Upper Buford Circle, St. Paul, Minnesota 55108-6097.

Simon A. Levin is from the Department of Ecology and Evolutionary Biology at Princeton University, 203 Eno Hall, Princeton, New Jersey 08544-1003.

Mark A. Lewis is from the Department of Mathematics at the University of Utah, JWB 233, Salt Lake City, Utah 84112.

Robert M. May is from the Department of Zoology at Oxford University, Centre for the Epidemiology of Infectious Disease, South Parks Road, Oxford, OX1 3PS, United Kingdom.

Stephen W. Pacala is from the Department of Ecology and Evolutionary Biology at Princeton University, 101 Eno Hall, Princeton, New Jersey 08544-1003.

Jonathan Roughgarden is from the Department of Biological Sciences and the Department of Geophysics at Stanford University, Stanford, California 94305.

Eleanor K. Steinberg is from the Department of Zoology at the University of Washington, NJ-15, Box 351800, 24 Kincaid Hall, Seattle, Washington 98195-1800.

Peter H. Thrall is from the Department of Botany at Duke University, Durham, North Carolina 27708.

David Tilman is from the Department of Ecology, Evolution, and Behavior at the University of Minnesota, 100 Ecology Building, 1987 Upper Buford Circle, St. Paul, Minnesota 55108-6097.

Howard B. Wilson is from the Department of Biology at Imperial College, Silwood Park, Ascot, Berks, SL5 7PY, United Kingdom.

PART I

Single Species Dynamics in Spatial Habitats

A first step in understanding the implications of space is to examine its impact on the dynamics of a single species. Although this might seem like a simple task, it is actually a complex issue that is far from resolved. Already, however, a variety of conceptual approaches and mathematical tools have provided major insights. The next three chapters present three major theoretical avenues: metapopulation models, cellular automata, and reaction-diffusion models. These are used, along with other related approaches, throughout the rest of this book.

Chapter 1 introduces the range of modeling detail used when building spatial theory. This introductory chapter is also designed to provide background needed to better understand later chapters and to illustrate some of the more novel phenomena that can occur in spatial habitats.

This book is a study in various approaches to spatial ecology, and a major contrast is provided by comparison of Chapters 1 and 2. Whereas Chapter 1 highlights simple, general models, Chapter 2 does the opposite. Here Ilkka Hanski proposes a mathematically more complex description, the incidence function approach, that adds information on the sizes and locations of each site to simple metapopulation models. He suggests that this added detail is critically important for understanding and managing any real population, such as the species of threatened butterfly that he studies. It highlights a critical process in ecology—the flow of ideas from theory, to field tests, and back.

Spatial systems are understood by combining knowledge of the rules governing interactions among individuals at sites with the linkage of these sites via dispersal. As a discipline, ecology has focused much effort on interactions among organisms, and remarkably little on dispersal, despite its great importance. Mark Lewis helps reverse this in Chapter 3 by considering both the complexities, and the underlying patterns and simplicity, of dispersal. The critical issue turns out to be the shape of the probability distribution describing dispersal distances. Many species have longer, "fatter" tails on these distributions than models have traditionally assumed, and these fat tails have an immense impact on the spread of organisms into novel habitats.

In total, this initial part introduces a variety of concepts and tools used to better understand spatial ecology and offers a glimpse of an oft-repeated debate between those advocating simple, elegant models versus those advocating more complex and realistic approaches. It should be clear to the reader that there is no one best approach—rather, the tack taken depends on the type of questions being asked. These early chapters also lay the conceptual groundwork for the remaining parts of the book.

CHAPTER ONE

Population Dynamics in Spatial Habitats

David Tilman, Clarence L. Lehman, and Peter Kareiva

All organisms are discrete entities that mainly interact with neighboring individuals of their own or other species. This discrete nature and spatial confinement is most evident for sessile organisms such as terrestrial plants, marine macrophytes, corals, and other organisms that live attached to surfaces. However, even motile organisms have their greatest impacts in a rather confined region—the region through which they move. These simple observations have profound implications for the dynamics and outcome of both intraspecific and interspecific interactions. In particular, local interactions and local movement/dispersal mean that population densities do not change in response to average conditions across a large habitat, as is assumed in classical nonspatial models, but rather in response to the local conditions experienced by each individual. This chapter presents three simple approaches for dealing with this spatial aspect of species interactions and population biology. The three approaches that we discuss are Levins's metapopulation-like model, cellular-automaton-like models, and reaction-diffusion models. The first and last approaches can be analytically tractable, whereas the middle one is best approached via computer simulation. The first two consider interactions in a spatially subdivided habitat, whereas the last considers interactions in a spatially continuous habitat. These approaches are the building blocks upon which much of this book is based. Our goal is to abstract the essential features of space that come from the discrete nature of individual organisms. There are many additional complexities to space that are not considered in this chapter but are discussed in subsequent chapters.

LEVINS'S MODEL FOR A SINGLE SPECIES

Levins's (1969) model provides a simple description of the dynamics of a single species living in a habitat composed of distinct sites. Although it has given many insights into communities that are subdivided into local populations, its use as a literal metapopulation model has been criticized (e.g., Hanski 1991, chap. 2; Harrison, Thomas, and Lewinsohn 1995; Harrison and Taylor 1997). For instance, it is viewed as a poor descriptor of metapopulations because it assumes that a single propagule can instantly transform an empty site into a local population that is at its carrying capacity, because local sites cannot differ in their carrying capacities or other measures of their quality, and because there is no possibility of local conditions "rescuing" a local site from extinction.

An alternative use of Levins's framework is to view it as a model for the occupancy of sites by single individuals as opposed to populations. To understand this interpretation of Levins's model, consider a habitat that is divided into sites that are just large enough to contain a single adult individual of a sessile species. A propagule of this species, upon entering an empty site, would occupy it. Individuals occupying sites could produce propagules that would be dispersed throughout the habitat. The rate of propagule production by a site is c (for colonization rate). In addition, there is some mortality rate, m, that is the chance of a currently occupied site becoming empty. To keep track of the dynamics of this species, it is necessary to know p, the proportion of all possible sites that are occupied at a given instant in time. If it is assumed that propagules disperse randomly among all possible sites, these assumptions can be converted into a simple model:

$$\frac{dp}{dt} = cp(1 - p) - mp. \tag{1.1}$$

This states that the rate of change in site occupancy (dp/dt) depends on the rate of propagule production (cp) multiplied by the proportion of currently open sites ($1 - p$). The rate of mortality (mp) is subtracted from this to get the net change in site occupancy.

This simple model of site occupancy has several interesting features. First, a species can persist in a habitat if $c > m$. Populations grow in a logistic fashion. When $c > m$, the proportion of occupied sites will approach an equilibrium, $\hat{p}$ (at which $dp/dt = 0$), with

$$\hat{p} = 1 - \frac{m}{c}. \tag{1.2}$$

This equilibrial density is globally stable (Hastings 1980), meaning that p will approach $\hat{p}$ from any starting density, and any perturbation, as long as $p > 0$. The most important and interesting aspect of this model is that no species is capable of completely filling its habitat at equilibrium. The proportion of all viable sites that are left unoccupied at equilibrium in a habitat occupied by a single species is $\hat{s}$, where

$$\hat{s} = 1 - \hat{p} = \frac{m}{c}. \tag{1.3}$$

Equation 1.3 shows that an unavoidable result of living in a spatial habitat is that a proportion of sites will be empty. The greater the mortality rate of a species relative to its colonization rate, the greater would be the amount of open space.

This model is mathematically simple and analytically tractable because of the simplifying assumption that it makes about dispersal. By assuming that all propagules are randomly dispersed across the entire habitat, this model eliminates the effects of local dispersal. However, by assuming that the habitat is subdivided into sites the size of an adult, the model implicitly addresses the discreteness of individuals and the idea of occupying space. Even though space is not treated explicitly, this model and its derivatives have been remarkably versatile in uncovering certain effects of space both on single species and on multispecies interactions (e.g., Skellam 1951; Levins and Culver 1971; Horn and MacArthur 1972; Armstrong 1976; Hastings 1980; Shmida and Ellner 1984; Bengtsson 1989, 1991; Tilman 1994; Chapters 8 and 10).

CELLULAR AUTOMATA AND INDIVIDUAL-BASED MODELS

The crucial simplifying assumption above was that of global, random dispersal of propagules. To model local dispersal we still envision a physically homogeneous habitat that is subdivided into sites, each of the size just capable of supporting a single adult. Each individual has a probability, m, of mortality per unit time. Each produces propagules at a rate of c. However, the propagules produced by each individual disperse locally. Local dispersal can be as simple as equiprobable movement to all adjacent cells (four adjacent cells for a habitat divided into a square grid, or six for a hexagonal grid) or a more complex pattern of dispersal to a greater number of nearby sites. A model in which propagules are randomly dispersed to adjacent sites is called a stochastic cellular automaton model. In the limit, as the number of sites that form the neighborhood of an individual is increased, the behavior of this model approaches that of Levins's model (Eq. 1.1).

In our simulations, individual organisms are distributed across a two-dimensional array of hexagonal sites (Figure 1.1). At any instant of time (of length dt), each occupied site sends propagules to a set of neighboring sites with a probability of $c\,dt$. If a propagule lands on an occupied site, the propagule is lost. If it lands on an empty site, the site is colonized and is now fully occupied. This process is repeated every time step, dt, to determine the dynamics of growth of the species.

Because this is a simulation model, there is no closed-form mathematical solution for the general case. However, the model shares many features with Levins's model and has the advantage of being easily modified. This allows its underlying assumptions to be manipulated in an exploration of the importance of these assumptions to the dynamics, equilibrium densities, and spatial patterns predicted by the model. Below

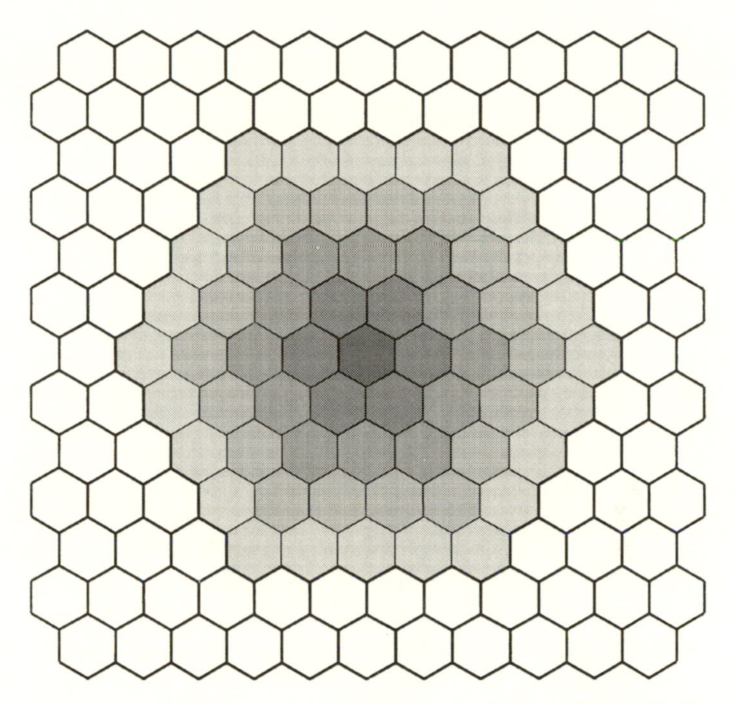

FIGURE 1.1. The hexagonal grid of sites used in our simulations of a spatially explicit habitat (see also Chapters 8 and 10). The shading illustrates the various rings of neighbors to which the propagules produced by the blackened site may be dispersed. The lightest shading is the fourth ring. Each hexagonal cell is the size of the area occupied by a single adult individual of this species.

we highlight some of the unique features that arise when Levins's model is modified into a spatial simulation that includes neighborhood dispersal.

An unavoidable outcome of local interactions (for a single species the interaction is that an occupied site cannot be invaded by a propagule) and local dispersal is clumping (see Durret and Levin 1994a,b for a thorough analysis). This is easily shown in cellular automaton models (Figure 1.2). Each of the six cases we illustrate is for a physically uniform habitat containing 10^4 sites (i.e., a 100×100 habitat). In each case the species becomes clumped, but the extent of this clumping, the

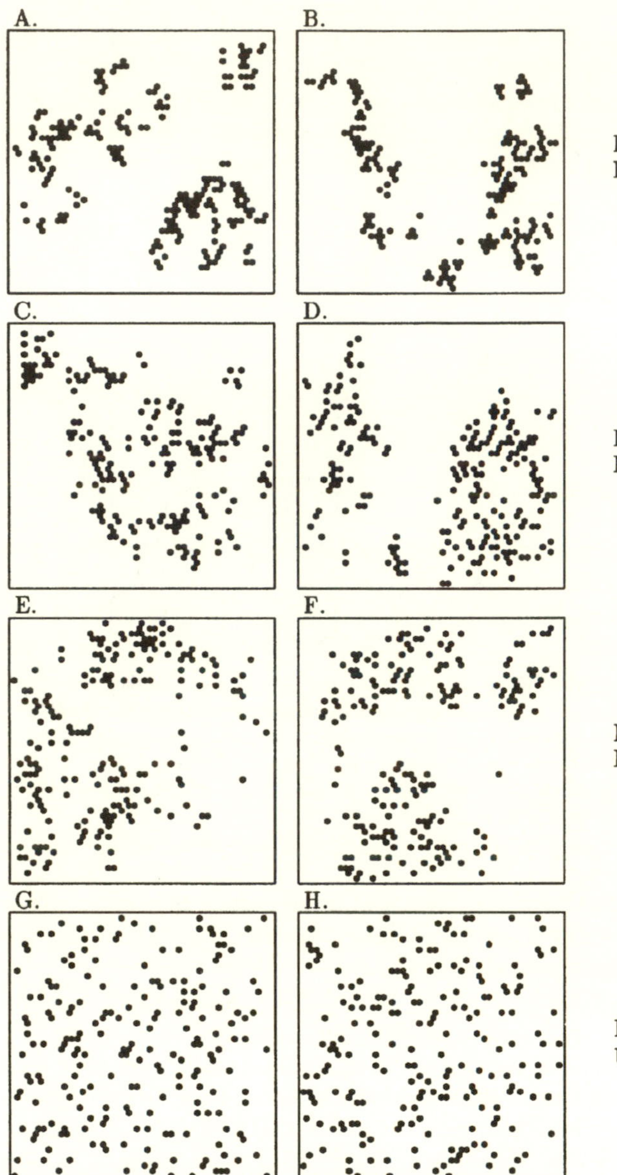
A.
B.
Dispersal
Range = 1
C.
D.
Dispersal
Range = 2
E.
F.
Dispersal
Range = 3
G.
H.
Dispersal
Uniform

size of the clumps, and the average distance between clumps depend on the mortality rate, the colonization rate, and the distance (number of adjacent rings) over which dispersal occurs. A field biologist encountering a clump of individuals of a species in nature, and noting a nearby habitat in which the species was rare or absent, might immediately wonder what environmental characteristics caused such clumping. These results show that there need not be any environmental cause. Because clumping is predicted to be an unavoidable result of local interaction and local movement in any spatial habitat—even if it is homogeneous—the appropriate null model for organism distributions might be a clumped dispersion (as opposed to the often-used random or Poisson distribution).

As discussed by Holmes (Chapter 5), another unavoidable result of local dispersal is decreased equilibrial abundances compared to the analytical Levins model. This occurs because local dispersal reduces the effective dispersal rate. With local dispersal, there is a greater chance of propagules falling on the same site and thus of all but the first propagule being ineffective. This decrease in the effective dispersal rate with local dispersal is further magnified by clumping. If most individuals of a species live in clumps, most of their propagules will fall in or near the clump and thus experience higher rates of site occupancy than occur on average, across the full habitat. This effect is illustrated in Figure 1.3, which shows how equilibrial population density (proportion of sites occupied) depends on the distance over which propagules are dispersed. Note that propagules have to be dispersed over a range greater than the

---

FIGURE 1.2. Clustering as a function of dispersal range. Parts A–F are snapshots of typical arrangements of individual organisms after three thousand simulated years, starting from an initially random arrangement, for dispersal ranges of one, two, or three rings around the parent. Parts G and H show a random arrangement, as occurs if there is global dispersal. Occupied cells are shown as black hexagons, and empty cells are not shown. In all cases, parameters were chosen to give equilibrial proportional site occupancy of about 0.08. For A, C, and E, $m = 0.05$, with c of 0.084, 0.068, and 0.065, respectively. For B, D, and E, $m = 0.03$ with c of 0.049, 0.041, and 0.03, respectively.

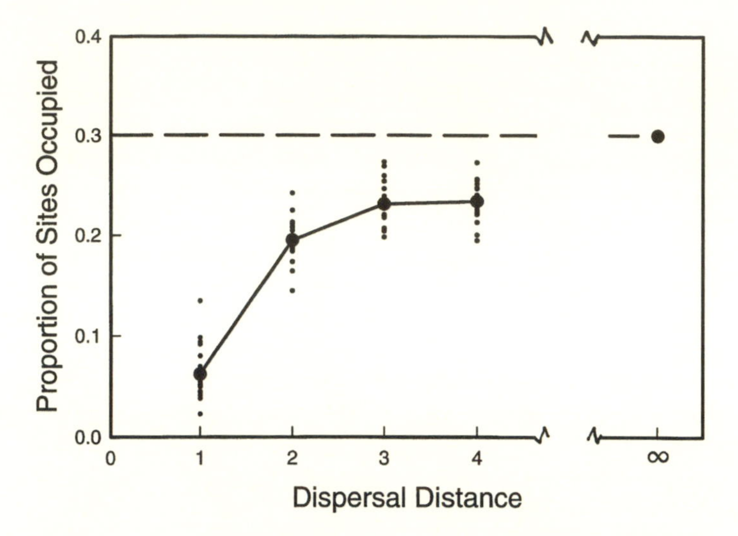

FIGURE 1.3. A comparison of different ranges of neighborhood dispersal with the global dispersal of Levins's model. Holding $m = 0.05$ and $c = 0.0714$ gives an equilibrial abundance of this species of 0.3 in the Levins model, but local dispersal of range 1 (dispersal to adjacent six cells) leads to much lower equilibrial abundances in the spatial simulator. As the dispersal range increases, the equilibrial abundance increases, approaching that of the Levins model when dispersal is across all possible sites.

adjacent four rings of neighbors for equilibrial density to approach that of the Levins analytical model.

Another feature of many spatially explicit models is called percolation. Percolation refers to how something responds to "clogs" in its habitat. For instance, consider water draining through a bed of gravel. The spaces between pieces of gravel form channels through which water may flow. If items are added that clog these channels, flow will eventually cease. However, the probability that any water will flow is not linearly dependent on the number of clogs added but is more like a step function. The average level of clogging at which flow ceases is called the percolation threshold. Populations living in spatial habitats can also exhibit a percolation threshold, but the dynamics are more complex because the "flow" is caused

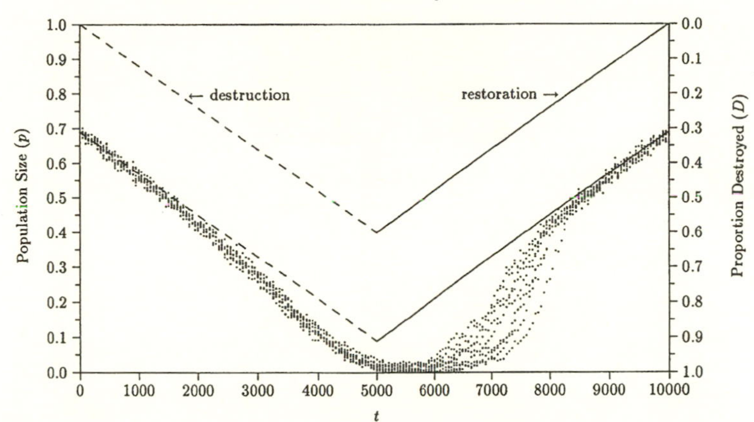

FIGURE 1.4. A composite of twelve replicate runs of random habitat destruction and restoration. The upper dashed line shows the amount of habitat destroyed, and the lower dashed line shows the decrease in the population size of this species analytically predicted by the Tilman et al. (1994) model. The solid lines similarly show habitat restoration and the recovery predicted by the Tilman et al. (1994) model. Dots show the results of runs in our spatial simulator, with simulations being sufficiently slow that each dot approximates the equilibrial outcome of those conditions.

by a combination of dispersal and reproduction, and individuals can survive, even if not reproducing, in completely isolated sites.

Consider a large, physically uniform spatial habitat occupied by a single species. For this habitat, let there be the potential destruction of individual sites. Destruction of a site means that no organisms can live there and that all propagules that enter the site die. The analytical Levins model predicts that a species would have its equilibrial abundance decrease, in a simple linear manner, as the amount of habitat destruction is increased (dashed line in Figure 1.4; see Tilman et al. 1994 and Chapter 10) and that its equilibrial abundance similarly would increase linearly as the habitat is restored to various degrees (solid line in Figure 1.4). Does this also occur in an explicitly spatial habitat? To find out, we imposed random destruction of

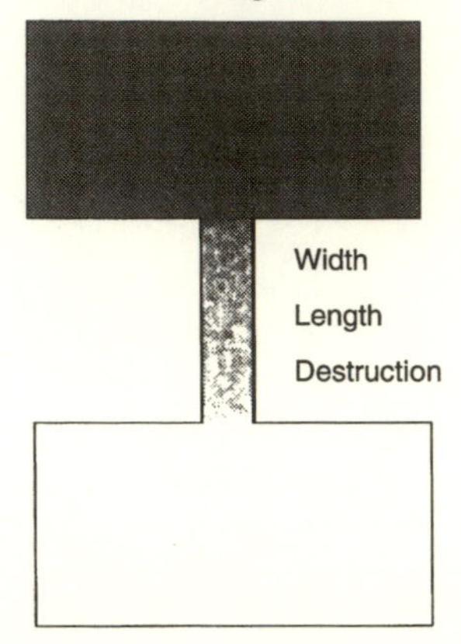

FIGURE 1.5. The ability of a species to move from a source area into a new viable habitat may depend on the length, width, and extent of destruction of a corridor, as illustrated here. See Figure 1.6 for results.

individual sites in our spatial simulator and kept all other model assumptions identical to those of the Levins analytical model. For each case, we ran the simulator while slowly increasing the number of sites destroyed, keeping the system always near equilibrium and continuing the process until the full range of habitat destruction had been covered. Then we started to slowly restore the area by randomly choosing destroyed sites and making them viable again, in similarly sized steps. We repeated these analyses many times, with each replicate simulation being separately randomized.

The resulting pattern differs markedly from the predictions of the analytical model (Figure 1.4). Each dot in Figure 1.4 shows the predicted quasi-equilibrium population abundance for a particular amount of habitat destruction or restoration. Abundance declined with destruction much as predicted analytically, but it did not increase as predicted by the Levins model following restoration. There was a marked "lag" between habitat restoration and the restoration of the population. This was not a time lag, because each point on the graph is close to the long-term equilibrial abundance of the species for that particular habitat condition. Rather, this was caused by the actual spatial pattern of destroyed and viable sites created by each run of random habitat destruction. As a habitat is restored, random restoration can create many viable sites that are totally inaccessible to current residents because there are no corridors connecting them with occupied sites. As restoration progresses, corridors are eventually created, allowing the species to occupy some formerly inaccessible sites. As this occurs, occupancy of the habitat comes closer to that predicted by the analytical model. This hysteresis results from what is called a percolation effect (Djordjevic 1992; O'Neill, Gardner, and Turner 1992). Percolation of living organisms differs from that of fluids because the percolation threshold depends on the biology of the organisms as well as the pattern of destruction. The variance observed in the responses depends on the actual spatial patterning of the random habitat destruction.

As anticipated by percolation theory, we have found a critical proportion of a habitat that can be destroyed, called D_c, at which point the habitat undergoes an abrupt change. When destruction is less than D_c there are paths that allow a species to migrate through the habitat; when destruction is greater than D_c there is almost no chance of migration from one side to the other of the habitat. The ability to migrate is of little importance as a habitat is being increasingly fragmented because there are few new open sites to which a species can move. However, when a habitat is being restored, the lack of corridors can prevent recolonization of viable sites. The nonlinear effect shown in Figure 1.4 thus is an insight provided by spatially explicit models that was not anticipated by the simpler analytical portraits of spatial processes.

Percolation effects point out some potential problems with corridors that link together the remnant fragments of once continuous ecosystem types. For a corridor to be effective, a species has to be able to migrate through it. This may not be a major problem, even for narrow corridors, if an organism can adjust its behavior to stay within the corridor (as may occur for some animals). However, plants move through a corridor by reproducing at one site and sending propagules out to adjacent sites. If a corridor is too narrow, a plant may lose too many propagules across its edges and be unable to spread down the corridor. If the corridor is too fragmented, passage may also be stopped. We used our spatial simulator to determine the effects of corridor length, width, and degree of destruction on the probability of movement through a corridor (Figure 1.5; Tilman, Lehman, and Yin 1997). To do this, we assumed we had a large source pool of propagules and investigated the probability that propagules would make it through a corridor to populate an empty intact habitat. We defined the critical amount of destruction for a given corridor as the amount of destruction, D_c, at which propagules were successful in half of the cases (where each case is a try in an independently randomly destroyed corridor). Corridor width (Figure 1.6A; Tilman et al. 1997) had a much greater effect on D_c than did corridor length (Figure 1.6B). We also found an interesting and alarming result: A habitat must be more pristine to serve as a corridor than to serve as a viable long-term refuge. This occurs because it is easier to maintain a species within a highly fragmented habitat than it is to facilitate migration through the same corridor. For example, in simulations reported in Figure 1.6 we found that species could survive in a habitat that had suffered up to 50% habitat destruction. However, the species could not migrate through a corridor that had suffered this much destruction. This means that virgin habitats, such as nature reserves and national parks, in addition to being reserves, might be considered to be viable corridors for repopulating surrounding areas that have suffered some habitat destruction, whereas habitats that have suffered destruction, fragmentation, or both might be considered to be most useful as supplemental reserves but *not* as corridors.

REACTION-DIFFUSION MODELS: SPATIALLY EXPLICIT MODELS FOR CONTINUOUS SPACE AND TIME

The traditional mathematical approach to spatial population models has involved the analysis of so-called reaction-diffusion models. In an ecological context "reaction" pertains to the process of population change or species interaction in the absence of any dispersal. Thus a reaction term in a reaction-

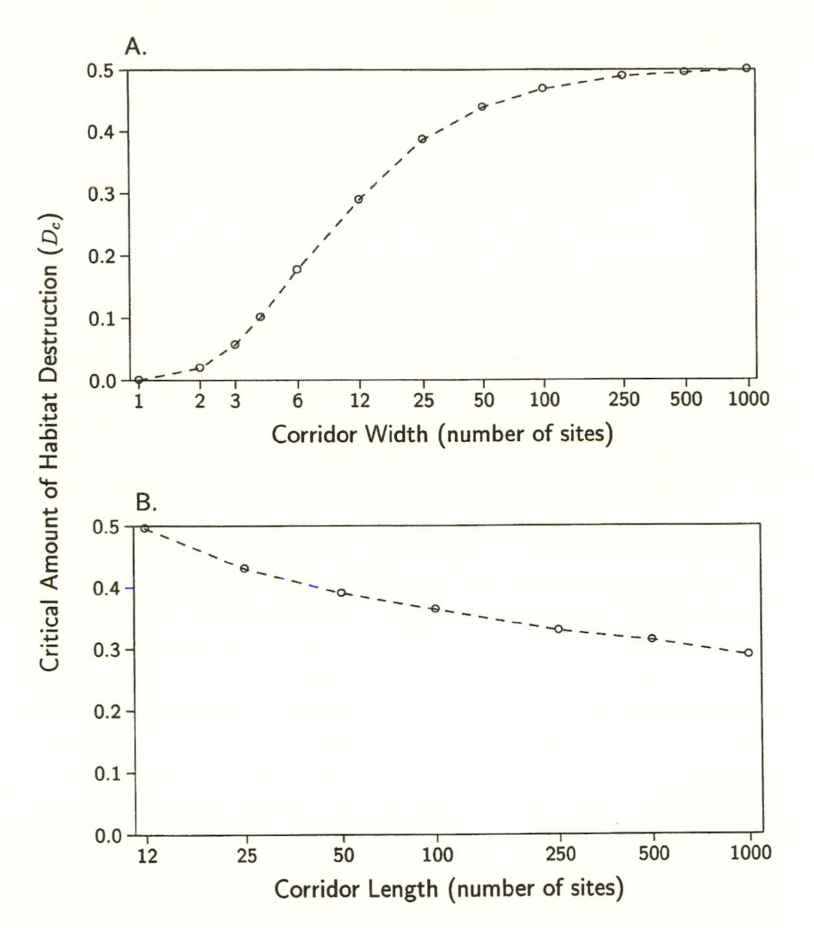

FIGURE 1.6. Numerous runs of a spatial simulator approximating the conditions of Figure 1.6 were performed to determine the critical amount of habitat destruction beyond which a population could not move, on average, through a corridor. (A) Even a strip of virgin habitat cannot function as a corridor if it is too narrow, and it can never function as a corridor, for this species, if it is more than 50% destroyed. (B) Corridor length also influences the extent of habitat destruction for which a corridor can remain viable.

diffusion model could be the logistic equation, or an exponential growth function, or any of ecology's many permutations on models of interspecific interactions. The "diffusion" term in these equations describes the movement of individuals and is usually the mathematical term corresponding to simple passive diffusion, that is,

$$D \frac{\partial^2 N(x,t)}{\partial x^2}, \tag{1.4}$$

for a one-dimensional model, where D is the diffusion coefficient (or dispersal rate) and $N(x,t)$ is the density of individuals at position x and time t. Notice that the entire term equals zero if N is uniform in space or if the spatial variation in N is described by a straight line. Reaction-diffusion equations are widely used in ecology, but their origin is rarely explained, leaving the mistaken impression that they require advanced mathematics to be understood. In fact, Equation 1.4 is easy to appreciate intuitively if we think about a one-dimensional environment—for example, that approximated by nearly sedentary organisms confined to the shoreline of a pond. Consider the shoreline to be divided into many discrete sites along its length, and suppose for the moment that there is no reproduction within the sites—all changes in population come from migration among sites. Concentrate on a given site and suppose that migration is purely random; that is, suppose each individual has a fixed probability of leaving the site in any given interval of time and leaves by moving either to the east or to the west with equal probability. Thus in each small unit of time, part of the population of the given site is lost when a fixed proportion of its occupants move west and an equal proportion move east. However, this loss is counteracted by individuals immigrating from neighboring sites. Because the same processes are assumed to operate in all sites concurrently, a fixed proportion of the number of individuals in the site to the west enters from the west, and similarly for the east. To put these words into mathematical symbols, let N_e be the number of individuals in the site to the east, N_w the number in the site to the west, N_c the number in the given site at the center, and

D the fixed proportion of individuals that leave in each direction per unit time. Then the change in population of the given site in a small unit of time is

$$\frac{\Delta N}{\Delta t} = DN_e + DN_w - 2DN_c. \tag{1.5}$$

Now notice that we can think of N_c as the density at some position x, N_w as the density at $x - (\Delta x/2)$, and N_e as the density at $x + (\Delta x/2)$. This means Equation 1.5 can be rewritten as

$$\frac{\Delta N}{\Delta t} = D\left\{\left[N\left(x + \frac{\Delta x}{2}\right) - N(x)\right] - \left[N(x) - N\left(x - \frac{\Delta x}{2}\right)\right]\right\}. \tag{1.6}$$

After some algebra this is equivalent to

$$\frac{\Delta N}{\Delta t} = D\,\frac{\Delta(\Delta N)}{\Delta x^2}. \tag{1.7}$$

This is a discrete approximation to

$$\frac{\partial N}{\partial t} = D\,\frac{\partial^2 N}{\partial x^2}. \tag{1.8}$$

Although this is clearly not a rigorous argument, it provides some intuition to the simple diffusion equation (see Levin and Pacala, Chapter 12, for rigorous mathematical derivations).

A simple reaction-diffusion model that includes logistic growth at any given site (the last term in the equation) and passive diffusive spread is

$$\frac{\partial N}{\partial t} = D\,\frac{\partial^2 N}{\partial x^2} + rN\left(1 - \frac{N}{K}\right), \tag{1.9}$$

with the x, t notation for each N dropped in order to "unclutter" the equations. Indeed, Equation 1.9 is the "Fisher equation," which has played a prominent role in the theory of ecological invasions. The population grows as a wave that spreads through a habitat at a speed determined by both D and r (see Lewis, Chapter 3). Although the vast majority of reaction-diffusion spatial theory assumes simple passive diffusion, there are important papers that explore more complex diffusive processes such as density-dependent diffusion, taxis, and convection (see Okubo 1980 for a review). Reaction-diffusion models gain their power from the fact that they treat space as a continuous variable and from the arsenal of analytical and numerical tools that can be brought to bear on partial differential equations. They represent the most compact description of spatially explicit population dynamics. For a biological audience, the main drawback of reaction-diffusion models is their unfamiliarity and the mistaken impression that they are "sophisticated mathematics." A more justified drawback is that analytical results using these equations are hard to come by, except for the simplest models and formulations. When reaction-diffusion models cannot be handled analytically, numerical methods end up effectively treating space as a series of coupled points (or discrete cells) linked by diffusion, with ordinary differential equations describing the dynamics at each point (or cell). In other words, we discretize partial differential equations to solve them numerically and thus should not avoid discrete spatial models on principle alone.

Two themes stand out from the theory of reaction-diffusion models in the context of spatial ecology. First, they have been used with great success to identify how patterns corresponding to predictable spatial variation in population density can be formed in uniform environments. In single species models, these patterns are typically in the form of traveling waves associated with invasions, whereas for interacting species the patterns can end up being fixed in space, so that there are standing waves or standing checkerboards, and so forth. The pattern formation revealed by reaction-diffusion equations has spawned a rich theoretical literature (e.g., see Murray 1989) that provides an excellent guide to similar processes of pattern formation in discrete cellular automata models (Hassell and Wilson, Chapter 4).

The second theme is the illustration of a tight connection between the ability of a population to persist as a function of the size and geometry of habitat islands, and the similar situation for isolated habitat fragments that find themselves surrounded by hostile lands. The mechanism underlying this influence of habitat shape and size is the effect of perimeter-to-area ratios on processes governed by dispersal across habitat boundaries. For instance, the simplest prediction is that if the radius of a circular habitat gets too small, diffusive losses to the surrounding world will outweigh population growth and the population declines deterministically to zero. This phenomenon has been hypothesized as a mechanism explaining plankton patchiness in marine systems and the absence of plankton patches beneath some finite small threshold size—the so-called *critical patch size* (Kierstead and Slobodkin 1953). Recent extensions of this theory involve more complex behaviors than diffusion (such as mixed boundary conditions) and less arbitrary external conditions than absolute instant mortality (Cantrell and Cosner 1993). These more sophisticated models obviously yield different predictions, but it is still variation in perimeter-to-area ratios that underlies the major effects of habitat size and shape.

SUMMARY

All organisms are discrete entities that interact and disperse locally. Space unavoidably causes individuals to differ in both the intra- and interspecific interactions and the resource levels they experience. Although detailed spatially explicit models of population growth and interaction can be cumbersome, three simple approaches can abstract different aspects of the essence of spatial ecology. The Levins model is only implicitly spatial because it assumes global dispersal, but this makes it analytically tractable. It often provides a reasonable approximation to explicitly spatial simulators, such as cellular automata. One of its interesting predictions is that no species living in a spatial habitat can occupy all sites at equilibrium, which has profound implications for biodiversity (Tilman 1994; Lehman and Tilman, Chapter 8). Cellular automata are created when

Levins-like models are made one step more realistic by including local dispersal in a habitat divided into distinct cells or sites. A range of new phenomena appear, chief among them spatial patterning. A species living in a homogeneous but spatial habitat will take on a spatially clumped distribution. The degree of local dispersal influences the equilibrial population size of species (including ability to persist). Local dispersal also leads to percolation effects that may mean that habitats need to be of higher quality to function as corridors than as preserves. An alternative approach is to consider space a continuous variable and to approximate dispersal as a diffusion process. Such reaction-diffusion models can be analytically tractable and also provide a rigorous framework within which to numerically explore the implications of different assumptions about the nature of species interactions, the dispersal process, and habitat boundaries and conditions. These three approaches are the major tools with which spatial ecology is explored throughout this book.

CHAPTER TWO

Predictive and Practical Metapopulation Models: The Incidence Function Approach

Ilkka Hanski

INTRODUCTION

For more than twenty years, the Levins model (Levins 1969, 1970) has served as a cornerstone of metapopulation studies, providing a conceptual foundation for much of the theory and inspiration for empirical studies (Gilpin and Hanski 1991; Hastings and Harrison 1994; Hanski 1996; Hanski and Gilpin 1997). By metapopulation studies I refer to a particular perception of space. Not only is space assumed to be discrete, but it is divided into suitable and unsuitable habitat, with the suitable habitat occurring in relatively small and well-defined habitat patches. In the Levins model, this abstraction is taken to its ultimate limit, with suitable habitat occurring in infinitely many, identical, and pointlike "patches" in a vacuum-like matrix, through which model individuals cruise fast and effortlessly (the Levins model is discussed in Hanski 1991, 1997). Such gross simplifications of reality have evoked critical comments about the value of the Levins model and thereby about the value of the classical metapopulation concept in general (Harrison 1991, 1994; Harrison and Taylor 1997). Field ecologists are keen to point out that in real fragmented landscapes, there is always great variation in the sizes and isolations of habitat patches, as shown by the example in Figure 2.1, and that migration distances are typically restricted. However, I suggest that these complexities violate only the letter, not the spirit, of the Levins view of metapopulations, and as it turns out, variance in patch sizes and isolation provides extra information on which a simple yet powerful modeling approach can be constructed.

The purpose of this chapter is to describe an extension of Levins's approach to metapopulation studies, which I have labeled the incidence function (IF) approach (Hanski 1994b). While retaining the notion of a metapopulation as a population of extinction-prone local populations, the IF model is different from the Levins model in the following three respects: The number of habitat patches and hence the number of local populations is finite; the patches may have different sizes; and each patch has unique spatial coordinates, with interactions among the patches being localized in space. These assumptions greatly complicate the mathematical analysis of the model (Gyllenberg and Silvestrov 1994), but they open up a useful and practical link between models and field studies (Hanski 1994b).

THE INCIDENCE FUNCTION (IF) APPROACH TO METAPOPULATION DYNAMICS

Figure 2.1 illustrates the type of scenarios in the real world for which the IF model has been developed (Hanski 1994b). This map shows the spatial locations and areas of fifty habitat patches within an area of twenty-five square kilometers. The patches are dry meadows with the plants *Plantago lanceolata* and *Veronica spicata*, the two host plants of the Glanville fritillary butterfly, *Melitaea cinxia* (Hanski, Kuussaari, and Nieminen 1994). The essential elements in this fragmented landscape that make the IF approach applicable are the following. First, the patches are large enough to support local breeding populations, but not so large that local extinctions would be very rare, nor so isolated that reestablishment following extinction would be very unlikely. Second, the habitat patches are discrete, well delimited from the rest of the environment, and cover only a fraction of the total area, which allows us to ignore the shapes of the patches (but not their areas) and makes it

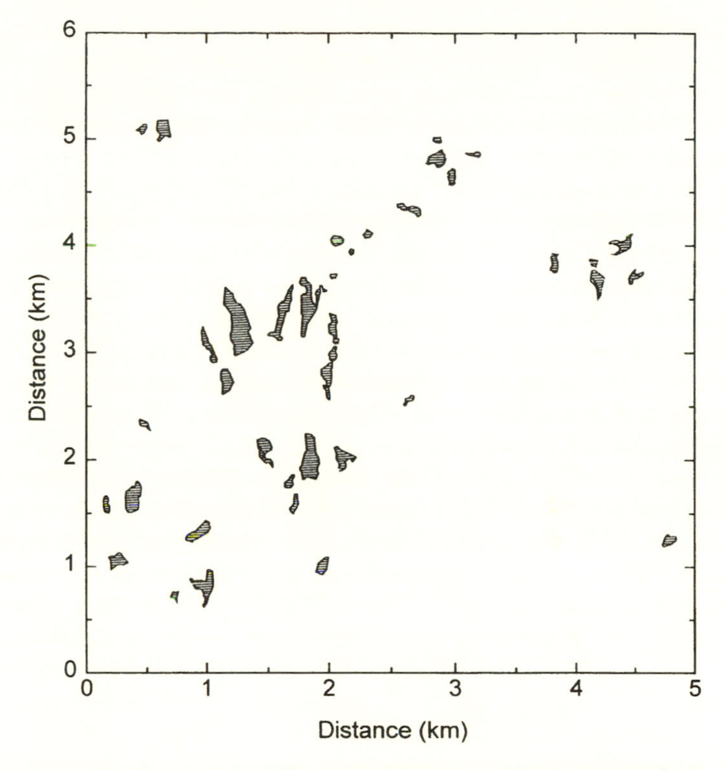

FIGURE 2.1. An example of the types of fragmented landscapes to which the IF approach can be applied. The habitat patches are dry meadows, suitable for the Glanville fritillary butterfly, *Melitaea cinxia* (Hanski et al. 1994).

reasonable to characterize isolation by the Euclidean distance calculated between the center points of two patches. The key idea is to construct a simple model that includes the first-order effects of patch area and isolation on local extinction and colonization and thereby on the pattern of patch occupancy. The aim is also to construct a model that can be parameterized with data that are widely available, snapshots of patch occupancies in landscapes such as the one shown in Figure 2.1. Finally, the aim is to construct a model with which one can make specific predictions as to how the occurrence of the species is affected by specific changes in the network, for instance, by the elimination of the three largest patches in Figure 2.1.

I call the effects of patch area and isolation on metapopulation dynamics first-order effects because these are the most general effects on extinction and colonization, these are the effects that ecologists have most frequently explored in field studies, and there are good reasons to expect that these are indeed the dominant effects. To start with patch area, numerous studies have reported on how the probability of population extinction, say, E, increases with decreasing patch size (Williamson 1981; Diamond 1984; Schoener and Spiller 1987; Hanski 1994b), evidently because small patches tend to have small and extinction-prone local populations (Schoener and Spiller 1987; Kindvall and Ahlén 1992; Hanski et al. 1995a). All extinction models tell us that small populations have a higher risk of extinction than do large ones (Lande 1993; Foley 1994, 1997 and references therein). Similarly, numerous field studies have reported on how the probability of colonization of empty patches, say, C, decreases with increasing isolation from existing local populations (Thomas et al. 1992; Hanski et al. 1995a). The explanation here is that an individual is more likely to move to a nearby than a faraway patch, for obvious reasons. Considering the dynamics of patch occupancy in terms of stochastic process, the long-term probability of occupancy of a particular patch increases with C and decreases with E, and hence large and little isolated patches have a greater chance of being occupied, at any given time, than do small and isolated patches.

Of course, there is no guarantee that a patch area effect, and especially an isolation effect, will be detected in a metapopulation. It may be that spatial heterogeneity does not allow a neat distinction between habitat and nonhabitat, or that there remains, within habitat patches, significant variation in habitat quality and hence in expected population density, eroding the basis for expecting a relationship between patch area and E. The nonhabitat may allow survival though perhaps not reproduction (source-sink dynamics; Pulliam 1988), further complicating the implications of patch area and population isolation. Patches may also be located so close to each other that isolation is no problem: All patches are occupied. If, however, a practical distinction between habitat and nonhabitat can be made, and if patch area and isolation effects on population extinction and reestablishment can be demonstrated, there are simple ways of modeling the dynamics as described below.

FIELD STUDIES

The IF approach involves a number of distinct tasks for field studies. Although most of the existing empirical metapopulation studies have been conducted without an explicit reference to this conceptual framework, many of these studies have nonetheless focused on the measurement of the parameters needed for an IF approach. First, a key initial task is to make the distinction between habitat and nonhabitat and to delimit all the habitat patches in the study region. Often habitat is defined subjectively, based on the ecologist's knowledge of the specific requirements of the study organism. More formally, one may use statistical methods to define suitable habitat (Lawton and Woodroffe 1991). The increasingly popular GAP analyses (Scott, Csuti, and Caicco 1991) deal with the definition of suitable habitat, but as the GAP approach is based on the assumption that species occur wherever environmental conditions are suitable, it is fundamentally in conflict with the metapopulation notion, where one assumes a priori that some suitable habitat is unoccupied at any given time. An experimental approach may be used to test the accuracy of an existing habitat classification: Experimental introduction to empty habitat should succeed (Harrison 1989; Oates and Warren 1990; Thomas 1992; Massot et al. 1994), introductions to nonhabitat should fail. My feeling, though, is that the IF approach is likely to be most helpful in those cases where the distinction between habitat and nonhabitat is conspicuous and poses no real problems. If the distinction is difficult to make, one should probably use some other approach. Note also that being fundamentally focused on the dynamics of single species, the IF approach cannot be easily applied to communities of many strongly interacting species (e.g., Pacala and Levin, Chapter 9).

Having successfully defined what is a suitable habitat patch, the next step is to map the occurrence of suitable patches within the study region and to survey the presence or absence of the focal species in these patches. Figure 2.2 shows the results of such surveys for four species of butterflies, a taxon that has provided many good examples of metapopulation dynamics (Thomas and Harrison 1992; Thomas and Jones 1993; Hanski and Thomas 1994; Hanski and Kuussaari 1995; Hanski et al. 1995a,b; Thomas and Hanski 1997). A few points are worth stressing. Making the discrete-population (presence/absence) assumption is often necessary to make large-scale field studies feasible at all (Steinberg and Kareiva, Chapter 14). All patches within the study region should be surveyed because we are ultimately interested in interactions among all populations in the metapopulation; if some populations interact with unknown populations, we are liable to draw misleading conclusions. The survey involves the measurement of patch areas and their spatial locations. Area measurement is critical, as area is a key variable both in the statistical data analysis and in modeling. The ability to measure patch area is tantamount to the ability to delineate habitat patches in the first place.

With data on patch areas and their spatial locations, and a snapshot of patch occupancies (Figure 2.2), one may analyze whether patch area and some measure of population isolation affect patch occupancy (Figure 2.3). Many empirical studies have used an overly simple measure of isolation: the distance to the nearest occupied patch. It is preferable to use a measure that takes into account the distances to all nearby populations, weighed by the sizes of, and the distances to, these populations (see Eq. 2.6). Furthermore, the shortest distance between two patches is not necessarily the most appropriate measure of distance, and one may use information about the species' biology to correct for the viscosity of different matrix (non-habitat) types (Wiens 1997). Typically, one is also interested in analyzing whether any other factors apart from patch area and isolation affect occupancy, and often they do (Sjögren 1991; Sjögren Gulve 1994; Eber and Brandl 1994). The population dynamic consequences of such other factors can be analyzed in the IF framework (see next section).

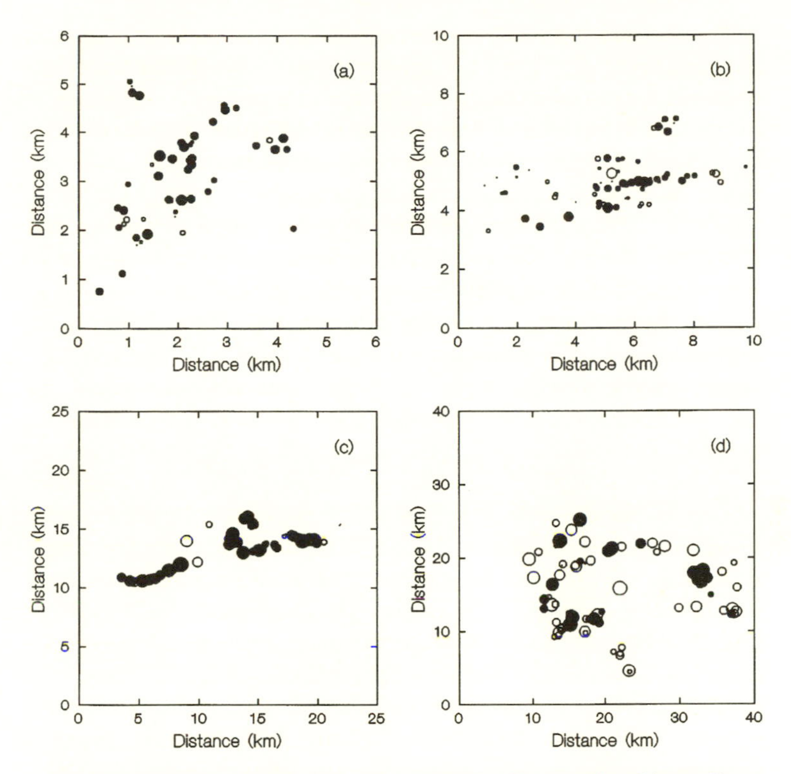

FIGURE 2.2. Maps of habitat patch networks for four species of butterfly: (a) *Melitaea cinxia* (Hanski et al. 1994), (b) *Scolitantides orion* (Saarinen 1993), (c) *Hesperia comma* (Thomas and Jones 1993; Hanski and Thomas 1994), and (d) *Melitaea diamina* (Wahlberg et al. 1996). In each case, the symbol shows the relative size of the habitat patch, with occupied patches shown by filled symbols.

If surveys have been conducted at two or more points in time, it becomes possible to analyze the effects of patch area, isolation, and possibly other factors on the rates of extinction and colonization (Figure 2.3). This is clearly a very valuable addition to the study of the pattern of patch occupancy, but often the turnover rate is so low that it takes great effort to accumulate sufficient data on extinctions and colonizations. The most appropriate study systems are the ones for which this can be accomplished (Sjögren 1991; Eber and Brandl 1994; Hanski et al. 1995a; Antonovics et al., Chapter 7).

An important assumption of the IF approach is that habitat patches have local breeding populations. Even if this were not the case, but rather the patch network would be occupied by a more or less panmictic population, we might expect to see the effects of patch area on (apparent) extinctions and colonizations, and on the instantaneous pattern of patch occupancy (Thomas 1994a). It would be more unexpected to observe an isolation effect if the patches are occupied by a single well-mixed population rather than by an assemblage of local populations. In any case, it is critical to have a good understanding of the population structure and migration behavior of the species under study.

THE INCIDENCE FUNCTION MODEL

The IF model is an extension of a simple Markov chain model of patch occupancy for a single patch to a metapopulation in many connected patches. Considering first a single patch, we assume it to have two possible states, occupied or empty. Assuming that the state of the patch may change only once in unit time, and assuming temporally constant but patch-specific extinction and colonization probabilities E_i and C_i, respectively, we find the long-term probability of patch i being occupied, J_i, which I call the incidence, given by

$$J_i = \frac{C_i}{C_i + E_i}. \tag{2.1}$$

I have modified this by adding the "rescue effect" (Brown and Kodric-Brown 1977) in the following manner (Hanski 1994b):

$$J_i = \frac{C_i}{C_i + E_i - C_i E_i}. \tag{2.2}$$

Equation 2.2 assumes that the probability of extinction in unit time is reduced from E_i to $E_i(1 - C_i)$ by the rescue effect (I will return to the biological interpretation of this assumption below).

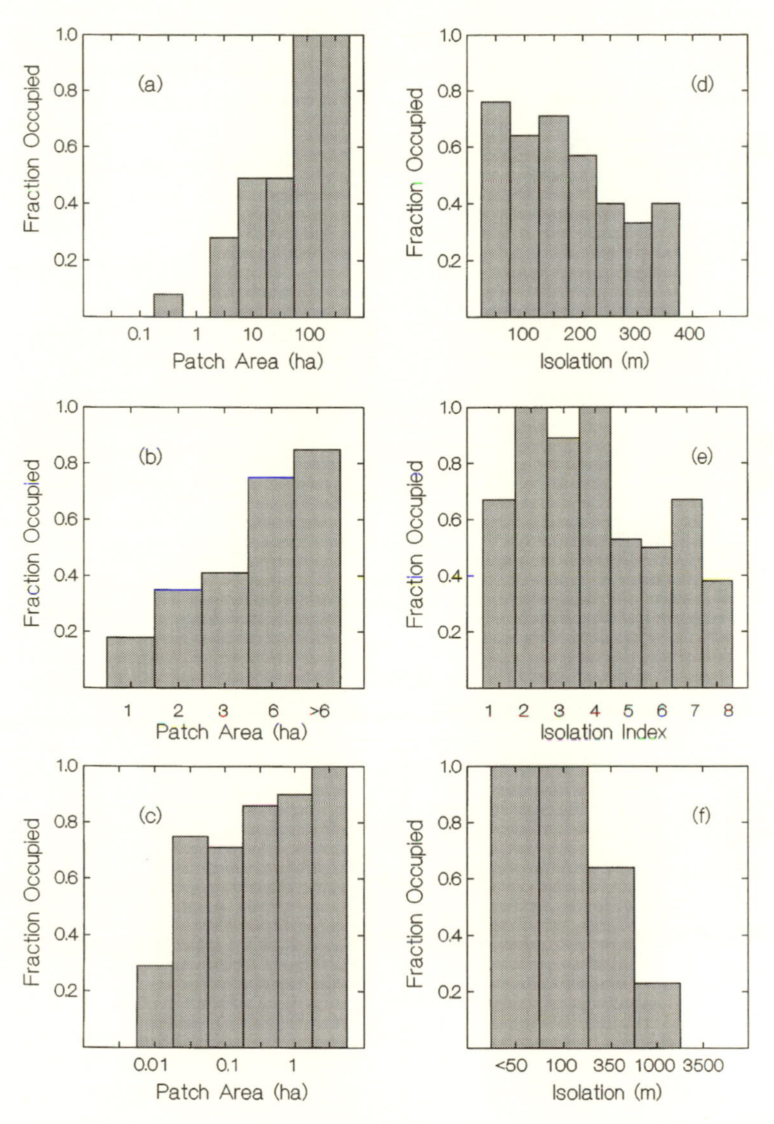

(a)
Fraction Occupied
1.0
0.8
0.6
0.4
0.2
0.1
1
10
100
Patch Area (ha)
(d)
Fraction Occupied
100
200
300
400
Isolation (m)
(b)
Fraction Occupied
1
2
3
6
>6
Patch Area (ha)
(e)
Fraction Occupied
1
2
3
4
5
6
7
8
Isolation Index
(c)
Fraction Occupied
0.01
0.1
1
Patch Area (ha)
(f)
Fraction Occupied
<50
100
350
1000
3500
Isolation (m)

FIGURE 2.3.

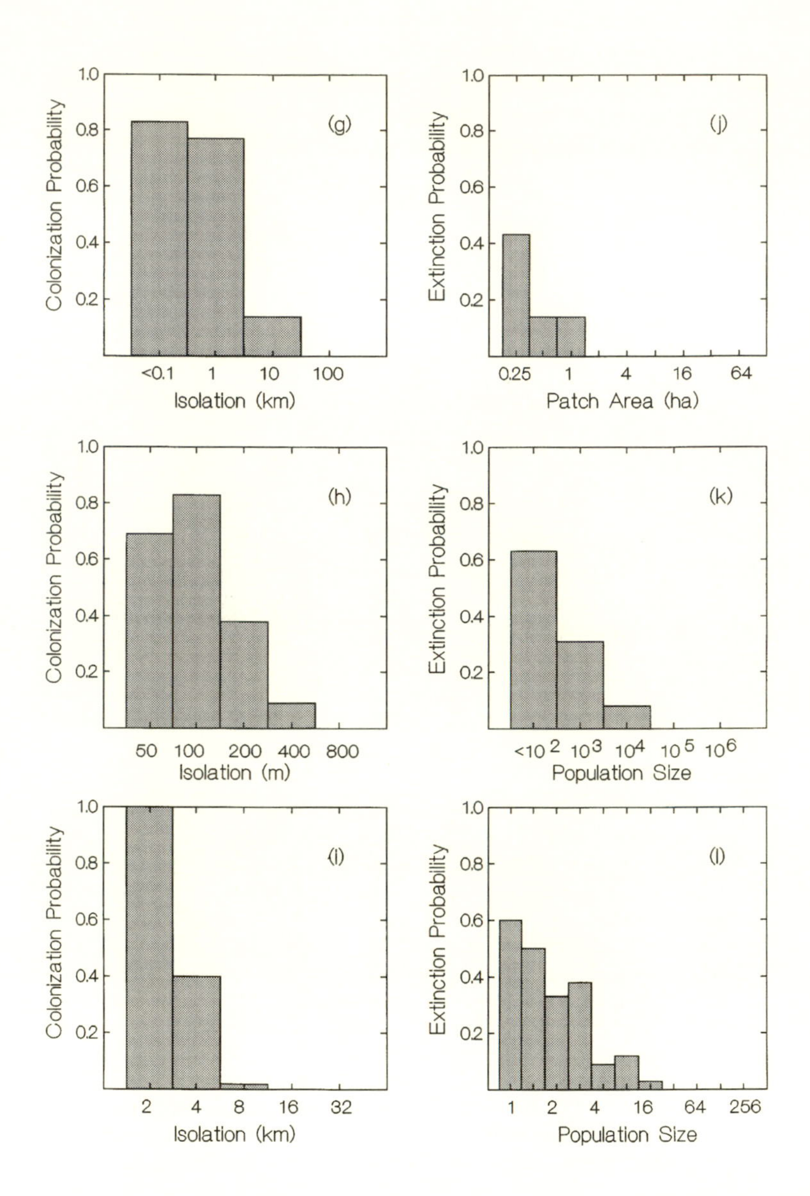

Colonization Probability
(g)
<0.1
1
10
100
Isolation (km)
Extinction Probability
(j)
0.25
1
4
16
64
Patch Area (ha)
Colonization Probability
(h)
50
100
200
400
800
Isolation (m)
Extinction Probability
(k)
<10 2
10 3
10 4
10 5
10 6
Population Size
Colonization Probability
(i)
2
4
8
16
32
Isolation (km)
Extinction Probability
(l)
1
2
4
16
64
256
Population Size

The next step is to relate the extinction and colonization probabilities to the structure of the fragmented landscape. To keep the model simple, I consider only the first-order effects of patch area and isolation. Patch area A_i will primarily affect the extinction probability E_i, and I assume that

$$E_i = \min\left(1, \frac{e}{A_i^x}\right), \tag{2.3}$$

where e and x are two parameters. Some other reasonable expression could be used instead of this formula. The colonization probability C_i must depend on the expected number of immigrants M_i arriving at patch i in unit time, for instance,

$$C_i = \frac{M_i^2}{y^2 + M_i^2}, \tag{2.4}$$

where y is a parameter. Equation 2.4 is justified when the per-immigrant probability of successful establishment depends on propagule size (Schoener and Schoener 1983; Ebenhard 1991). If there is no interaction among the immigrants at colonization, C_i increases exponentially with M_i,

$$C_i = 1 - \exp(-yM_i). \tag{2.5}$$

FIGURE 2.3. Examples of the effects of habitat patch area and isolation on patch occupancy and the rates of extinction and colonization. Patch area affects occupancy: (a) the masked shrew, *Sorex caecutiens* (Hanski 1992); (b) the nuthatch, *Sitta europaea* (Opdam and Schotman 1987); (c) the silver-spotted skipper butterfly, *Hesperia comma* (C. D. Thomas, unpubl.); patch isolation affects occupancy: (d) the pika, *Ochotona princeps* (Smith 1980); (e) the checkered blue butterfly, *Scolitantides orion* (Saarinen 1993); (f) the common shrew, *Sorex cinereus* (Lomolino 1993; Crowell 1986); patch isolation affects colonization: (g) the silver-spotted skipper butterfly, *Hesperia comma* (Thomas and Jones 1993); (h) the cricket *Metrioptera bicolor* (Kindvall and Ahlén 1992); (i) the Edith's checkerspot butterfly, *Euphydryas editha* (Harrison, Murphy, and Ehrlich 1988); patch area (or population size) affects extinction: (j) the cricket *M. bicolor* (Kindvall and Ahlén 1992); (k) pooled data for four species of spiders (Schoener and Spiller 1987); (l) the silver-studded blue butterfly, *Plebejus argus* (Thomas 1994b). (Originally published in Hanski 1994b with more details on the datasets.)

The value of M_i depends on the sizes of and distances to the existing populations from patch i. A natural choice for metapopulations without a mainland (a permanent large source of immigrants) is

$$M_i = \beta S_i = \beta \sum p_j \exp(-\alpha d_{ij}) A_j, \qquad (2.6)$$

where the sum is taken over all patches apart from the focal patch i, p_j equals 0 for empty and 1 for occupied patches, and d_{ij} is the distance between patches i and j. Here α and β are model parameters.

The relationship between the set of occupied patches and the colonization probability of patch i, via the number of immigrants M_i, lifts the model from the level of a single patch to the level of the metapopulation. The model is linear at the patch level but nonlinear at the metapopulation level. The key assumption here is that M_i is constant at steady state. This is clearly only an approximation, because the set of occupied patches (the p_i values in Eq. 2.6) will change at steady state, but numerical studies have shown that this approximation works well for realistic patch networks and metapopulations (Hanski 1994b). Notice also the implications of Equation 2.2 at the metapopulation level. The assumptions behind this equation may appear illogical, as patches that went extinct will nonetheless contribute to the "rescue" of other patches. The rationale here is that such patches have previously sent immigrants to patch i, and the immigrants have increased the population size in patch i and thereby decreased its probability of extinction: the rescue effect. Equation 2.2 is an attempt to incorporate the effect of migration on local dynamics in a patch occupancy model.

Plugging the above assumptions into Equation 2.2 gives the following expression for the incidence of patch i:

$$J_i = \left(1 + \frac{ey'}{S_i^2 A_i^x}\right)^{-1}, \qquad (2.7)$$

where $y' = (y/\beta)^2$.

PARAMETER ESTIMATION

Equation 2.7 may be written as

$$J_i = \frac{1}{1 + \exp[\ln(ey') - 2\ln(S_i) - x\ln(A_i)]}, \qquad (2.8)$$

and this equation may be further linearized to

$$\ln\left(\frac{J_i}{1 - J_i}\right) = -\ln(ey') + 2\ln(S_i) + x\ln(A_i). \qquad (2.9)$$

Here is a very concrete link between models and field studies. Equation 2.8, and therefore Equation 2.7, is a logistic regression model, which has become the standard method of data analysis in empirical metapopulation studies (Sjögren 1991; Verboom et al. 1991; Thomas et al. 1992; Eber and Brandl 1994; Hanski et al. 1995b). The IF model suggests that in logistic regression, area should be log-transformed (as is usually done) and that isolation should be measured by Equation 2.6. One could estimate the coefficient of $\ln(S)$ from data instead of using the constant 2, which I have assumed on biological grounds (Hanski 1994b). Instead of logistic regression, one may use nonlinear regression to estimate the parameters of Equation 2.7. This is necessary for the version of the model without the rescue effect (Hanski 1994b), because that model cannot be linearized with the logit-transformation. Figure 2.4 gives plots of $\ln(A)$ versus $\ln(S)$ for the four butterfly species with survey data shown in Figure 2.2. Figure 2.4 also shows the fitted $J = 0.1$, 0.5, and 0.9 incidence lines for these data.

An important assumption in parameter estimation is that the metapopulation from which the data come occurs at a stochastic steady state. If the size of the metapopulation (number of local populations) shows a long-term increasing or decreasing trend, the metapopulation is not at a steady state, and parameter estimation would produce misleading results. It is, of course, very difficult to formally test the steady state assumption, especially because the steady state is stochastic and may involve

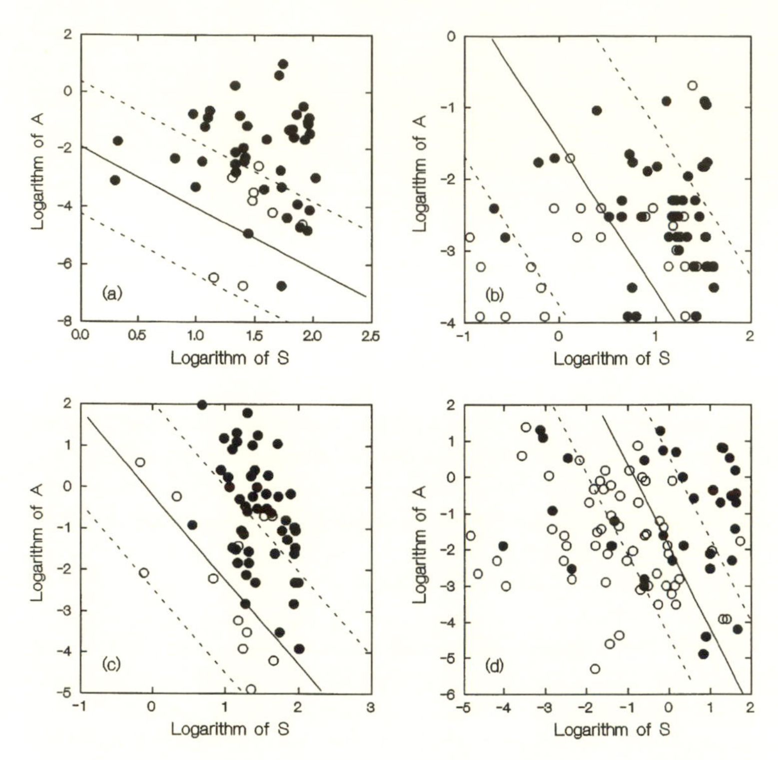

FIGURE 2.4. Plot of the logarithm of patch area against the logarithm of patch isolation, measured by S (Eq. 2.6; isolation increases to the left), for the four butterfly species in Figure 2.2 (same order of species as in Figure 2.2). The four panels show the $J = 0.1$, 0.5, and 0.9 incidence lines, that is, the fit of Equation 2.7 to these data.

substantial oscillations in the number of extant populations (Hanski et al. 1996b). In practice, one has to use general information about the environment and the species to judge whether this assumption is justified.

One apparent difficulty with Equation 2.7 is that the values of parameters e and y' cannot be estimated independently. In biological terms this means that we cannot distinguish between patterns of patch occupancy produced by high rates of extinction and colonization from patterns produced by low rates.

Teasing apart the values of e and y' is nonetheless essential if we want to iterate the extinction-colonization dynamics of the species in a patch network. The simplest approach is to estimate the threshold patch area for which $E = eA_0^{-x} = 1$, from which one may calculate the value of e knowing the value of x. If no other data are available apart from one snapshot of patch occupancies, one could use the area of the smallest occupied patch as an estimate of A_0. The predicted rates of extinction and colonization are affected by the choice of A_0, but the equilibrium fraction of occupied patches (metapopulation size) is not. Alternatively, one may use observed turnover rate to fix the values of e and y' in the product ey' as explained in Hanski (1994b). The amount of turnover data required for this purpose is modest, as we are not attempting to model the factors affecting the probabilities of extinction and colonization (as in transition models based on logistic regression; Sjögren Gulve and Ray 1996).

ITERATING METAPOPULATION DYNAMICS WITH THE IF MODEL

When the values of the model parameters have been estimated from empirical data, we may proceed to iterate metapopulation dynamics in any patch network, given the knowledge about patch areas and spatial locations and the initial state of occupancies (the p_i values). The dynamics are iterated by assigning, at discrete time intervals, extinction and colonization events to individual patches according to the probabilities E_i and C_i. Thus, if patch i is empty at time t, we draw a random number uniformly distributed between 0 and 1 and allow patch i to become colonized if that number is less than

$$C_i = \frac{M_i^2}{y^2 + M_i^2} = \frac{1}{1 + \left(\dfrac{y}{M_i}\right)^2} = \frac{1}{1 + \left(\dfrac{y'}{S_i}\right)^2}, \quad (2.10)$$

where S_i is given by Equation 2.6. Similarly, if patch i is occupied, we allow it to go extinct with probability

$$E_i(1 - C_i) = \min\left(1, \frac{e}{A_i^x}\right)(1 - C_i). \qquad (2.11)$$

Equation 2.11 includes the rescue effect. Following the evaluation of extinction and colonization events in each patch at time t, the occupancies p_i are updated and the next round of extinctions and colonizations at time $t + 1$ is assessed. It is worth repeating that the model can be used to iterate metapopulation dynamics in any patch network, as long as the patch areas and spatial locations are known, and the iteration may be started from any initial set of occupied patches. The IF model may therefore be used to answer very practical questions about metapopulation dynamics as illustrated by the examples below and in Hanski (1994a,b; Hanski et al. 1996b).

MODEL PREDICTIONS FOR REAL METAPOPULATIONS

Figure 2.5 shows again the area-isolation plot for one of the butterfly species, *Melitaea diamina*, with the $J = 0.5$ incidence lines for all of the four butterfly species estimated from their respective patch networks (Figure 2.4). In three species, the effects of patch area and isolation on occupancy are very similar, whereas *Hesperia comma* has a lower incidence than the other species in small and isolated patches. Most simply, this could reflect a lower density of *H. comma* than the other species, but there is no such difference (Hanski and Thomas 1994, table 4). Apparently, *H. comma* differs in some other significant manner from the rest of the species, giving it a low colonization rate, high extinction rate, or both. The analysis in Hanski and Thomas (1994) suggested that *H. comma* has higher demographic stochasticity and a lower intrinsic rate of population increase than *M. cinxia*, which are both expected to increase the extinction rate (Lande 1993; Foley 1994).

The lower panel in Figure 2.5 shows the predicted metapopulation trajectories of the four species when iterated in the *M.*

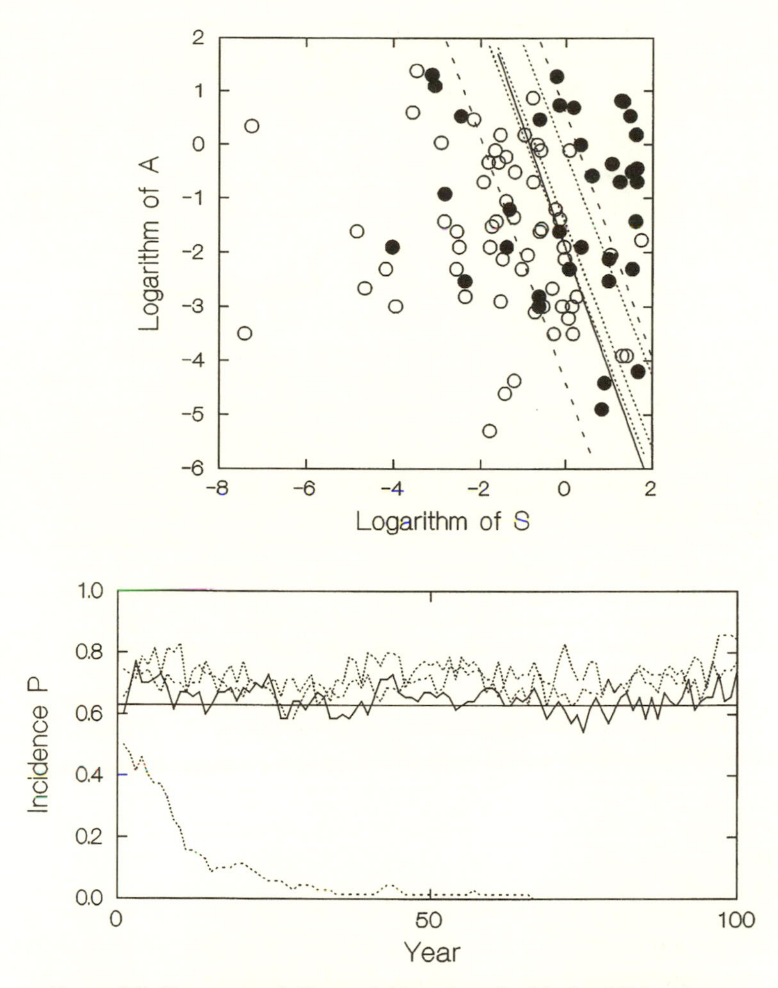

FIGURE 2.5. Upper panel, Figure 2.4d replotted, with $J = 0.5$ incidence lines (dotted lines) for all the four species in Figure 2.4, estimated from their respective patch networks. The species with the incidence line above the others is *Hesperia comma*. Lower panel, predicted fraction of occupied patches in the *M. diamina* network (Figure 2.2d), using the four sets of parameter values estimated for the four butterfly species in Figure 2.4. Each run was started with half of the patches occupied. The *H. comma* metapopulation is predicted to go rapidly extinct; the other species fluctuate around the observed fraction of occupied patches in the *M. diamina* metapopulation (solid line).

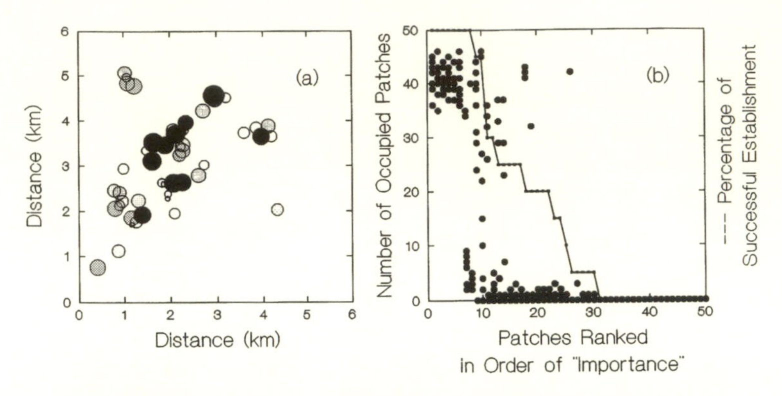

FIGURE 2.6. An example of how the IF model may be used to assess the significance of individual habitat patches for metapopulation dynamics. (a) The metapopulation of the Glanville fritillary butterfly, *Melitaea cinxia*, in a fifty-patch network in Finland, within an area of 5 × 5 km^2 (Hanski et al. 1994). Using the parameter values estimated in Figure 2.4a, I numerically evaluated the probability of successful reinvasion into the patch network from each patch, following extinction from all other patches. Dark shading indicates large probabilities of reinvasion. (b) The patches have been arranged in an order of decreasing "importance" on the horizontal axis. The vertical axis shows the fraction of ten independent model iterations in which only the focal patch was initially occupied and that persisted for more than twenty generations ("successful invasions," line), and the number of patches occupied after twenty generations (dots). (From Hanski 1994b.)

diamina network, started with half of the patches occupied. *Hesperia comma* is predicted to go extinct in such a network, but the other spcies occupy about equal fractions of patches in stochastic steady state, as was expected based on the results in the upper panel in Figure 2.5.

The above comparison should make it clear that, once the IF model has been parameterized and (hopefully) tested, it can be used to answer very practical questions about transient and steady state metapopulation dynamics in constant or changing patch networks. Figure 2.6 gives an example in which I have evaluated the "importance" of individual patches for metapopulation persistence in the fifty-patch network occupied by the Glanville fritillary butterfly (see Figures 2.2a and 2.4a).

The IF model may be generalized to include the effects of other factors apart from patch area and isolation on the rates of extinction and colonization. Moilanen and Hanski (1997) modified Equations 2.6 and 2.7 to include the effect of patch quality as a factor producing an "effective" patch area or isolation. The model was applied to the large metapopulation of the Glanville fritillary butterfly on the Åland Islands in Finland. Several factors were shown to influence the "effective" patch area or isolation, including the amount of the larval host plant *Plantago lanceolata* in the habitat patch (positive effect on "effective" patch area), abundance of flowers (decreases emigration but increases immigration), grazing (negative effect on patch area), and percentage of low vegetation (negative effect on patch area). These results demonstrate how, given sufficient data, the IF model may be used to formally test the population dynamic consequences of variation in patch quality and in the surrounding environment. In our study, the main conclusion nonetheless was that patch area and isolation had by far the strongest effects (Moilanen and Hanski 1997), suggesting that it is often sufficient to use the basic IF model for predictive purposes.

COMPLEX SPATIAL DYNAMICS

The Levins model predicts a threshold rate of colonization for metapopulation persistence; beyond the threshold, increasing colonization rate leads to a monotonically increasing fraction of occupied patches (Tilman and Lehman, Chapter 10). Structured metapopulation models that include the effect of migration on local dynamics (the rescue effect) predict the possibility of alternative stable states for a range of population parameters, landscape structures, or both (Hanski 1985; Gyllenberg and Hanski 1992; Gyllenberg, Hanski, and Hastings 1997; Figure 2.7a). The intuitive explanation of alternative stable states is the increasing fitness of immigrants with increas-

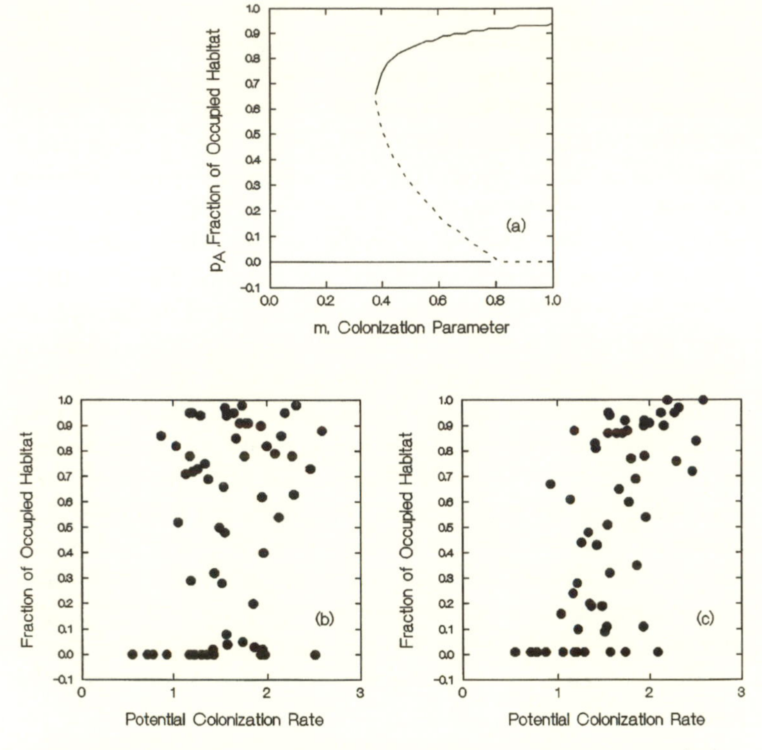

FIGURE 2.7. Bifurcation diagrams for the fraction of occupied habitat (P_A) with increasing colonization rate (m). (a) A theoretical result based on the model described in Hanski and Gyllenberg (1993; using parameter values in their figure 3d). Solid lines represent stable equilibria; broken lines are unstable equilibria. (b) The empirical result for sixty-five semi-independent patch networks of the Glanville fritillary butterfly, *Melitaea cinxia*. The occupied fraction of the pooled patch area, P_A, was used as the dependent variable. The potential colonization rate, M, gives the average value of the expected immigration rate to the patches, calculated on the assumption that all patches are occupied (because M has to measure the potential, not actual, rate of colonization). For more details see Hanski et al. (1995b). (c) The same information as (b) plotted in exactly the same way, but now the values were predicted by the IF model parameterized for this species and iterated for two thousand generations in the sixty-five networks. Each point represents one snapshot from the model-predicted trajectory.

DISCUSSION

One purpose of this volume is to explore the kinds of modeling approaches that are most helpful for better understanding of spatial population dynamics. Even the simplest patch models, such as the Levins model, are useful for the purpose of elucidating the general concepts of metapopulation persistence in extinction-colonization equilibrium and the threshold condition for persistence, implying that not all suitable habitat is occupied at equilibrium (May 1991; Nee and May 1992; Nee 1994; Lawton et al. 1994; Tilman et al. 1994; Tilman and Lehman, Chapter 10). The Levins approach is very simplified, however, with a focus on just the density of habitat patches within a large area. The IF approach brings the extinction-colonization dynamics closer to real metapopulations by adding three ingredients to the Levins approach: the actual number of habitat patches, the area of each habitat patch, and the spatial location of each patch. The IF approach represents a practical abstraction because it allows the model to be parameterized, realistically, for species living in fragmented landscapes. The IF model is a spatially explicit metapopulation model and hence an alternative to often very complex simulation models that have been advocated for management purposes (Akcakaya 1994). Clearly, with only four parameters, the IF model cannot deal with all the complexities that the simulation models are supposed to handle, but before one has actually demonstrated that all these complexities really matter, one might be better off with a simpler model. The successful applications of the IF model to a range of examples (Hanski 1994b; Hanski et al. 1995b, 1996b; Wahlberg, Moilanen, and Hanski 1996; Moilanen, Smith, and Hanski 1997) suggest that often the effects of patch area and isolation on population extinction and colonization explain as much of extinction-colonization dynamics as one could hope to explain. This is most likely for species with relatively rapid turnover rate in highly fragmented landscapes (Figure 2.1). It remains an important task for further research to find out under which circumstances and for which kinds of species it is sufficient to focus only on the patch area and isolation effects in metapopulation modeling.

The IF concept is focused on the effects of patch area and isolation on metapopulation dynamics, but this modeling framework also provides a natural way to assess the population dynamic consequences of patch quality and the quality of the surrounding environment, thus providing a link between metapopulation dynamics and landscape ecology (Wiens 1997). For instance, one may include in the model a parameter of patch quality that modifies patch areas; or one may modify distances among pairs of patches depending on the structure of the intervening landscape. The significance of such factors on metapopulation dynamics can be tested by observing whether the model with the extra parameter better describes observed patterns of patch occupancies (Moilanen and Hanski 1997) or predicts population turnover more accurately. The IF approach can also be extended to two-species interactions, by making assumptions about how interspecific interactions affect colonization and extinction rates (Moilanen and Hanski 1995; Lei and Hanski 1997).

CONCLUSIONS

The Levins model and related patch occupancy models have provided a useful conceptual and modeling framework for metapopulation studies for the past twenty years. The Levins approach assumes an infinite number of identical habitat patches, which are all equally connected to each other. I have described here a simple alternative approach, the incidence function (IF) model, with a finite number of habitat patches of different sizes and of different degrees of connectance to each other. Unlike the highly abstract Levins model, the IF model can be closely linked with field studies. The model can be parameterized with a snapshot of patch occupancies to make quantitative predictions about transient and steady state metapopulation dynamics in particular fragmented landscapes. Although the IF model is a patch occupancy model, like the

Levins model, it is possible to investigate in this framework how the quality of the surrounding landscape and how various patch attributes apart from area affect metapopulation dynamics. A range of examples in this chapter illustrates the use of the IF model in empirical metapopulation studies. Successful applications of the IF model point to the conclusion that metapopulation dynamics in highly fragmented landscapes are often largely controlled by the first-order effects of patch area and isolation on local extinction and recolonization.

CHAPTER THREE

Variability, Patchiness, and Jump Dispersal in the Spread of an Invading Population

Mark A. Lewis

INTRODUCTION

Biological invasions can be dramatic events. They have received much attention in recent years (Mooney and Drake 1986; Drake et al. 1989; Hengeveld 1989), largely because virtually every ecosystem has been invaded by exotic organisms with often drastic consequences for the native fauna or flora. Historical records of biological invasions abound, with data from a variety of diverse sources such as Christmas Bird Counts (Veit and Lewis 1996), agricultural records of insect pests (Andow et al. 1990), lynx trap line data (Hengeveld 1989), and medieval bills of mortality for the black death.

Parallel with the empirical interest in invasions has been an explosion of mathematical models that attempt to describe or predict the fate of particular invasions (Williamson 1989). One especially common way of mathematically analyzing invasions is to investigate the form and rate of a population's spread in a new environment. The usefulness of these models arises from the fact that they provide testable analytical predictions about the asymptotic rate of spread (ARS) of a population, based upon parameters that can be measured at the level of an individual, such as fecundity and mean squared spatial displacement per unit time. Furthermore, the rate of spread, expressed in terms of the rate of change of effective range radius (proportional to $\sqrt{\text{area invaded}}$), is readily available from maps of an expanding range.

A large portion of the mathematical literature on invasions is couched in terms of deterministic partial differential equations, which often yield appealingly tractable and compact models of invasion (Skellam 1951). The mathematical approach can be traced back to R. A. Fisher, who was interested in the spatial spread of new genes that appeared in a population (Fisher 1937). The simplest possible model of this form is the reaction-diffusion model

Rate of change in density = dispersal rate + growth rate

$$\frac{\partial u}{\partial t} = D\nabla^2 u + f(u), \tag{3.1}$$

where dispersal enters through ∇^2, which denotes the diffusion operator, t is time, $u(x, y, t)$ is the local population density, D is the coefficient of diffusion, and $f(u)$ is a nonlinear function describing the net population change from birth and death.

First analyzed in one spatial dimension by Fisher (1937) and Kolmogorov, Petrovsky, and Piscunov (1937), the case where $f(u)$ describes logistic population growth and a few individuals are initially introduced to a localized region has a wavelike solution that eventually expands in each direction at a speed (ARS) of $2\sqrt{rD}$, with $r = f'(0)$ or what is traditionally called the intrinsic rate of increase of population growth. In front of the wave is the uninvaded territory, and behind the wave the population is at the carrying capacity. In fact, this result is remarkably robust to changes in the shape of the nonlinear growth function f and holds true providing the maximum per-capita growth rate is highest at low population densities. In spite of its simplicity, the ARS formula derived from Equation 3.1 has had remarkable success in explaining the rates at which species have invaded new environments. Using well-documented invasions of the muskrat (*Ondatra zibethica*) in Europe and the cereal leaf beetle (*Oulema melanopus*) and cabbage white butterfly (*Pieris* [= *Arogeia*] *rapae*) in North America and using parameters taken from the measurement of individual behaviors, Andow et al. (1990) found that the theory matched the data well in all cases but the cereal leaf beetle.

Andow et al. suggested that this discrepancy might be due to the fact that the beetle was dispersing much faster than the microscale data indicated, perhaps by macroscale jumps provided by air currents or human transport.

Clearly Equation 3.1 does not describe the exact ecological situation. For example, it neglects stochastic aspects and assumes no long-range jumps by individuals. As we will show, these additional factors can play an important role in determining the rate and form of an invasion.

The purpose of this chapter is threefold: (1) to extend classical theory to incorporate long-distance dispersal events and to evaluate how this detail affects the ARS; (2) to introduce and develop the idea of underlying stochastic variability in invasion rates and patterns, even when parameters are known and fixed; and (3) to argue how the theory should be extended to explore the stochastic factors affecting patchiness and variability in spread rates of invaders.

LONG-DISTANCE DISPERSAL

Implicit in the diffusion approximation is the assumption that the distance that an individual disperses in a *fixed* length of time is normally distributed. However, dispersal patterns have been measured for a tremendous number of organisms either as seed shadows (Willson 1993), plant disease dispersal gradients (Fitt et al. 1987), or mark-recapture data (Taylor 1978), and most distributions of dispersal distances are leptokurtic.

To understand how leptokurtic distributions arise naturally, we can relax the assumption that individuals take a fixed length of time to disperse and, instead, assign the probability of settling as a function of the time t since dispersal started. The assumption of random movement, modeled by diffusion, a constant settling rate, and a sufficiently long time for dispersers to settle, gives rise to a leptokurtic dispersal function that drops off exponentially with distance (Okubo 1980).

By way of contrast, an empirical approach to deriving the dispersal distance function is simply to fit mark-recapture or dispersal data to a family of functions (Figure 3.1). Once a dispersal function has been determined, a model is needed to link the dispersal with the reproduction and death of individuals. The Fisher model (3.1) assumes individuals reproduce continuously and move via simple diffusion. This is clearly not the case for a large number of organisms including most birds, many mammals, and most plants for which reproduction is at discrete intervals and dispersal typically occurs once a year.

When reproduction and dispersal occur at discrete intervals an integro-difference equation is the relevant formulation. Here individuals first undergo reproduction and then redistribute their offspring according to the dispersal function before reproduction occurs once again. If generations are nonoverlapping, as is the case with annual plants and many insect species, then the process is described by

$$N_{t+1}(x) = \int_{\Omega} k(z - x) f[N_t(x)]\, dz, \tag{3.2}$$

where $N_t(x)$ is the density of individuals at point x and time t, $f[N]$ describes density-dependent fecundity, and $k(z - x)$ is the dispersal function, which depends upon the distance $|z - x|$ between the location of birth z and the location of settlement x so that

$$k(z - x)\Delta z\, \Delta x = \text{Probability of dispersing from } \Delta z \text{ to } \Delta x. \tag{3.3}$$

Equation 3.2 states that the density of offspring at point z (denoted by $f[N_t(z)]$) multiplied by the dispersal function (denoted by $k(z - x)$), and then integrated over all possible locations z in the study area Ω yields the density of individuals at the next time step $N_{t+1}(x)$. Historically integro-difference models have been used to predict changes in gene frequency (Slatkin 1973; Weinberger 1978, 1982; Lui 1982a,b, 1983), and only more recently have they been applied to ecological problems (Kot and Schaffer 1986; Hardin, Takac, and Webb 1988a,b, 1990; Kot 1989, 1992; Hastings and Higgins 1994; Neubert, Kot, and Lewis 1995). The advantage of using the

A)

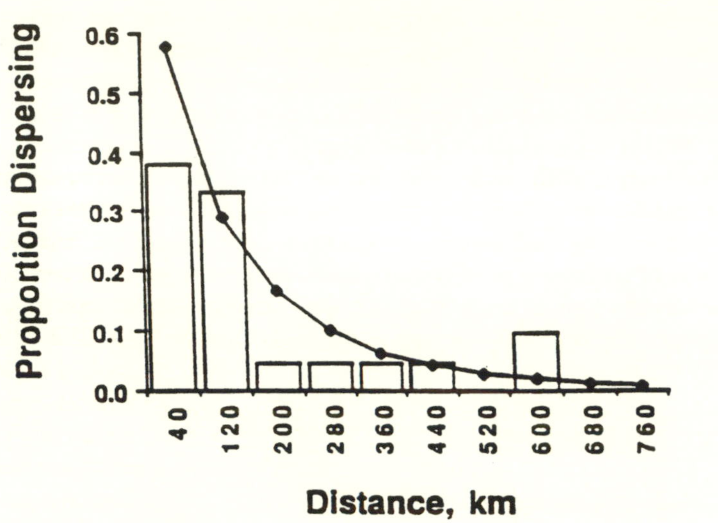
Proportion Dispersing
0.6
0.5
0.4
0.3
0.2
0.1
0.0
40
120
200
280
360
440
520
600
680
760
Distance, km

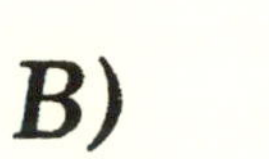
B)

Proportion Dispersing

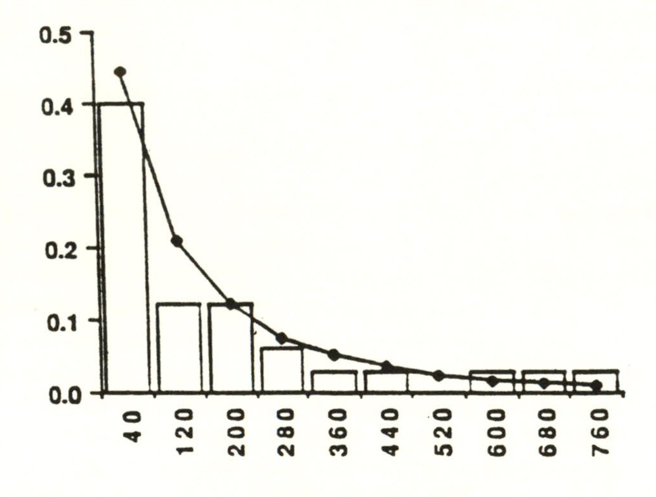
0.5
0.4
0.3
0.2
0.1
0.0
40
120
200
280
360
440
520
600
680
760
Distance, km

integro-difference formulation over a reaction-diffusion model (Eq. 3.1) is the ability to strictly correlate the model with biological observations of discrete reproduction and dispersal events. To illustrate this approach we consider the explosive invasion of eastern North America by house finches (*Carpodacus mexicanus*) after the birds were released in Long Island in 1940 following a ban on their sale as pets (see Veit and Lewis 1996 for details).

House finch reproduction occurs in the spring, and dispersal occurs in the fall. Birds born in the spring, surviving the summer, and dispersing in the fall then can form breeding pairs that may reproduce in the following spring. Rather than remaining at their old nesting sites, some adults also disperse in the fall. To model the population spread we need initially to distinguish between birds that are nine to twelve months old in the spring of year t (juveniles, denoted $J_t(x)$) and birds that are older than twelve months in the spring of year t (adults, denoted by $A_t(x)$). The density of potential breeders in year t is thus $N_t(x) = A_t(x) + J_t(x)$. Dispersal distance data and the fitted dispersal functions for juvenile and adult house finches are shown in Figure 3.1.

A crucial ingredient in any invasion model is a formula for the density of offspring as a function of the density of potential breeders. Veit and Lewis (1996) derived this formula for the house finch under assumptions of (1) a 1:1 sex ratio and random searching of males and females to form pairs at rate σ, (2) a fixed length of time T over which pair formation can occur, (3) a finite density δ of possible nesting sites, and (4) density-dependent competition for nesting sites. The formula is

$$f[N_t] = \frac{cN_t^2}{4/(\sigma T) + 2N_t + N_t^2/\delta}, \tag{3.4}$$

where c is the average number of offspring born to a breeding pair that survive the summer.

FIGURE 3.1. (A) Frequency distribution of dispersal distances for adult house finches from Massachusetts. (B) Dispersal distances for young-of-the-year house finches. (Redrawn from Veit and Lewis 1996.)

At the end of the first summer, a proportion p_J of the surviving offspring disperse, as does a proportion of the adult population p_A. The dispersal functions for fledglings and adults are shown in Figure 3.1 and are represented here as k_J and k_A. Equations describing reproduction, survival, and dispersal are then

$$J_{t+1}(x) = (1 - p_J)f[N_t(x)] + p_J \int_{-\infty}^{\infty} k_J(z - x)f[N_t(z)]\,dy, \tag{3.5}$$

$$A_{t+1}(x) = s(1 - p_A)N_t(x) + sp_A \int_{-\infty}^{\infty} k_A(z - x)N_t(z)\,dy, \tag{3.6}$$

where s is the probability that a juvenile or adult in year t will survive to year $t + 1$. Finally, adding Equations 3.5 and 3.6 gives a single equation for the density of adults plus juveniles prior to the spring breeding (N_{t+1}) in terms of the density of adults plus juveniles before the previous spring breeding (N_t):

$$\begin{aligned} N_{t+1}(x) = {} & s(1 - p_A)N_t(x) + (1 - p_J)f[N_t(x)] \\ & + \int_{-\infty}^{\infty} k_A(z - x)sp_A N_t(z)\,dy \\ & + \int_{-\infty}^{\infty} k_J(z - x)p_J f[N_t(z)]\,dy. \end{aligned} \tag{3.7}$$

Thus quantities $N_t(x)$ and $N_{t+1}(x)$ in Equation 3.7 are the expected density of birds at Christmas Bird Counts in successive years.

With explicit equations for bird densities over successive years, the next step is to estimate model parameters. Veit and Lewis (1996) estimated the life history parameters c and s from a variety of independent studies (Milby and Wright 1976; Hooge 1990; Hill 1993). The parameters σT and δ were calculated indirectly under assumptions that the environment would maintain an average carrying capacity of 3.8 individuals per square kilometer and that the finite rate of pair formation (proportional to σ) and the fixed length of time for pair formation T would mean that 4% of birds would not be able to find mates when the bird density was reduced to half that of the carrying capacity. Such inability to find mates at reduced densities is typically referred to as an Allee effect in recognition of the work by Allee (1938) on this subject. Last, proportions dispersing p_J and p_A were assumed to increase linearly with the density of potential breeders until the carrying capacity is reached.

Using the parameter estimates, numerical predictions from the model can be compared with the very richly detailed historical documentation of house finch spread (Figure 3.2). The model output reflects two key features of the data: (1) initially slow, then abruptly accelerating, rate of population spread, which eventually reaches an invasion velocity given by the slope of the line in the top right of Figure 3.2a: (2) strong correlation between rate of population spread and rate of population growth near the center of the range. Numerical simulations for a variety of parameter values show that even a very small proportion of birds not finding mates dramatically slows the population spread, especially the early spread, and that density-dependence for dispersal fractions p_J and p_A causes an increase in the length of time taken before the invasion achieves a constant spread rate (Veit and Lewis 1996). While the actual rate of spread of the house finch population is slightly underpredicted by the model for the first ten years after release, and slightly overpredicted thereafter, given the simplifications made in the modeling process, it can be argued that the fit of the data and model is as good as reasonably expected.

Rather than relying upon numerical simulations to give spread rate predictions, it is possible to use a formula for the asymptotic spread rate when the population obeys the simple form of the integro-difference equation (Eq. 3.2). The formula is analogous to the formula $2\sqrt{rD}$ for Fisher's equation (Eq. 3.1) and involves the geometric growth rate of the population

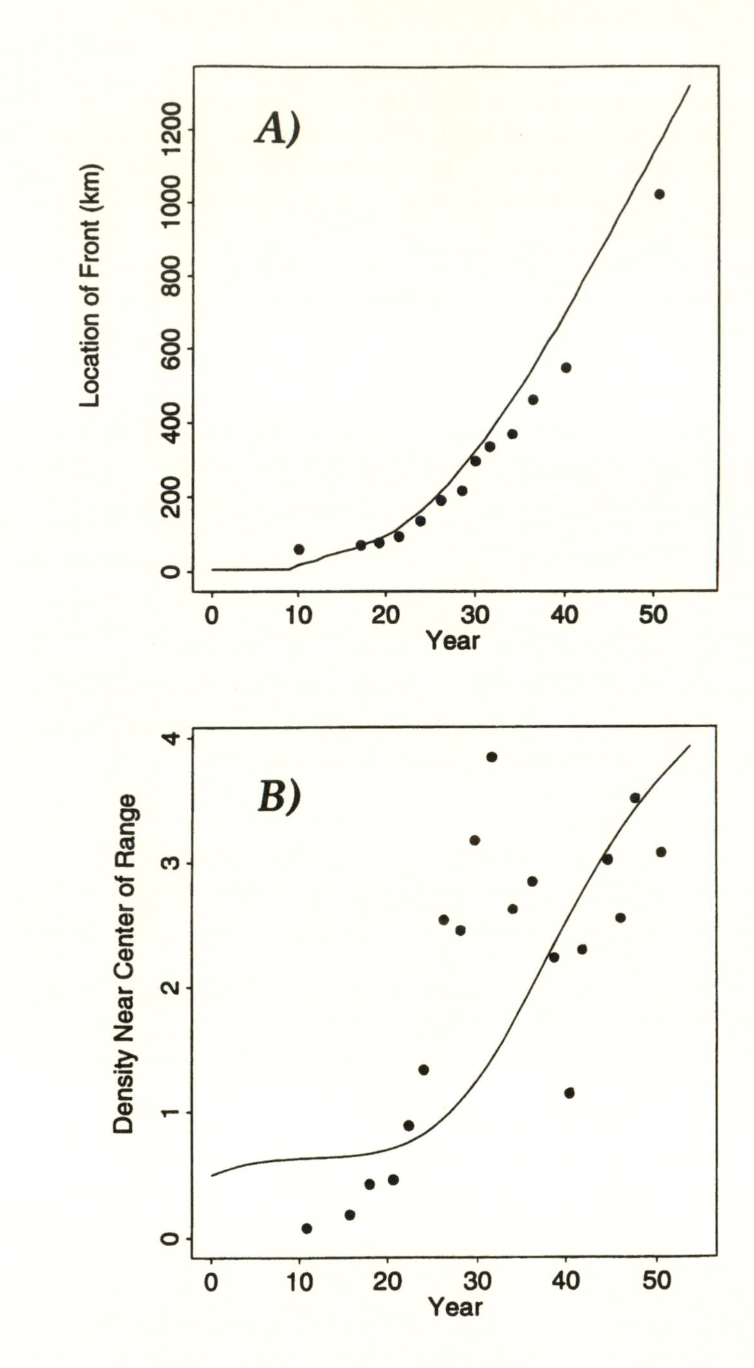
A)
Location of Front (km)
0
200
400
600
800
1000
1200
0
10
20
30
40
50
Year
B)
Density Near Center of Range
0
1
2
3
4
0
10
20
30
40
50
Year

and the shape of the spatial redistribution function. Kot et al. (1996) showed how the ARS formula depends critically upon the very long-distance dispersal events, that is, upon how "fat" the tails of the dispersal function are. Specifically, they showed that, in the absence of an Allee effect in growth dynamics, a constant ARS requires the tail of the dispersal function to drop off at least as fast as an exponential function (i.e., be exponentially bounded). When this is not true the ARS is no longer constant but can grow geometrically. The dramatic consequences of "fattening" the tail of the dispersal function and thus switching from a constant ARS to geometric growth of the ARS are illustrated in Figure 3.3. With this in mind, and reanalyzing dispersal data from Dobzhansky and Wright's study of genetically marked *Drosophila pseudoobscura*, Kot, Lewis, and van den Driessche (1996) showed that the data were insufficient to distinguish between an exponentially bounded and a very fat-tailed dispersal function that was not exponentially bounded, in spite of the fact that invasions for the two would proceed at very different rates. In other words, to distinguish between the two functions one needs data on the probability of rare, extremely long-distance dispersal events occurring—data that may not be available (Figure 3.4).

The result regarding the fat tails, however, depends crucially upon the assumption of no Allee effect; even a mild Allee effect drives very long-distance lone dispersers extinct, and invasion rates from very fat-tailed dispersal functions do not continue to grow. This helps to explain the house finch

FIGURE 3.2. (A) Range of population spread for the house finch based on the numerical solution of the Veit and Lewis (1996) age-structured integro-difference equation model, with parameters independently estimated from life history and banding data. The range radius, calculated as the radius of a semicircle covering the area invaded from the eastern seaboard, is superimposed. (B) Density at center of range ($x = 0$) for the numerical solution of the model described in (a). Mean density of house finches on five Christmas Bird Counts within the core area of their range is superimposed. (Redrawn from Veit and Lewis 1996.)

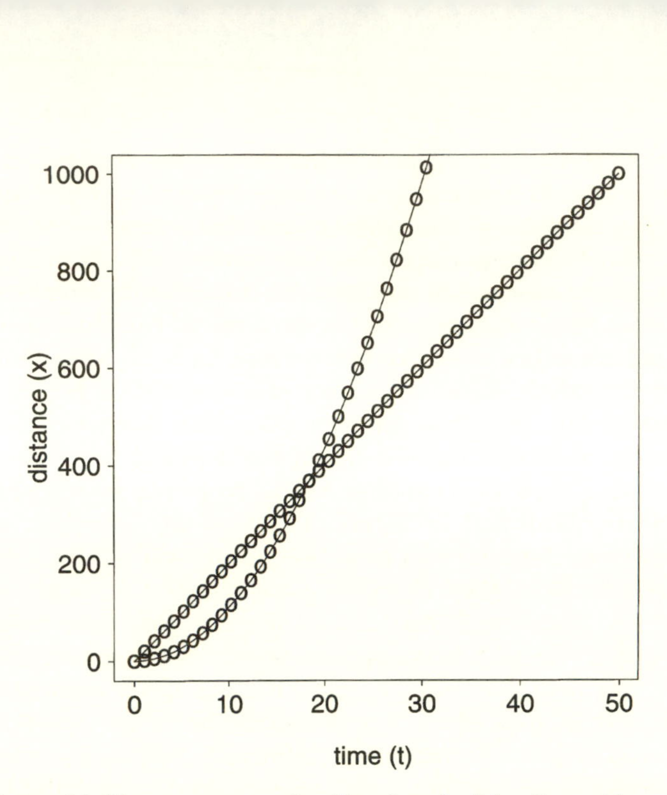

FIGURE 3.3. The consequence of making the tail of the dispersal kernel "fatter" yields a dramatic change in predictions for progression of the invasion. With R expected offspring per parent, the normal dispersal function

$$k(x) = \frac{\exp\left(-\frac{x^2}{4D}\right)}{\sqrt{4\pi D}}$$

yields constant spread with velocity (slope) $c = 2\sqrt{D \log(R)}$, whereas the "fat-tailed" dispersal function

$$k(x) = \frac{\alpha^2 \exp(-\alpha\sqrt{|x|})}{4}$$

yields an increasing spread rate given by

$$x_t \to \frac{1}{\alpha^2}\left[t \log(R) + \log\left(\frac{\alpha^2 N_0}{4N_c}\right)\right]^2,$$

where N_c is the threshold level for detection and N_0 is the number of individuals initially released from a point. Compare these results with Figure 3.4. Details of the calculations are given in Kot et al. (1996).

numerical results, which showed a dramatic slowing of spread rates in the presence of a mild Allee effect. Assumptions of sexual reproduction and pair formation after dispersal make it highly unlikely that extremely long-range dispersers will find each other and be able to mate successfully. By way of contrast, the chance long-distance dispersal of even a single microbe may be sufficient to start an entire colony.

STOCHASTIC ASPECTS OF SPREAD

Up to this point we have thought about the spread of invading organisms in terms of the expected density of organisms as a function of space and time. Actual data from biological invasions indicate a high degree of patchiness; the invasion, itself, is less often a wave of invasion than a series of invaded patches that spread, coalesce, and spawn new patches. Often such patchiness is ignored when comparing models with data; calculation of the effective range radius (proportional to $\sqrt{\text{area invaded}}$) averages out the explicit spatial patchiness.

Although some patchiness is clearly a function of environmental variability, stochasticity also plays a role here. This can be seen by modeling the behavior of individual particles in a homogeneous environment via a "branching process" where particles have certain probabilities of reproducing and dying per unit time and can spatially redistribute. Monte Carlo simulations of the branching processes show that when the dispersal is long-distance and occurs over multiple space scales "patchy" spread results (Figure 3.5a) and the locations of the patches vary from simulation to simulation (see also Minogue 1989). The stochastic nature of the patchy versus smooth structure of an invasion plays an important role in what is actually observed. A single observer with limited observation range would miss the leading edge of the invasion entirely unless one of the patches happened to "land" close by.

Stochastic factors also govern the precision of predictions for the speed of invasion. For example, defining the invasion front as the location of the farthest dispersing particle, one can plot

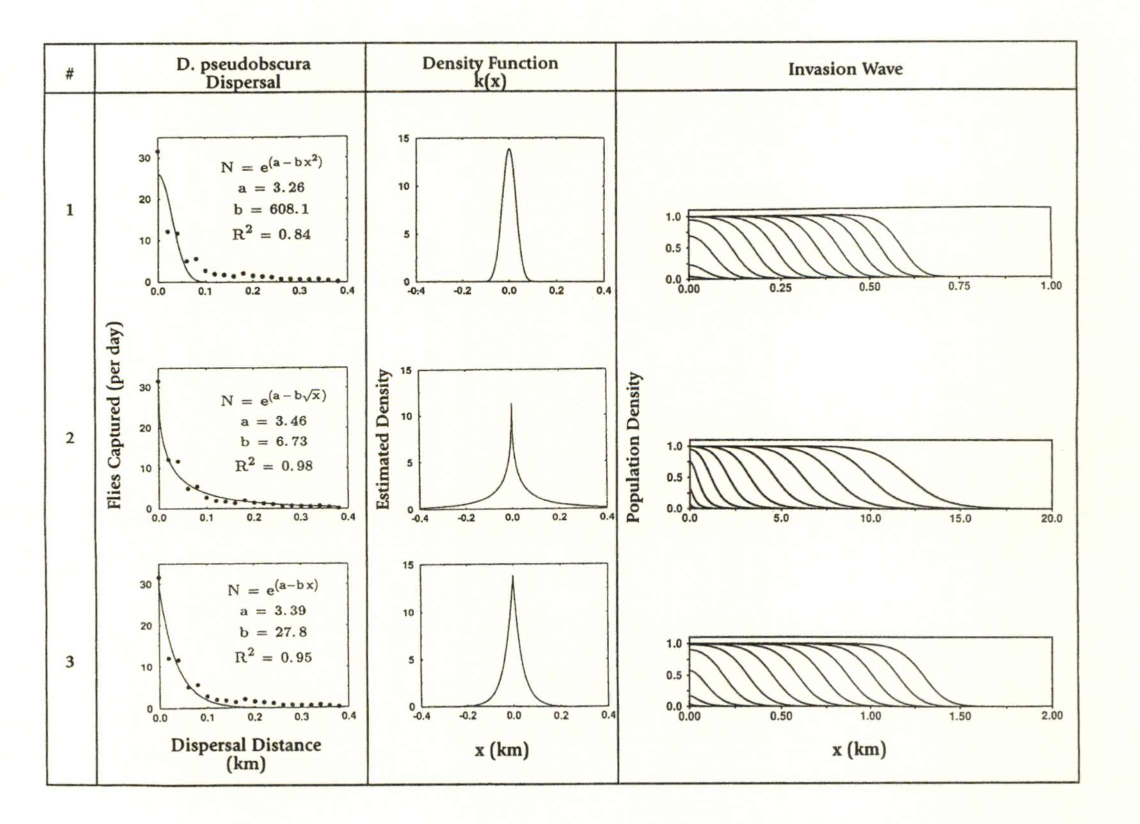
#
D. pseudobscura Dispersal
Density Function k(x)
Invasion Wave
1
2
3
Flies Captured (per day)
$N = e^{(a - bx^2)}$
a = 3.26
b = 608.1
R^2 = 0.84
$N = e^{(a - b\sqrt{x})}$
a = 3.46
b = 6.73
R^2 = 0.98
$N = e^{(a - bx)}$
a = 3.39
b = 27.8
R^2 = 0.95
Dispersal Distance (km)
Estimated Density
x (km)
Population Density
x (km)

location versus time for Monte Carlo simulations of a spreading population (Fig. 3.5c). While the simulation results vary about a line of fixed slope equal to the theoretical ARS, the dispersal function that is more leptokurtic has the higher variability, that is, this function yields a simulation result with locations, on average, farther from the line of fixed slope.

To understand invasion patchiness and variable invasion rates we need to come up with a way to measure the variability. One way to measure the intrinsic variability of an invasion process is in terms of statistics about the invasion process taken over an *ensemble*, or large number of different realizations of the invasion process under identical conditions. In defining an ensemble we have already run into a difference between models and the real world. The ensemble average makes sense for a computer model where one can run large numbers of Monte Carlo simulations and simply take the average over the large number of runs. However, a real-life invasion process cannot simply be repeated many times under identical conditions—most invasions only occur once under any given set of conditions. However, as shown below, ensemble averages still provide powerful tools with which to understand the intrinsic sources of variability.

FIGURE 3.4. Dispersal curves, redistribution kernels, and predicted invasions for *D. pseudoobscura*. The left column shows data for the number of *D. pseudoobscura* captured per trap day and a set of fitted functions. The middle column shows kernels resulting from the assumption that the direction of dispersal was chosen randomly, and the right column shows a numerically simulated solution to Equation 3.2 for growth governed by the Beverton-Holt curve with $R = 10$. The kernels have tails that range in fatness from the normal distribution (1) to the modified Weibull distribution (2) (whose tails are not exponentially bounded), with the Laplace kernel (3) as a leptokurtic kernel with tails of intermediate fatness. Note that although the kernels with exponential bounded tails (1 and 3) generate a wave with a constant ARS, the modified Weibull distribution (2) generates a wave that continues to accelerate as it becomes more and more shallow. (Based on Kot et al. 1996.)

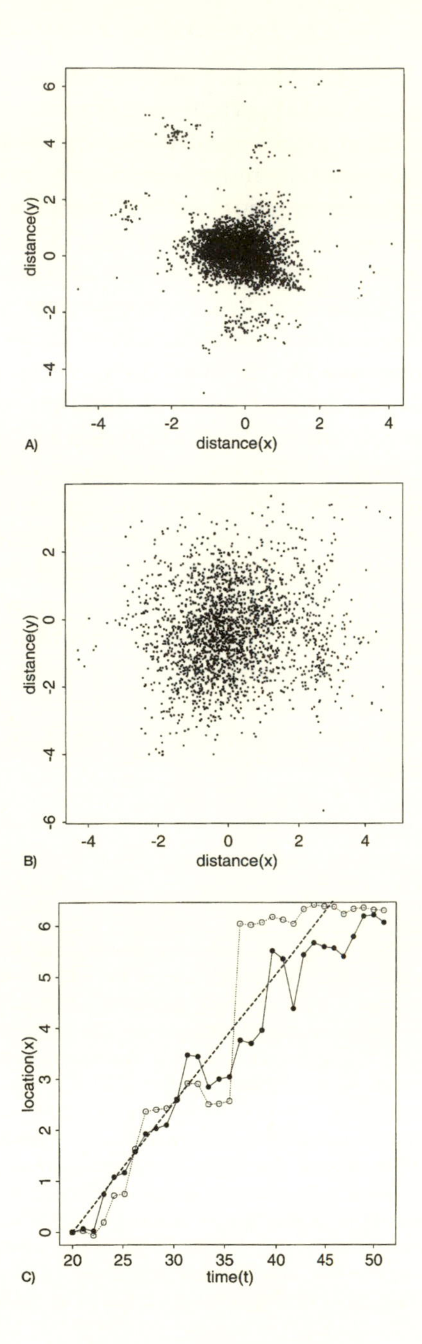
distance(y)
distance(x)
A)
distance(y)
distance(x)
B)
location(x)
time(t)
C)

Analyzing behavior in terms of the *ensemble*, we define the expected density of individuals $p_t(x)$ as the number of individuals per unit area, averaged over a large number of different realizations of the invasion process under identical conditions (ensemble average). The spatial variance is the squared deviation in the number of individuals per unit area away from the expected density, averaged over the ensemble. In other words, if the invasion process were entirely predictable, then every realization of the invasion process would be the same. Thus the number of individuals per unit area would not deviate from the expected density, and the spatial variance therefore would be zero. By way of contrast, a highly variable invasion process would have a large spatial variance. A related statistic, the spatial covariance between two points in space $c_t(x, y)$, is the deviation away from the expected density at the first point x times the deviation away from the expected density at the second point y, averaged over the ensemble. When the two points coincide, the spatial covariance simply becomes the spatial variance. The spatial covariance is thus a scaled measure

FIGURE 3.5. When dispersal events occur on markedly different spatial scales, populations spread in a patchy manner. Monte Carlo simulation of reproducing dispersing particles. Each time step: (1) particles have a Poisson number of offspring, with mean of 1.2; (2) offspring disperse with random distances drawn from the dispersal kernel, and angles are drawn from a uniform distribution; and (3) the parent particle then dies. Initially ten particles were released at $x = 0$. The distribution of particles is shown after thirty time steps. (A) Composite Laplace kernel

$$k(x) = \frac{p\sqrt{\delta}}{2}\exp(-\sqrt{\delta}\,|x|) + \frac{(1-p)\sqrt{\varepsilon}}{2}\exp(-\sqrt{\varepsilon}\,|x|)$$

with $\delta = 100$, $\varepsilon = 1$, and $p = 0.99$. (B) Composite Laplace kernel with $\delta = 12.8$, $\varepsilon = 4$, and $p = 0.99$. (C) When kernels have fatter tails there is more stochastic variation about the theoretically predicted ARS. The location of the particle dispersing farthest in the positive x-direction is plotted as a function of time. Open dots indicate case (A) in this figure, and solid dots indicate case (B). The dashed line has slope 0.2529, which is the theoretical ARS.

of spatial correlations within the invasion process. In other words, if deviations away from the expected density at the first point were entirely uncorrelated with deviations away from the expected density at the second point then the spatial covariance would be zero. Positive correlations in the deviations lead to a positive covariance, and negative correlations lead to a negative covariance.

With definitions of the spatial statistics in hand, our attempt to understand patchiness and variability of the invasion process initially focuses on following model predictions for the statistics in space and time as the invasion progresses.

The goal of following the spatial statistics over time and space can be achieved by solving dynamical equations that mathematically describe the progress of the statistics. Armed with assumptions of discrete reproduction and dispersal stages, and ignoring the effects of density-dependent interactions, a detailed derivation of the equations was made in Lewis and Pacala (in preparation). Inputs into the equations are the expected number of offspring per individual R, the variance in this number $\mathrm{Var}(R)$, and the dispersal function $k(x)$. Finally, outputs from the equations are expected density $p_t(x)$ and spatial covariance $c_t(x, y)$.

Numerical solutions to the equations are shown in Figures 3.6 and 3.7. Note that the invasion process is characterized by invasive waves of both expected density ($p_t(x)$) and spatial covariance ($c_t(x, y)$). These quantities can be translated into the mean m and variance σ^2 in the number of individuals in any arbitrary region of space and at any time step t. As we will show next, these are the very quantities needed to understand both patchiness within a wave and variability in spread rates.

A measure of patchiness is a comparison between the "mean crowding," or density of neighbors experienced by an individual, and the mean density, or expected number per unit area. With patchy spatial distributions the presence of one individual will increase the likelihood of nearby neighbors. The mean crowding index, given by Lloyd (1967), is

$$m^* = m + \left(\frac{\sigma^2}{m} - 1\right), \tag{3.8}$$

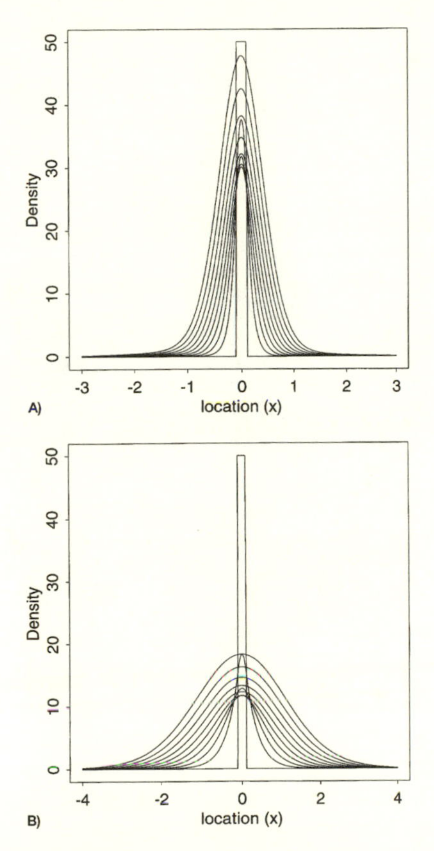

FIGURE 3.6. Expected density of individuals as a function of time. The equation governing the spread is simply a linear version of Equation 3.2. Initial conditions are consistent with ten individuals being released with random locations lying in the interval $-0.1 \leq x \leq 0.1$. Each adult has a Poisson number of offspring with mean $R = 1.2$. Solid lines show the expected density of individuals for the first ten time steps. (A) Composite Laplace kernel as in Figure 3.5a. (B) Composite Laplace kernel as in Figure 3.5b.

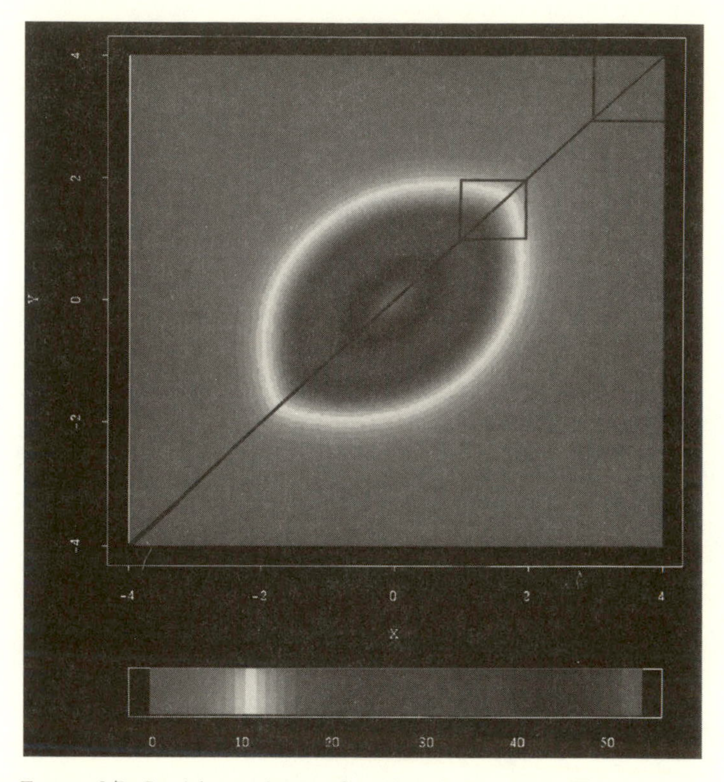

FIGURE 3.7. Spatial covariance of an invasion process. The equation governing the spread of spatial covariance has one contribution from siblings who had the same parents, and another contribution from more distantly related individuals who inherit spatial correlatives from their parents. Numerical solution of the spatial covariance equations for $t = 10$ with the kernel as described in Figure 3.5b. The initial covariance was taken to be identically zero. This is consistent with the initial spatial distribution of the individuals being Poisson with a mean density of fifty on interval $-0.1 \le x \le 0.1$ and being zero elsewhere. The spatial covariance function was integrated over $1 \le x, y \le 2$ to evaluate the patchiness index and over $2.98 \le x, y \le 4$ to evaluate the level of variation in the number of particles near the leading edge of the wave. The line $x = y$ indicates that the spatial covariance function is not defined at these values. See the text for further details of the calculations.

where m is the mean number of individuals in the region and σ^2 is the variance in the number. The ratio of mean crowding index to mean over a region $z_1 \leq x \leq z_2$,

$$\frac{m^*}{m} = 1 + \frac{\displaystyle\int_{\substack{z_1 \\ y \neq x}}^{z_2} \int_{z_1}^{z_2} c(x, y)\, dx\, dy}{\left[\displaystyle\int_{z_1}^{z_2} p(x)\, dx\right]^2}, \tag{3.9}$$

is an index of patchiness. Randomly distributed individuals have $c(x, y) = 0$ and thus $m^*/m = 1$. By way of illustration, choosing $z_1 = 1$ and $z_2 = 2$ yields $m^*/m = 1.719$ when the dispersal scales are strongly different (parameters as in Figure 3.6a) and $m^*/m = 1.158$ when the dispersal scales are closer (parameters as in Figure 3.6b). Thus, in this illustrative example, the patchiness index is higher when dispersal scales are strongly different.

Any given invasion will have a particular velocity of spread that differs from the expected "average" velocity. The amount of variation we can anticipate may be enormous—but fortunately this amount itself can be predicted. The variability in spread rate can be most easily understood by calculating the variance in the leading edge of the wave. Defining the leading edge of the wave as the region over which the expected number of particles is 0.5, by way of illustration, the leading edge for Figure 3.6a is given by $x > 1.48$, and the leading edge for Figure 3.6b is given by $x > 2.98$. The corresponding variances are 1.23 and 0.772, respectively, indicating a much higher variability for the case with markedly different spatial dispersal scales (compare with Figure 3.5). Although this approach does not directly calculate the variance in the *location* of the farthest dispersing particle (Figure 3.5c), it is of practical use; given a specified region that may soon be invaded, one can predict not only the expected number of invaders but also the variability in this estimate.

Thus the mathematical model eliminates the need to investigate stochastic variation by repeated Monte Carlo simulations for each dispersal function. It provides a means to directly calculate the patchiness index or variance in the leading edge of the wave for any given spatial redistribution function $k(x)$. The results illustrated here by Monte Carlo simulations (Figure 3.5) and sample calculations are also analyzed mathematically in detail by Lewis and Pacala (in preparation). Results indicate that for a given invasion speed, both patchiness of the invasion and variability in spread rates are governed by three factors:

1. Disparity in length scales present in the dispersal function $k(x)$. Leptokurtic dispersal functions give rise to highly variable spread rates and yield of patches on length scales that are many times the average dispersal distance.
2. Average number of offspring per individual R. If R is large then an individual has many offspring dispersing independently that tend to "average out" the stochastic nature of the dispersal process. This translates into less variability in spread rates and a reduced patchiness.
3. Variance in number of offspring per individual Var($\mathscr{R}$). If the number of offspring per individual is highly variable, then in the next generation spatial correlations are, on average, larger than if each individual had a fixed number of offspring. This translates into higher variability in the spread rates and an increased patchiness.

BIOLOGICAL SIGNIFICANCE OF STOCHASTIC ASPECTS

As illustrated in the previous section, stochastic factors play a role in the precision of predictions based on an ensemble mean (i.e., on the average behavior over a large number of invasions under identical conditions). For example, if one wishes to estimate the number of invaders arriving within a prescribed area and time, then a model couched in terms of the mean density of organisms can be integrated over the area to predict the mean number in the area but gives no indication of variance of that number. However, it is the variance that indicates how precise the prediction is: The square root of the variance $\sigma = \sqrt{\sigma^2}$ is the average deviation that a typical invasion process will have away from the expected value. In this case a stochastic integro-difference formulation yields estimates for both the mean and variance. When the variance is high, a large discrepancy between the observation of single invasion and the predictions of a model may be due to intrinsic stochastic variability in the invasion process rather than an inaccurate model. The precision with which one can estimate the number of invaders arriving within a prescribed area is also intimately tied to the Lloyd mean crowding index (Eq. 3.8), which, when compared to the mean density, leads to an index for patchiness (Eq. 3.9).

The results provide a new approach to understanding mechanisms governing the intrinsic variation in invasion rates. Leptokurtic dispersal functions lead to higher spatial covariance in the leading edge of the invasion, which in turn yields a larger variance in the number of individuals in the leading edge of the wave. In other words, the possibility of a very few individuals dispersing a very long distance will increase the variance of the number of individuals having dispersed past a certain point. Note that this variance is with respect to an ensemble of stochastic invasion processes occurring under identical conditions, whereas typical field observations are for a single invasion. The result of high variance is thus a concomitant decrease in the ability to precisely predict the yearly spread rate for a given invasion. Therefore it is the dispersal function that plays a crucial role in determining whether or not invasion rates can be precisely predicted. For example, the case where most propagules travel a short distance, but a few may be carried a very long distance (say, seeds carried by birds or by wind currents), leads to a low likelihood of estimating the true spread rate. This relates back to the failure of Fisher's model to predict the spread of the cereal leaf beetle discussed in the introduction. Andow et al. (1990) suggested that the reason for the failure might arise from possible macroscale jumps provided by air transport or human transport. The results here indicate that inclusion of these factors, while crucial for accuracy, would intrinsically lead to highly variable invasion rates.

SUMMARY

Returning to the general ecological issue of invasions, there are several key insights that have emerged from our analyses. First, invasion rates depend crucially upon rare, long-distance dispersal events, that is, upon the tails of the redistribution function. Thus when a disperser has different mechanisms for dispersal, the key mechanism for understanding invasion rates is the one that takes the disperser the farthest, even when only a small fraction move in this manner. For example, blue jays can transport acorns for distances of up to several kilometers, whereas other modes of dispersal, such as squirrels, transport the seed distances on the order of several hundred meters. It now appears it was the blue jay "factor" that led to the rapid recolonization of postglacial eastern North America by oak trees (Johnson and Webb 1989). However, to quote Johnson and Webb, "Despite these recent advances [in understanding dispersal mechanisms], Darwin's (1859) statement that 'we are as yet profoundly ignorant of many occasional means of transport' [of seeds and organisms] is still true." The challenge to biologists is to focus our attention where it can yield the most effect—on understanding those occasional, but crucial, long-distance means of transport.

Second, although classical invasion models may be able to accurately predict average rates of spread, a single invasion process will be very hard to predict with any precision if there is a stochastic rare long-distance component to the dispersal process. Here the stochastic models are crucial for quantitatively determining variability in the predictions. The basic ingredients needed for the model are the reproductive rate R, the variance in this rate, and the spatial redistribution function. The simplest output from the model is a prediction of the expected number of organisms in a specified region and the variance in that number. Such analyses demonstrate when even the "right" model is likely to do a poor job of estimating spread rates for a particular invasion.

Third, the spatial patterning of an invasion can be linked to the dispersal function in a way that allows one to predict patterns such as patchiness. It may even be possible to work backwards, estimating properties of the dispersal function given data on spatial patterning. This is an ongoing area of theoretical research. Whether the theory can actually provide useful information under field conditions remains to be tested.

The focus of this chapter has been quantitative analysis of the interplay between dispersal and chance in biological invasions. One key message is that if long-distance chance dispersal events dominate, then exact predictions of spread rate may be very difficult or impossible to obtain. This difficulty may not simply stem from problems in getting long-distance dispersal data; inexactness also comes from variability intrinsic to the actual invasion process. The heart of the problem lies in distinguishing between these sources of variability so as to deduce whether effort spent on measuring dispersal distances will pay off sufficiently in improved spread rate estimates. The alternative is that the stochastic factors dominate a process which no amount of sampling effort can resolve.

ACKNOWLEDGMENTS

The work on stochastic aspects of spread was started during a visit to the Levin and Pacala labs at Princeton. The visit was funded by the Alfred P. Sloan Foundation. Steve Pacala and George Hurtt are collaborators on this work. I thank Fred Adler, Julian Cook, and Mike Neubert for helpful discussions and feedback, and Peter Kareiva for insisting that I write this is English and for his help with the translation. This work was funded in part by a grant from the National Science Foundation (DMS-9457816).

PART II

Parasites, Pathogens, and Predators in a Spatially Complex World

Predators, parasites, and pathogens are always at risk of overexploiting their hosts or prey and in the process dooming themselves to starvation. Indeed, cycles and oscillations are an undercurrent in all consumer-resource interactions; this realization has led theoretical ecologists to ask why hosts or prey are not more commonly driven to extinction. The answer to this question has often been "space." In particular, spatial heterogeneity, spatial refugia, or simply the existence of a spatial dimension are thought to stabilize many inherently oscillatory systems. Although the importance of space to predator-prey systems is ingrained in ecological folklore, careful mathematical resolution of exactly how space alters these antagonistic interactions has been a relatively recent development. This part includes the major themes surrounding space and consumer-victim interactions but offers no easy answers. Although the ensuing examples are drawn primarily from host-parasitoid and host-pathogen systems, the concepts apply equally well to plant-herbivore and predator-prey interactions.

Hassell and Wilson summarize nearly two decades of progress on the dynamics of spatially distributed host-parasitoid systems. At the level of individual hosts within a population, spatial variation in exposure to risk is consistently stabilizing. However, when one connects subpopulations of hosts and parasitoids via dispersal, a staggering variety of patterns can emerge. One especially important finding is that even though the environment is homogeneous, locally unstable host-parasitoid interactions can be stabilized by dispersal. As one builds up multispecies models of hosts and parasitoids, the spatial patterning that develops can become even more intricate. Biologically the key result is the emergence of spatial segregation between highly mobile parasitoids and less mobile parasitoids. This spatial segregation effectively produces the "niche separation" required for coexistence, assuming there is a trade-off between competitive ability and dispersal ability in the parasitoids.

Holmes's chapter on disease dynamics is primarily concerned with how subtleties of space alter the threshold density of hosts, above which an epidemic will sweep through a population. Her approach involves lattice models in which the movement and fate of discrete individuals are tracked at fixed positions. She finds that when transmission is localized among discrete individuals occupying particular positions in space, epidemics die out more readily than they do for equivalent nonspatial models—because by chance a diseased individual may find itself isolated from susceptible hosts. Less obvious is the effect of discrete spatial neighborhoods on the period of disease cycles—periods are dramatically lengthened in spatial epidemics compared to well-mixed populations. One of the most intriguing aspects of Holmes's exploration of spatial effects is the challenge of identifying what is the fair nonspatial comparison for her lattice disease model. To answer clearly whether "space matters," one must have a nonspatial "control" model that isolates the effect of space from other differences in model formulation.

Ferguson, May, and Anderson provide a striking counterpart to the simplicity of Holmes's approach, imbuing a model of measles epidemics with several layers of spatial detail and age-structure. In contrast to Holmes, they follow populations of infected and susceptible individuals rather than track the fate of discrete individuals. The fascinating aspect of measles is the quality of data available—there is no better dataset on spatiotemporal dynamics for a disease than that which exists for measles in human populations, which means this modeling exercise probes the limits of what can be done using the best possible spatially structured data. Two striking points emerge:

1. Although nonspatial models approximate well the measle cycles in large populations, spatial models are clearly needed to reproduce the irregularities evident in small populations.
2. Even for this best-case scenario with respect to data availability, truly realistic spatial models are hindered by the difficulty of parameter estimation, suggesting they may not be a worthy goal for diseases in plant and animal populations, which are bound to be much more data-poor.

Finally, Antonovics, Thrall, and Jarosz describe the best empirical study to date of a spatially distributed plant-pathogen interaction and use a nested set of models to ask how consideration of space improves our understanding of this system. Unlike the other chapters in this part, this examination of a smut fungus that attacks *Silene* is first and foremost a field study, with models being a tool rather than a centerpiece. The Antonovics et al. contribution is also the only chapter in the book to deal explicitly with genetics and the interplay of ecology and genetics. We learn that genetic variation for disease resistance is much more likely to persist in a metapopulation than in a nonspatial *Silene*-smut interaction. Thus, both genetics and space matter, and in a way that depends on the details of the interaction between these two oft-neglected "complications." Although it might seem daunting that the main message of this contribution is that we must attend to both genetics and the spatial dimension, Antonovics et al. offer the good news that the field study of genetics and metapopulations for plants can be surprisingly tractable, given the right choice of study systems.

Several themes emerge from the four chapters in this part that will reappear throughout this book. First, patchiness and spatial patterning are readily generated in a homogeneous environment simply because dispersal is localized and species interactions are nonlinear. Second, the spatial patterning that emerges in a wide variety of lattice or cellular automata models often involves a segregation of species that affords the opportunity for a type of coexistence that would be impossible in a nonspatial world. Third, one does not always need spatially explicit models to capture or understand key effects of space (such as the stabilizing effect of heterogeneity in risk). Fourth, trade-offs between dispersal and competitive ability can yield a type of coexistence that requires a spatial expanse in which species segregate into separate portions of the landscape. Finally, a major challenge for theoreticians is the identification of what levels of detail (or conversely what degree of simplification) are needed in spatial models, and what the nonspatial "mean field" analogues are to spatially explicit models.

CHAPTER FOUR

The Dynamics of Spatially Distributed Host-Parasitoid Systems

Michael P. Hassell and Howard B. Wilson

INTRODUCTION

Insect parasitoids make up some 20% to 25% of all insect species (Godfray 1994) and are often a major cause of mortality to their insect hosts. That parasitoids have at least the *potential* to regulate their host populations is uncontested. Empirical observations in the laboratory have shown hosts maintained well below their carrying capacity by introduced parasitoids, biological pest control programs have shown the same effect in the field (e.g., Embree 1966; Waage and Hassell 1982), and the ecological literature is full of examples of potentially stable host-parasitoid models. But how often parasitoids *actually* regulate their hosts in the field, and what precise mechanisms are involved, is still widely debated. Ever since the classic experiments of Huffaker (1958), spatial heterogeneity in host-parasitoid and predator-prey interactions has been a prime candidate for counteracting the inherent instabilities predicted by the early models of Thompson (1924) and Nicholson and Bailey (1935) and the discrete-time equivalents of the Lotka-Volterra models (May 1973b). In this chapter, we explore the effects of spatial heterogeneity on the dynamics of host-parasitoid systems at a variety of different scales. In particular, we distinguish between *individuals* occurring in spatially distributed patches within their habitat (these will often be discrete food plants for the host insects), *local populations* characterized by more or less complete mixing of individuals at some point during the generation period, and *metapopulations* formed by a number of these local populations connected by *restricted* dispersal.

A GENERAL FRAMEWORK

In this chapter we shall follow the premise that the populations have discrete and synchronized generations. This is a tradition initiated mainly by entomologists (Thompson 1924; Nicholson 1933; Varley 1947) with insect hosts and their parasitoids in mind, both of which often have more or less discrete generations. We being by assuming that a host population with discrete generations is attacked by a population of specialist parasitoids in which the searching adult females are perfectly synchronized with the host stage that they attack. This leads to the following general framework:

$$N_{t+1} = \lambda N_t f(N_t, P_t), \tag{4.1a}$$

$$P_{t+1} = cN_t[1 - f(N_t, P_t)] \tag{4.1b}$$

where P and N are the population sizes of the adult female parasitoids and the susceptible host stage, respectively, in generations t and $t + 1$. In the host Equation 4.1a, the parameter λ is the *net* finite rate of increase of hosts in the absence of the parasitoids, which may be a constant or some density-dependent function of N_t. The unspecified function $f(N_t, P_t)$ gives the fraction of the N_t hosts escaping parasitism, so that $[1 - f(N_t, P_t)]$ therefore gives the fraction of hosts parasitized. Finally, c is the average number of adult female parasitoids emerging from each host parasitized. Clearly, these simple equations subsume a large amount of host and parasitoid biology, and the apparent simplicity of model (1) is deceptive. To be properly parameterized for a particular host-parasitoid system would require extensive life table information on both populations of a detail that has rarely been obtained.

The Nicholson-Bailey model is the best-known specific model of the form of Equation 4.1 (Nicholson 1933; Nicholson and Bailey 1935). The number of hosts encountered per female parasitoid per generation is linearly related to host density, where the slope of the relationship is the per capita searching efficiency of the parasitoids, a. These encounters are then distributed randomly among the hosts, all of which are equally susceptible, so that host survival is given by the zero term of the Poisson distribution,

$$f(P_t) = \exp(-aP_t). \tag{4.2a}$$

Notice that the per-capita searching efficiency is now given by

$$a = \frac{1}{P_t} \ln\left(\frac{N_t}{S_t}\right) \tag{4.2b}$$

where S_t are the number of surviving hosts in generation t and that it only requires estimates of the searching parasitoid density and the proportion of hosts they parasitize to be calculated in a particular situation. The properties of the Nicholson-Bailey model are familiar: There is a locally unstable host-parasitoid equilibrium given by $N^* = \lambda \ln \lambda/[c(\lambda - 1)a]$ and $P^* = \ln \lambda/a$, the slightest perturbation from which leads to oscillations of rapidly increasing amplitude. Although observed in a few laboratory experiments (e.g., Gause 1935; Burnett 1956; Huffaker 1958), such instability is hard to reconcile with the long-term persistence of natural systems, and much of the work on the dynamics of host-parasitoid systems over the past three decades has resolved around the identification of possible mechanisms promoting stability.

TWO-SPECIES LOCAL POPULATIONS

What Is Heterogeneity?

The term "heterogeneity" has become something of a "catchall" in ecology. Its use in this chapter, however, will be quite specific. Following Chesson and Murdoch (1986), we define "heterogeneity" in terms of the variation in risk of parasitism between different individuals in the host population.

In deriving a precise measure of this variability, Chesson and Murdoch (1986) first noted that the instantaneous risk of parasitism from Equation 4.2b is given by $aP\,(=\ln[N/S])$. Because the Nicholson-Bailey model assumes both homogeneous populations and a constant searching efficiency, all host individuals are exposed to the same number, P, of parasitoids and share the same risk of parasitism. If a were to vary between hosts (or groups of hosts), however, or some hosts were exposed to more searching parasitoids than were others, this risk of parasitism would be bound to vary within the host population. Let us consider the example of a host population divided into j groups, each within a distinct patch, where the host individuals within each group share the same risk of parasitism, $a_j P_j = \ln(N_j/S_j)$, but this risk may vary between groups. We now define the *relative risk of parasitism*, ρ, by comparing this risk *within* a group with the risk assuming a homogeneous Nicholson-Bailey interaction:

$$\rho = \frac{a_j P_j}{a\bar{P}} = \frac{\ln(N_j/S_j)}{\ln(\bar{N}/\bar{S})} \qquad (4.3)$$

where $\bar{P}$, $\bar{N}$, and $\bar{S}$ are averages per patch. The distribution of ρ values thus represents the variability in the risk of parasitism between hosts, and it is this variability that is the key to understanding the stabilizing effect of spatial or other forms of heterogeneity. For example, the Nicholson-Bailey model is recovered if ρ is uniformly one per patch, whereas the negative binomial model of May (1978) is obtained when ρ is gamma distributed (see below).

Spatial Patterns of Parasitism

In this chapter, and irrespective of the spatial scale, we consider heterogeneity from only one source—arising from aggregation of individuals within a spatially structured environment. We commence with a habitat in which spatially distinct patches of plants serve as food for a local population of an insect herbivore species, which in turn is host for a specialist parasitoid. The scale of this environment is such that both populations mix thoroughly during periods of dispersal so that individuals within patches have no degree of autonomous dynamics over a period of generations. Both hosts and parasitoids have discrete and synchronized generations. On emergence, the adult hosts disperse from their natal patch, and, having mated, the females then move among the patches ovipositing. The resulting immature hosts are confined to their respective patches until the next generation of adults emerge. Like the adult hosts, the emerging adult parasitoids mix thoroughly prior to foraging for host eggs, larvae, or pupae within the patches.

Suppose that the foraging parasitoids divide themselves evenly between the different patches, and that within any patch they have a linear functional response and a constant searching efficiency and encounter hosts at random. The fraction of hosts parasitized in each patch is now constant, and the Nicholson-Bailey model is exactly recovered. Such uniform risk of parasitism across patches, although convenient mathematically, is contrary to a wealth of evidence that has accrued from both laboratory and field studies. For example, in a total of 194 different examples reviewed by Lessells (1985), Stiling (1987), and Walde and Murdoch (1988), fifty-eight show patterns of parasitism that are positively density dependent (Figure 4.1a), fifty show inverse patterns (Figure 4.1b), and eighty-six show parasitism that is uncorrelated with host density per patch (Figure 4.1c,d). Pacala, Hassell, and May (1990) and Hassell, Pacala, May, and Chesson (1991) proposed that the direct and inverse patterns stem from what they called *host-density-dependent heterogeneity* (HDD) and the density-independent patterns from *host-density-independent heterogeneity* (HDI). This terminology corresponds to, but is more broadly descriptive than, Chesson and Murdoch's (1986) dichotomy between *pure regression models* in which the distribution of parasitoids (and hence of the resulting parasitism) is a deterministic function of local host density and *pure error models* in which the density of searching parasitoids in each patch is a random variable independent of local host density.

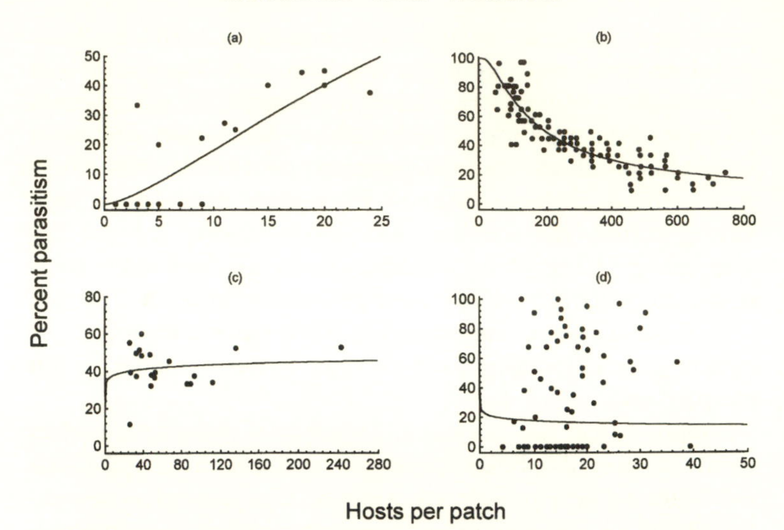

FIGURE 4.1. Examples of field studies showing percent parasitism as a function of host density from patch to patch. (a) Direct density-dependent parasitism of the scale insect *Fiorinia externa* by the eulophid parasitoid *Aspidiotiphagus citrinus* (McClure 1977). (b) Inverse density-dependent parasitism of gypsy moth (*Lymantria dispar*) eggs by the encyrtid parasitoid *Ooencyrtus kuwanai* (Brown and Cameron 1979). (c) Density-independent parasitism of the scale insect *Parlatoria oleae* by the aphelinid parasitoid *Coccophagoides utilis* (Murdoch et al. 1984). (d) Density-independent parasitism of the gall midge *Rhopalomyia californica* by the eulophid parasitoid *Testrasticus* sp. (Ehler 1987). The solid lines are means fitted using the maximum likelihood procedure in Pacala and Hassell (1991).

Quite different claims have been made for the potential importance to population dynamics of the different patterns in Figure 4.1. Some of the earlier theoretical literature emphasized the primacy of the density-dependent patterns in promoting stability (e.g., Hassell and May 1973, 1974; Murdoch and Oaten 1975), and this in turn influenced the interpretation of empirical data from the field (e.g., Cronin and Strong 1990; Hails and Crawley 1992; Ehler 1987). Later work, however, has argued that both the inverse and the density-independent patterns can be just as important as the density-dependent ones (Hassell 1984; Chesson and Murdoch 1986; Hassell and May

1988; Walde and Murdoch 1988; Pacala et al. 1990). Recently, Murdoch and Stewart-Oaten (1989) have challenged the whole basis of these studies, concluding from models in continuous time that density-independent patterns have no effect on stability and that density-dependent ones are most likely to be destabilizing. In the following sections, we make some attempt to unravel this rather convoluted literature.

Spatially Explicit Host-Parasitoid Models

The general model (Eq. 4.1) above can be applied to discrete interactions in a patchy environment, as long as the function $f(N_t, P_t)$ represents the *average*, across all n patches, of the fraction of hosts escaping parasitism. In any explicit representation of patchiness, therefore, $f(N_t, P_t)$ must depend on both survival from parasitism *within* patches as well as the spatial distributions of hosts and parasitoids *between* patches. We commence with the simplest scenario—a habitat with n patches, within each of which parasitism follows the Nicholson-Bailey model. The distribution of hosts and parasitoids from patch to patch can follow any distribution defined by α_i and β_i, the fractions of total hosts and total searching parasitoids, respectively, in the ith patch, giving

$$N_{t+1} = \lambda N_t \sum_{i=1}^{n} [\alpha_i \exp(-a_i \beta_i P_t)] \tag{4.4a}$$

$$P_{t+1} = cN_t \sum_{i=1}^{n} \{\alpha_i[1 - \exp(-a_i \beta_i P_t)]\} \tag{4.4b}$$

where a_i is the searching efficiency in patch i, and λ and c are as defined in Equation 4.1 (Hassell and May 1973, 1974). Notice that the Nicholson-Bailey model is recovered in the limit that the parasitoids treat all patches equally (i.e., where there are P_t/n parasitoids in each patch). However, once the parasitoids are unevenly distributed, the risk of parasitism will vary between patches. Stability in this model hinges on both host and parasitoid distributions as well as the host rate of increase. Two specific examples illustrate this.

A MODEL WITH HDD HETEROGENEITY. Hassell and May (1973) explored a particular case with simple descriptions of host and parasitoid distributions. The hosts were assumed to have a single high-density patch with the remainder equally divided between the $n - 1$ other patches. The searching parasitoids were assumed to respond to local host densities according to a simple power function

$$\beta_i = w\alpha_i^{\mu} \quad (4.5)$$

where α_i and β_i are as defined above, μ is a measure of how much the searching adult parasitoids aggregate in patches of high host density, and w is a normalization constant such that the β_i values sum to unity. The searching parasitoids are thus distributed evenly across patches if $\mu = 0$ and increasingly tend to aggregate in patches of high host density as μ rises until, in the limit $\mu \to \infty$, they all congregate in the single patch of highest host density leaving the remainder as complete refuges. Following the terminology of MacArthur and Levins (1964) and MacArthur (1968), the parasitoids are perfectly "fine-grained" in their distribution when $\mu = 1$ (i.e., equal fractions of hosts and parasitoids in each patch) and increasingly "coarse-grained" as μ increases or decreases. Finally, if $\mu < 0$ the parasitoid distribution is reversed, and their local abundance is now inversely correlated with host density per patch.

The dynamics of model 4.4 with β_i and α_i given by the function 4.5 above may be summarized as follows:

1. Local stability does *not* require the distribution of parasitism to be directly density dependent; inverse density dependence can be just as powerful a mechanism promoting population stability. Which contributes more depends on the details of the host distribution (Hassell 1984).
2. The equilibrium host population, N^*, tends to be at its lowest when host and parasitoid distributions are most closely correlated (i.e., $\mu \approx 1$) and to increase markedly as μ gets greater or smaller (Hassell 1984; Walde and Murdoch 1988).

3. Outside the locally stable regions, at both high and low host rates of increase, a range of interesting cyclic and chaotic dynamics occurs (Rohani, Godfray, and Hassell 1994).

A MODEL WITH HDI HETEROGENEITY. There has been much more emphasis on understanding and modeling HDD than HDI heterogeneity, partly by population ecologists who for a long time have had a much clearer view of how spatial density dependence promotes population stability (e.g., Murdoch and Oaten 1975; Beddington, Free, and Lawton 1978; Hassell 1978; Comins and Hassell 1979) and partly by behavioral ecologists wishing to make the link between foraging behavior and population-level processes (e.g., Goss-Custard et al. 1995; Bernstein, Kacelnik, and Krebs 1988, 1991). In this section we consider a specific example generating HDI.

Let us consider the example where (1) hosts are distributed across patches, according to any distribution whatsoever; (2) the distribution of searching parasitoids in patches is *unrelated* to this host distribution but is *aggregated* from patch to patch, in this case following a gamma distribution; and (3) parasitism within patches is again random with a constant per-capita searching efficiency. The fraction of hosts escaping parasitism is now given by

$$f(P_t) = \int_0^\infty g(\varepsilon)\exp(-aP_t\varepsilon)\,d\varepsilon. \tag{4.6}$$

Here $g(\varepsilon)$ is the gamma probability density function for parasitoids per patch with unit mean and variance $1/\theta$ (θ is a positive constant governing the shape of the density function) and a is the usual per-capita searching efficiency of the parasitoid. In each patch, therefore, host survival by $P_t\varepsilon$ randomly searching parasitoids is given by the zero term of a Poisson distribution with mean $aP_t\varepsilon$ (Murdoch et al. 1984; Chesson and Murdoch 1986). Exactly the same derivation would apply were the parasitoids uniformly distributed across patches with searching efficiency, a, varying according to a gamma distribution (Bailey, Nicholson, and Williams 1962). In both cases the effect is a gamma distribution in the host likelihood of parasitism (Kakehashi, Suzukin, and Iwasa 1984). Model 4.1 with f defined by model 4.6 is stable provided the HDI aggregation of adults is sufficient for $\theta < 1$.

A convenient feature of Equation 4.6 is the way it reduces exactly to May's (1978) familiar negative binomial model for parasitism, where the host survival is given by

$$f(P_t) = \left(1 + \frac{aP_t}{k}\right)^{-k} \tag{4.7}$$

and where the exponent k defines the degree of aggregation from the negative binomial distribution. By identifying k with θ above, Equations 4.6 and 4.7 become identical (Chesson and Murdoch 1986; Pacala et al. 1990), and hence the local stability criterion becomes $k < 1$. Thus, as well as being a purely phenomenological means of introducing stabilizing heterogeneity into host-parasitoid models (Godfray and Hassell 1989; Godfray, Hassell, and Holt 1994), the negative binomial model for parasitism also has a detailed and precise origin from a particular spatial model.

In short, density-independent patterns of parasitism can be a powerful means of promoting stability in these interactions. We now turn to a more unified way of representing the impact of HDD and HDI on population stability.

A More Unified Approach

An alternative way of expressing the stability criterion $k < 1$ from the HDI model above makes use of the square of the coefficient of variation (CV^2 = variance/mean^2), which in this case is simply given by $1/k$. Hence, the stability condition $k < 1$ is identical to the condition $CV^2 > 1$ (May 1978; Chesson and Murdoch 1986; Hassell and May 1988); in other words, the host-parasitoid interaction is stable if the distribution of searching parasitoids per patch in each generation, measured as the square of the coefficient of variation (CV^2), is greater than one (the reason for expressing this as the CV^2 rather than just the CV is apparent from Equation 4.8 below).

With this specific example in mind, Pacala et al. (1990) and Hassell, Pacala, May, and Chesson (1991) set out to determine whether such a simple criterion for stability (with CV^2 made less scale dependent by weighting the number of searching parasitoids in a patch by host density) might apply more generally across the spectrum of discrete-generation host-parasitoid models. They analyzed five modes of the general form of model 4.1 to see how well the stability criterion, $CV^2 > 1$, applies. For example, their Model I is the HDI example above, in which $CV^2 > 1$ exactly represents the condition for stability. From the completely uncorrelated densities of hosts and searching parasitoids in the HDI Model I, their Model II goes to the other extreme of a pure HDD model in which there is a perfect correlation between the two. The parasitoids can thus deterministically track the patch-to-patch variation in local host density. Specifically, they assumed an aggregated host distribution given by the gamma distribution, and parasitoid aggregation per patch described by Equation 4.5. The stability criterion, $CV^2 > 1$, that was exact for Model I, is now only approximate but still a good approximation if values of λ are low and the hosts highly aggregated. Models I and II make specific assumptions about the host and parasitoid distributions and represent end points on the continuum between HDI and HDD models. Model III is much more general: (1) The distribution of relative host numbers in patches can take any form whatsoever as long as it does not change with total host density, and (2) the parasitoid distribution is defined by an arbitrary function that combines HDD and HDI aggregation. In this case, the $CV^2 > 1$ rule applies well if on average there are many parasitoid visits per patch but not so well if the average number of visits per patch is small (see Hassell, Pacala, May, and Chesson 1991 for full details).

Applications to Field Data

A useful feature of the $CV^2 > 1$ rule is that it is readily decomposed into its constituent parts of HDD and HDI heterogeneity, both of which can be quantified directly from empirical data collected in the field. Pacala and Hassell (1991) began by assuming that the distribution of searching parasitoids is described by the same regression function (Eq. 4.5), to which is added the gamma distributed residual, ε, as in the HDI model. HDD heterogeneity is thus governed by the magnitude of μ (direct density dependence if $\mu > 0$ and inverse density dependence if $\mu < 0$), and HDI heterogeneity depends on the magnitude of the variance of ε. Full details of how this leads to the survival function $f(N_t, P_t)$ and the analysis of the resulting model are given in Pacala and Hassell (1991). The key conclusion to emerge is that the $CV^2 > 1$ rule may be approximated as

$$CV^2 \approx C_I C_D - 1 \tag{4.8}$$

where C_I is the HDI component given by $C_I = 1 + \sigma^2$ in which σ^2 is the variance of ε, and C_D is the HDD component given by $C_D = 1 + V^2\mu^2$ in which V is the weighted coefficient of variation of the host density per patch. The $CV^2 > 1$ rule therefore applies if $C_I C_D > 2$, which can arise from HDI alone if $C_I > 2$, HDD alone if $C_D > 2$, or some combination of the two.

Applying the $CV^2 > 1$ rule to field data would be straightforward if there were data on the actual distribution of searching parasitoids (but see Waage 1983; Casas 1988, 1989; Driessen and Hemerick 1991; Jones, Godfray, and Hassell 1996). For example, Driessen and Hemerick (1991) were able to observe the foraging of *Leptopilina clavipes*, an eucoilid parasitoid of *Drosophila* larvae in stinkhorn fungi, and estimated the CV^2 directly from the mean and variance of the distribution of parasitoid observations to fall between 0.23 and 0.53. Reliable observations on the distribution of foraging effort by parasitoids are, however, very difficult to make in the field; it is much easier to record the levels of parasitism across a number of patches, merely by sampling hosts, and dissecting or rearing through the samples. At a level "once removed" from the searching adult parasitoids, however, such data can still be used to reflect the distribution of the adults as long as the parasitoid functional response within a patch is known. Pacala and Hassell (1991) show how this step can be made if a type I functional response is assumed. They developed a maximum likelihood method for estimating the key parameters of C_I and C_D from data on levels of parasitism and host density per patch, from which the estimated CV^2 is then obtained (see Ives 1992 for the analysis assuming type II functional responses). Some examples are given in Figure 4.1.

From an analysis of sixty-five examples in the literature where spatial patterns of parasitism have been recorded, the following results emerge. In eighteen of the cases $CV^2 > 1$, indicating levels of heterogeneity that, were they consistent from generation to generation, should be sufficient to stabilize the host-parasitoid interaction. Furthermore, in fourteen of these cases $C_I > C_D$ and $C_I > 2$ indicating that HDI heterogeneity alone is sufficient to make $CV^2 > 1$. Finally, in a further eight cases, there were values of CV^2 that, although less than one, could be promoting stability to some significant degree ($CV^2 > 0.6$). Many caveats need to be attached to this analysis (see Taylor 1993 for an excellent review). For instance, most of the examples above come from single-site studies carried out within a single generation. The extrapolation to population dynamics is made by assuming that this "snapshot" of estimates of CV^2 and host distributions is typical of patterns from generation to generation. Thus, to be applied to any particular system, the CV^2 needs to be estimated over a period of time (e.g., Jones, Hassell, and Pacala 1993; Reeve, Cronin, and Strong 1994).

The most interesting suggestion to emerge from this body of work, and contrary to what is still the popular view, is that although both density-dependent and density-independent patterns of parasitism can contribute to the stability of host-parasitoid interactions, it is the density-independent ones that are far the more important.

An Interesting Debate

This conclusion, however, is very different from that drawn recently by Murdoch and Stewart-Oaten (1989). They model a host-parasitoid interaction in continuous time with perfectly overlapping generations and with infinite dispersal rates. Both

HDI and HDD aggregation are modeled in terms of the variance in the parasitoid distribution and covariance in the distributions of hosts and parasitoids. Within this framework, HDI heterogeneity has *no effect* on stability at all, whereas HDD heterogeneity contributes little to stability or, more likely, is *destabilizing*. The stark contrast between these two sets of conclusions has made for an interesting debate (Godfray and Pacala 1992; Ives 1992; Murdoch et al. 1992).

In an attempt to clarify how much it is the presence or absence of within-generation redistribution that is at the core of these very different predictions, Rohani et al. (1994) started with a discrete-generation model, but then allowed more or less continuous redistribution of the parasitoids (which could be HDD or HDI) during the generation period. In this respect the model is in the same spirit as the optimal foraging model of Comins and Hassell (1979) but differs in having explicit age structure of the populations (see Godfray et al. 1994 for full details) and in allowing greater flexibility of parasitoid foraging behavior ranging from density-dependent to density-independent aggregation among the patches. The juvenile hosts, on the other hand, do not redisperse within a generation; as in the majority of insect species, they are relatively immobile and remain in their natal patches.

The conclusions from this model are relatively straightforward: (1) The stabilizing effect of HDI aggregation is not influenced at all by within-generation parasitoid movement; the aggregated distribution of parasitoids across patches is preserved as the within-season interaction proceeds. (2) This is not the case, however, with HDD aggregation, where within-generation redistribution certainly reduces the impact of a density-dependent parasitoid foraging strategy, compared to a situation with a single parasitoid dispersal event in each generation followed by no subsequent movement. Just as with the optimal foraging model of Comins and Hassell (1979), the differences between patches in their risk of parasitism at the start of the season (due to parasitoids aggregating in the high-host-density patches) are progressively reduced by within-generation movement. Having said that, there remain many examples in the literature where the spatial distribution of parasitism is still strongly density dependent when measured after all redistribution of the foraging parasitoids has occurred (e.g., Figures 4.1a,b), and in these cases the CV^2 analysis remains appropriate.

This analysis suggests, therefore, that HDI aggregation is not at all influenced by within-season movement. The importance of HDD, however, depends on the strength of the covariance between parasitism and host density per patch at the end of the season after all parasitism has occurred. This is likely to be greatest in species in which individual parasitoids make relatively few patch visits during their searching lifetime (e.g., Driessen and Hemerik 1991). These conclusions also have interesting implications for the spatial scale at which heterogeneity may be exerting its maximal impact. Density-dependent aggregation to patches of high host density, by its very nature, operates on a spatial scale determined by the dispersal and foraging behavior of the individuals. Although strongly favored by natural selection, its impact on dynamics may well turn out to be slight. Density-independent aggregation, on the other hand, is likely to operate at larger spatial scales reflecting environmental processes operating on a larger spatial scale than that of local parasitoid and host dispersal. In this case, the effects of heterogeneity on dynamics may need to be explored much more at the metapopulation level, as advocated by Murdoch et al. (1992). It is to this metapopulation scale that we now turn.

TWO-SPECIES METAPOPULATIONS

Metapopulation models for predator-prey and host-parasitoid systems have developed in quite different ways (see reviews by Taylor 1988, 1990, 1991; Gilpin and Hanski 1991; Nee, May, and Hassell 1997). Some are based on the original formulation of Levins (1968); all areas of suitable habitat are equally in contact with each other and are either empty or are occupied by one or the other, or both, of the species. The models record the rate of change of these states as a function of colonization and extinction parameters. Others are spatially explicit so that each habitat containing a local population has specific neighbors. These "cellular" models may also record different states of habitat occupancy (e.g., Maynard Smith 1974; Wolfram 1984; Wilson, De Roos, and McCauley 1993; Durrett and Levin 1994a,b; Rand, Keeling, and Wilson 1995; Rand and Wilson 1995), or they may deal explicitly with the dynamics of the local populations (e.g., Reeve 1988; Hastings 1990; Taylor 1990; Comins, Hassell, and May 1992; Rohani et al. 1994; Swinton and Anderson 1994). This section deals primarily with the latter. Our main aim is to explore how limited dispersal linking local populations can introduce metapopulation dynamics that differ from the dynamics of the individual local populations. We then extend this to examine a range of three-species systems, looking in particular at how the invasion and persistence properties change in moving from isolated local populations to metapopulations.

We begin with the following scenario: an environment subdivided into local habitats that contain the food plant for an herbivorous insect with discrete generations. The emerging adult insects either remain to oviposit in their natal habitat or move to one of the neighboring areas and oviposit there. Dispersal therefore only occurs between adjacent local populations. The remainder of the life cycle is spent within the areas where the individuals were born. The larvae of these insects are hosts for a specialist parasitoid species whose life cycle is synchronized with that of its host. Like the adult hosts, the adult female parasitoids either remain where they emerged or disperse to one of the surrounding habitats. Such a scenario therefore differs importantly from the patchy system described earlier in that only some hosts and parasitoids disperse in any one generation, and those that do move outwards in a diffusive kind of way rather than mixing thoroughly and dispersing throughout the habitat.

This environment is represented here in a very simple way: as a two-dimensional square grid of cells of arena width n, each representing a habitat for a local population. Such models with discrete time and space but continuous population states have been dubbed "Coupled Map Lattices" (Kaneko 1992; Solé, Bascompte, and Valls 1992). Within each generation there are two distinct phases: (1) a dispersal phase and (2) a phase when the local population reproduces and matures within its patch. Reproduction can thus occur following, or prior to, migration, but not at the same time; this would require a different and detailed formulation of the model to keep track of those individuals that migrated and those that remained developing within the patch (Hassell et al. 1995; Bascompte and Solé 1995; Rohani et al. 1994). In the dispersal phase, a certain fraction of adult hosts, μ_N, and a fraction of adult female parasitoids, μ_P, leave the patch from which they emerged, while the remainder stay behind to reproduce in their original patch. The dispersing individuals, rather than dispersing throughout the area, spread outwards to colonize equally the eight nearest neighbors of the patch from which they emerge. Longer-range dispersal can only occur through repetition of these single-patch movements over multiple generations. The equations for the dispersal stage in each patch are

$$N'_{i,t} = (1 - \mu_N)N_{i,t} + \mu_N \overline{N}_i \tag{4.9a}$$

$$P'_{i,t} = (1 - \mu_P)P_{i,t} + \mu_P \overline{P}_i. \tag{4.9b}$$

Here $N_{i,t}$ and $P_{i,t}$ are the predispersal host and parasitoid population densities in patch i at time t, $N'_{i,t}$ and $P'_{i,t}$ are densities after dispersal, and $\overline{N}_{i,t}$, $\overline{P}_{i,t}$ are the average host and parasitoid populations over the eight nearest neighboring patches. Slightly different definitions for $\overline{N}_{i,t}$ and $\overline{P}_{i,t}$ apply for patches along the boundary of the arena, depending on whether one assumes cyclic, absorbing, or reflective boundary conditions. Cyclic boundaries have opposite edges of the arena effectively joined together. This is obviously unrealistic but has the advantage that all patches are dynamically equivalent, with no edge effects. With absorbing boundaries, individuals dispersing across the boundary are assumed to be lost, and with reflective boundaries, dispersing individuals are prevented from crossing the boundary and remain in the edge patches. In all the results presented here, the boundary conditions make little difference, except that simulations with cyclic boundaries tend to produce more symmetrical spatial patterns of abundance.

To lay bare the effects of limited dispersal alone, we keep the within-habitat phase of the interaction at its simplest. All habitats are assumed to be identical in "quality," the hosts have a constant rate of increase, and local parasitism is given by the negative binomial model:

$$N_{i,t+1} = \lambda N'_{i,t} f(P'_{i,t}) \tag{4.10a}$$

$$P_{i,t+1} = c N'_{i,t} [1 - f(P'_{i,t})] \tag{4.10b}$$

where $f = (1 + aP'_{i,t}/k)^{-k}$. First we note that the local (as opposed to the global) stability properties of this model are broadly the same as in the homogeneous equivalent, although the stability properties can differ, at least for the one-dimensional version of the spatial model, when the host and parasitoid dispersal rates are strongly asymmetrical (Hassell et al. 1995; Rohani, May, and Hassell 1996). In other words, the introduction of metapopulation structure has little or no effect on the local stability properties of the system, and thus model 4.10 has the same local stability properties as the negative binomial HDI model discussed above (i.e., locally stable if $k < 1$). This result makes sense intuitively: Provided that the environment is uniform, all local populations at equilibrium will have the same density, and movement to and from local populations will be in balance. A number of factors will, of course, confound this simple conclusion. Most obviously, varying habitat "quality" affecting the demographic rates of the local populations is bound to introduce different dynamics depending on the nature of the heterogeneity that is imposed. But even in homogeneous environments, moving to a metapopulation may change the stability properties under some conditions, for instance, if the hosts have a density-dependent rate of increase (Reeve 1988).

It is outside the region of local stability (i.e., $k > 1$) that such metapopulations show their most markedly different dynamics (Hassell, Comins, and May 1991; Solé et al. 1992; Bascompte and Solé 1994). Broadly, these are characterized by the unstable local populations tending to fluctuate asynchronously, enabling the metapopulation as a whole to persist much more readily than the comparable spatially homogeneous model (Taylor 1988, 1990, 1991; Comins et al. 1992; Swinton and Anderson 1994; Hassell, Comins, and May 1994). Arena size is clearly an important factor in this persistence, and we return to this later. Associated with this persistence are also some striking and varied spatial patterns of local population abundances (Figure 4.2). These have been labeled as "spatial chaos," "spirals," and "crystal lattices" (Hassell, Comins, and May 1991; Comins et al. 1992). Figure 4.3 gives an example of the approximate boundaries for these different patterns in relation to the fractions of hosts and parasitoids dispersing. The *spiral structures* are characterized by the local population densities forming spiral waves that rotate in either direction around relatively immobile focal points, while the combined metapopulation exhibits what appear to be stable limit cycles (Figure 4.4a). A time series taken from a particular local population would be characterized by fairly regular cycles produced by the regular "passage" of the arm of the spiral through the habitat, unless at the foci of the spiral where the population densities remain relatively constant. Local populations "swept" by the trailing arms of the spiral fluctuate with increasing amplitude the farther they are away from the focus. Despite the order in these patterns, they are still apparently chaotic since the position and number of focal points vary slowly with time in nonrepeating patterns. The region of *spatial chaos* is characterized by much more complex sets of intersecting wave fronts, such that the host and parasitoid populations fluctuate from cell to cell with no long-term spatial organization. Randomly oriented wave fronts are observed, but each persists only briefly. The total metapopulation generally remains within narrow bounds, but occasional large excursions are observed (Figure 4.4b). Finally, the rather extreme combination of very low host dispersal and very high parasitoid dispersal gives a small region of persistent *crystal lattice-like* structures in which relatively high-density cells occur at a spacing of approximately two grid units, and the metapopulation as a whole is stable (Figure 4.4c).

Alternatively, the entire metapopulation may go extinct in this model for several reasons. First, extinctions become increasingly likely as the arena width n gets smaller (see Figure 4.5). In the limit of very small arena widths ($n \leq 3$) and all individuals dispersing from their respective patches ($\mu_N = \mu_P = 1$), local and metapopulation models become the same; for example, with random parasitism ($k \rightarrow \infty$) within local populations, the Nicholson-Bailey model is recovered for the metapopulation as a whole, and the global dynamics are thus always unstable. The host and parasitoid dispersal rates also influence the probability of extinction for a given arena size. Generally, as dispersal increases, adjacent patches become more synchronized, and the host and parasitoid become more autocorrelated in space. The characteristic spatial scale of the dynamics increases, the overall variability between patches decreases, and so the metapopulation as a whole becomes less persistent. Failure to persist in small arenas is thus associated with insufficient space in which to fit a self-maintaining pattern. "Persistence" here is arbitrarily defined as the population size of each species remaining above a certain extinction threshold for at least two thousand generations. This, however, does not imply that the population will persist indefinitely; there is always a small, constant probability of the host or parasitoid going extinct at each time step. Increasing n (up to a threshold size) will decrease the probability of extinction but never eliminate it.

Fragmentation of the whole environment, in much the same vein as reducing the overall area, runs the risk of disrupting the whole metapopulation dynamics, either by reducing the number of local populations (cells) below some level required for the combined metapopulation to persist, or by interfering with the dispersal required to link the locally unstable local populations, often leading to population outbreaks as the ability of the parasitoid to regulate the host is disrupted (Hassell, Godfray, and Comins 1993). The extent of this depends to a large extent on the characteristic spatial scale of the dynamics. Thus parameter combinations producing large-scale spirals are especially vulnerable to shrinking arena sizes, whereas interactions producing chaotic spatial patterns are less vulnerable.

(a)

(b)

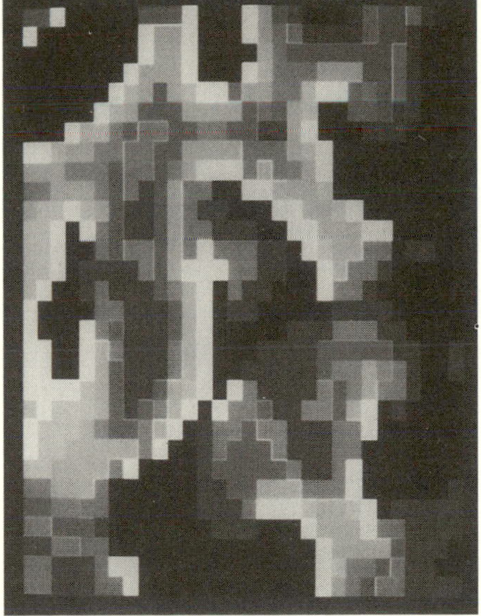

FIGURE 4.2.

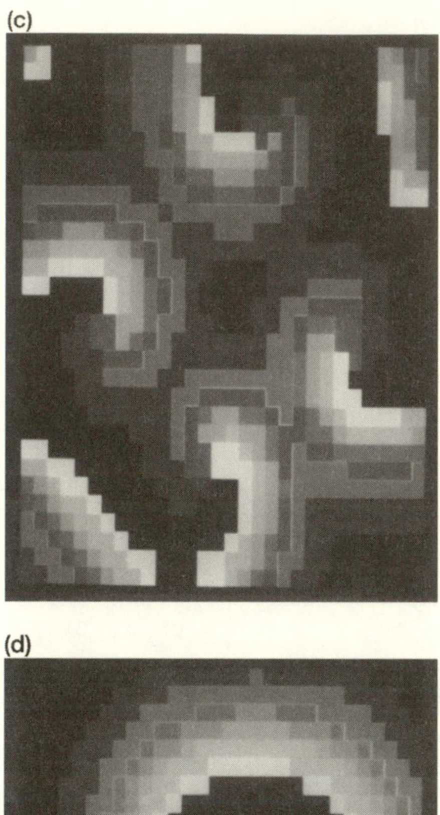
(c)

(d)

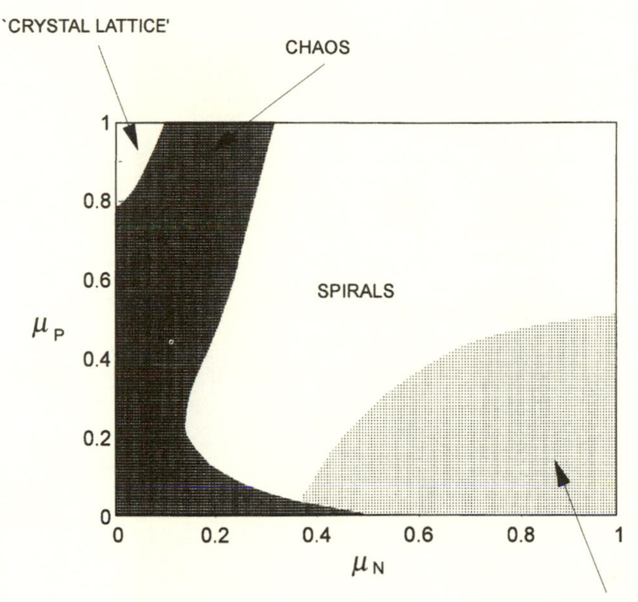

FIGURE 4.3. Diagram showing how the spatial dynamics in the host-parasitoid metapopulation models depends upon the relative host and parasitoid dispersal rates, μ_N and μ_P, respectively, for $n = 30$ and $\lambda = 2$. The boundaries are obtained by simulation and are approximate. The hatched areas indicate the regions in which the spatial pattern is chaotic. The shaded areas represent parameter combinations for which the persistent spatial pattern is unlikely to be established by starting the simulation with a single nonempty patch. (From Comins et al. 1992.)

FIGURE 4.2. Photographs illustrating the different patterns of spatial dynamics obtained from metapopulation model (4.10) with $k \to \infty$. Each photograph is a snapshot in time with the shading representing different relative abundances of hosts and parasitoids within a patch. (a) The "crystal lattice" pattern obtained for $\mu_P = 1$ and $\mu_N = 0.05$. (b) Spatial chaos obtained for $\mu_P = 0.89$ and $\mu_N = 0.2$. (c) Small spirals obtained for $\mu_P = 0.5$ and $\mu_N = 0.5$. (d) Large spirals obtained for $\mu_P = 0.89$ and $\mu_N = 1$ ($\lambda = 2$ throughout). (In part after Hassell, Comins, and May 1991.)

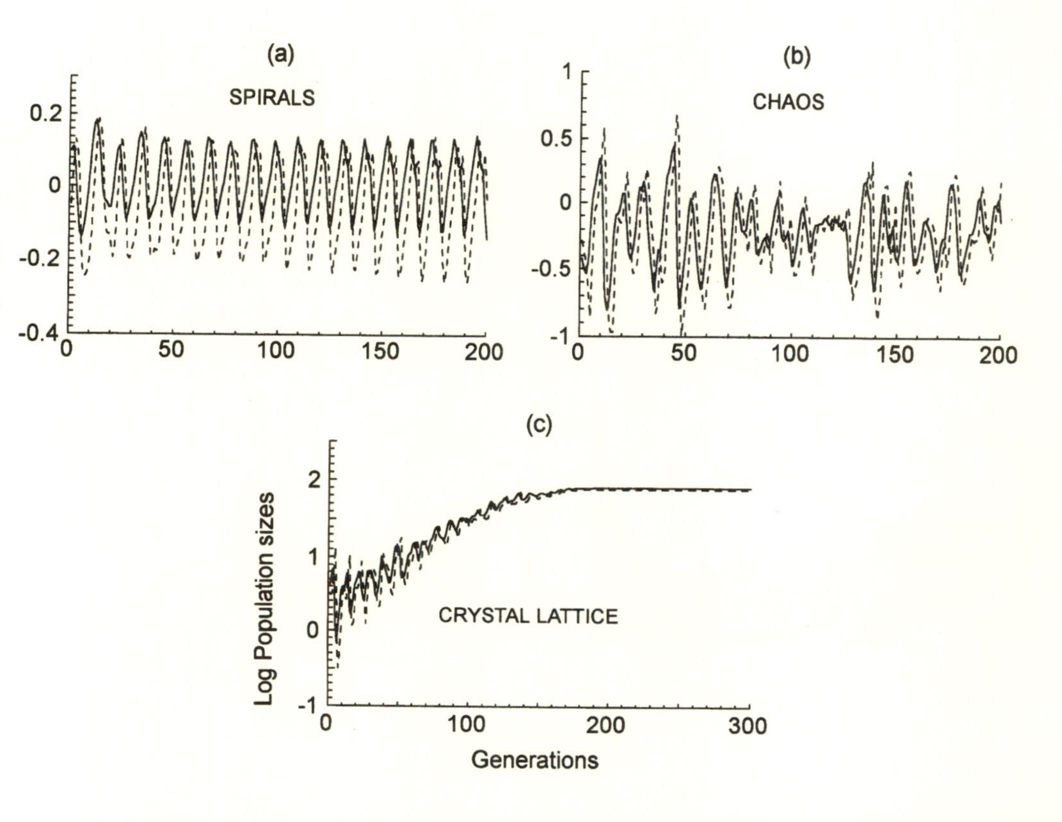
(a)
SPIRALS
Log Population sizes
0.2
0
-0.2
-0.4
0
50
100
150
200
(b)
CHAOS
1
0.5
0
-0.5
-1
0
50
100
150
200
(c)
CRYSTAL LATTICE
Log Population sizes
2
1
0
-1
0
100
200
300
Generations

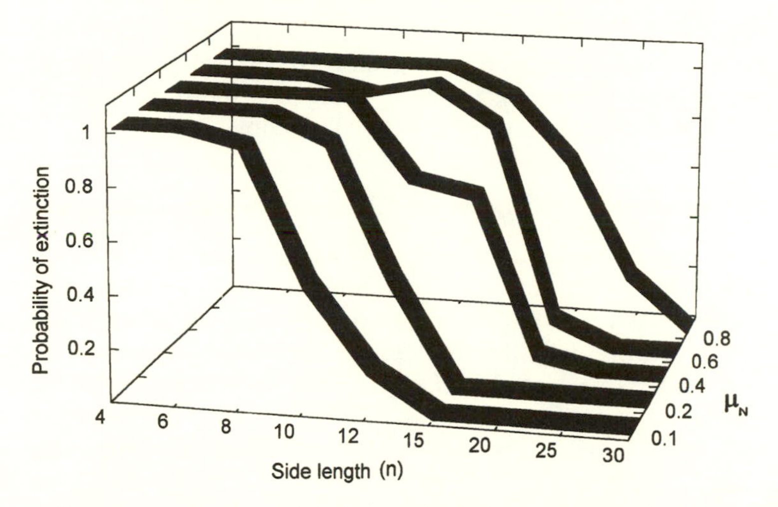

FIGURE 4.5. Extinction probabilities for simulations started in various sizes of arena width n, and varying host dispersal rate μ_N ($\mu_P = 0.89$, $\lambda = 2$). (From Hassell, Pacala, May, and Chesson 1991.)

A second reason for extinction is that the starting conditions for the simulation may be unsuitable. For example, in the region described as "hard-to-start spirals" in Figure 4.3, persistence is impossible if the simulations are started from a single nonempty cell. Once the populations are initiated, however, by simultaneous colonization of many of the cells, persistence always occurs.

In short, restricted local dispersal coupled with unstable within-patch dynamics strongly promotes population persistence, even in an environment of uniform habitats. But does this mechanism play an observable role in real systems? Unfortunately, the data to answer this are not available; there are very few studies on a spatial and temporal scale sufficient to test these metapopulation concepts (but see Hanski, Chapter 2). At present we can only resort to making the models more

---

FIGURE 4.4. Examples of time series of mean population densities illustrating the three regions in Figure 4.3. (a) Spirals ($\mu_N = 1$, $\mu_P = 0.89$), (b) spatial chaos ($\mu_N = 0.2$, $\mu_P = 0.89$), and (c) crystalline structures ($\mu_N = 0.05$, $\mu_P = 1$). (From Comins et al. 1992.)

robust—for example, by introducing demographic and other forms of stochasticity, by using more realistic descriptions of within-patch dynamics, by using more realistic dispersal patterns, and by including additional interacting species.

MULTISPECIES LOCAL POPULATION INTERACTIONS

Despite years of intensive study on host-parasitoid dynamics, there have been rather few attempts (but see Hochberg and Hawkins 1992, 1993) to examine the dynamics of model communities more complicated than two parasitoids attacking one host (May and Hassell 1981; Hogarth and Diamond 1984; Kakehashi et al. 1984; Godfray and Waage 1991; Briggs 1993), two hosts attacked by a common parasitoid (Comins and Hassell 1976, 1987; Holt 1977; Holt and Lawton 1994), or a community of a host, parasitoid, and hyperparasitoid (Beddington and Hammond 1977; Hassell 1979; May and Hassell 1981). And yet it has long been known that parasitoids and hosts are typically found as members of large communities of insects (e.g., Askew 1975; Memmott, Godfray, and Gauld 1994). Although models of such complex webs are likely to be analytically intractable, a mechanistic study of how the different individual links within them come together, or split apart, can shed light on the community dynamics as a whole. In particular, we can analyze the conditions for certain species being able or unable to invade a food web from which they are absent. If they can invade, we then need to analyze whether they can persist in the ensuing community. In this section we examine the dynamics of a community interacting in a single habitat and consisting of up to five species—two hosts attacked by two specialist parasitoids and one generalist parasitoid that is able to attack both host species. The aim is to determine which trophic configurations can most easily persist or be invaded. Questions of invasion and persistence in multispecies metapopulations are considered in the final section.

We consider a community of hosts and parasitoids with discrete and synchronized generations. In particular, there are two host species, H_1 and H_2, that do not interact directly but are attacked by three species of parasitoid: two specialist parasitoids, P_{S1} and P_{S2}, one on each host, and a third generalist species, P_G, capable of attacking both hosts. In real systems, specialist parasitoids are often endoparasitoids, whereas generalists tend to be competitively superior ectoparasitoids (Godfray 1994). We therefore assume that if a host is attacked by both the specialist and the generalist, only the generalist is successful. This leads to the following model:

$$N_{j,t+1} = \lambda_j N_{j,t} f(P_{Sj,t}) f(P_{G,t}) \qquad (j = 1,2) \quad (4.11a)$$

$$P_{Sj,t+1} = N_{j,t} f(P_{G,t})\left[1 - f(P_{Sj,t})\right] \qquad (j = 1,2) \quad (4.11b)$$

$$P_{G,t+1} = \sum_{j=1}^{2} N_{j,t}[1 - f(P_{G,t})] \qquad (4.11c)$$

where λ_j is the net reproductive rate of host j and $f(P)$ are the host survival terms. These survivals are given by the negative binomial function (Eq. 4.7) in which the distribution of attacks by the two parasitoids searching for the same host is independent. As we have already seen, in the context of a spatially patchy environment this can be interpreted specifically as the parasitoids aggregating from patch to patch in an HDI way. In addition to parasitism by the different species being independent of each other, we also assume equal values of k and equal fecundities of the two hosts. Both these assumptions can be relaxed and qualitatively similar results obtained (Wilson, Hassell, and Godfray 1996).

Whether or not an absent parasitoid species can invade a particular community can be determined from the growth rate of the invading parasitoid when its population size is very small. In such a situation we can linearize the equation for the parasitoid dynamics and require that its growth rate be greater than one. In particular we can show that it is only the relative searching efficiencies of the specialist and generalist parasitoid (a_S and a_G, respectively) that determine whether the absent parasitoid can invade (see Wilson et al. 1996 for full details). These invasion boundaries are plotted in Figure 4.6 as a function of the amount of parasitoid aggregation in the system, k, and the magnitude of the searching efficiency trade-off,

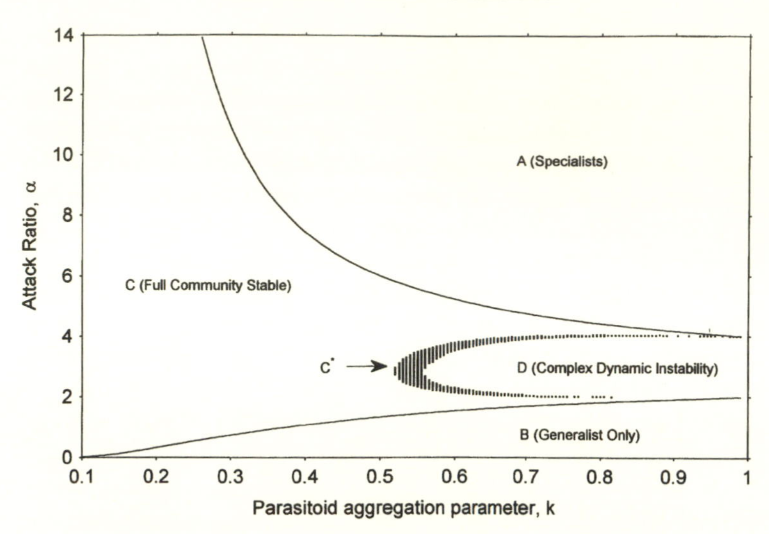

FIGURE 4.6. The regions, explained in the text, where different communities are observed for given parameter values. The x-axis reflects the spatial aggregation of the parasitoids, where all the parasitoids have the same value of k. Low values of k imply strong density dependence. The y-axis gives the relative attack rates, a_S/a_G, of the specialist and generalist parasitoids, respectively, where both specialists have $a_S = 0.3$, and with host rates of increase $\lambda_1 = \lambda_2 = 2$. The two solid lines mark the invasion boundaries of the generalist and specialist parasitoids invading a community from which they are absent. The area of persistence of the full five-species community was found numerically by iterating Equation 4.11. The persistence criterion is that all species should have a population density greater than $1.0e^{-6}$ after ten thousand iterates, all starting from the same initial density of 0.1. Whether the five species can coexist with complex dynamics, the region C*, is dependent on initial conditions. This region thus reflects the initial condition and persistence criterion chosen above. However, other criteria give qualitatively the same result.

a_S/a_G. The figure also shows whether the communities are stable or not as determined analytically or numerically. First, we note that no persistent communities can occur unless a threshold level of parasitoid aggregation is present ($k < 1$), in which case four separate regions of parameter space can be distinguished (A to D). In region A the specialists are much more efficient than the generalist, and we see only specialist parasitoids present. In region B the specialists are either less efficient, or only a little more efficient than the generalist, and so a community with only two hosts and one generalist occurs. In region C a stable full five-species community is possible, whereas in subregion C* all species can persist but limit cycles or chaotic dynamics are observed. Finally, in region D more complex dynamic behavior occurs, which Wilson et al. (1996) have dubbed a region of "complex dynamic instability." Here all three parasitoid species can invade a community in which they are absent, yet the full five-species community cannot persist: The densities of both host species show large amplitude fluctuations leading inevitably to the extinction of the two specialists. However, once the specialists have gone extinct, either or both specialists are able to reinvade, leading to a four- or five-species community with complex dynamics that ultimately breaks down again with the loss of the specialists. No one community can persist, and the community observed at any one time depends critically on the speed with which the five-species community collapses, and the frequency with which the absent specialists reinvade.

A crucial determinant of community structure in these models is thus the relative searching efficiencies of specialists and generalists, and the nature of the density dependence acting on the parasitoid population (captured here by the degree of aggregation, k). In some areas of parameter space there is no dynamically stable end point, and complex cycles occur where the food web structure is constantly changing, species being added through invasion and lost through direct and indirect competitive interactions. We conclude that some communities of hosts and parasitoids may display complex dynamic behavior determined by both local population interactions and the frequency of species invasions.

MULTISPECIES METAPOPULATIONS

Just as the criteria for persistence in two-species host-parasitoid systems are much increased in moving from local to metapopulations, so we may expect community persistence to be enhanced at the metapopulation level. In this section we examine the following range of three-species host-parasitoid metapopulations: (1) two host species with a shared parasitoid species (H-P-H), (2) a single host species attacked by two parasitoid species (P-H-P), and (3) a host-parasitoid-hyperparasitoid interaction (H-P-Q). All are modeled in much the same way as done earlier for the two-species metapopulations, and fuller details of the models may be found in Hassell et al. (1994) and Comins and Hassell (1996). Once again, the principal objective is to determine how the straightforward linking of neighboring local populations by limited dispersal affects invasion and persistence of the interacting populations.

We commence with the simple food chain of a host-parasitoid-hyperparasitoid (H-P-Q) community. As before, there are two phases in each model generation: (1) the dispersal of a fraction of the merging adult hosts and parasitoids and (2) the interaction of hosts and parasitoids within individual local populations. In the dispersal phase, a certain fraction of the adults in each habitat redistribute themselves to the eight immediate neighboring patches, while the remainder stay behind to reproduce in their natal patch. As before, the dispersal is spatially symmetrical for all species, but the fraction of dispersers is species dependent. In the second phase, the population densities of host, parasitoid, and hyperparasitoid population densities in the ith habitat are given by

$$N_{i,t+1} = \lambda N'_{i,t}\, f(P'_{i,t}) \tag{4.12a}$$

$$P_{i,t+1} = N'_{i,t}[1 - f(P'_{i,t})]f(Q'_{i,t}) \tag{4.12b}$$

$$Q_{i,t+1} = N'_{i,t}[1 - f(P'_{i,t})][1 - f(Q'_{i,t})] \tag{4.12c}$$

where λ is the usual host rate of increase term, the survival terms $f(P'_{i,t})$ and $f(Q'_{i,t})$ are given by the Nicholson-Bailey term (Eq. 4.2a), and the primes indicate the densities after dispersal.

We first consider if a hyperparasitoid can invade a host-parasitoid community and then whether the resultant community is persistent or not. We know from the five-species interactions in the previous section that a community can be invadable yet the resultant system fail to persist. On this local scale, persistence depends on the relative competitive advantage of the parasitoid species: The hyperparasitoids can only invade if their searching efficiency relative to the primary parasitoid ($a_Q/a_P = \alpha$) is high enough (Beddington and Hammond 1977; Hassell 1979; May and Hassell 1981). A similar condition is also true for hyperparasitoids invading a host-parasitoid metapopulation: Higher attack rates are always better able to invade (α approx. > 1.5). However, while invasion of a single local population patch case is a purely deterministic process, invasion of a metapopulation has a stochastic component that stems from the heterogeneity in the abundance of local populations in space and the particular local population where the invasion occurs. An invasion must start in a habitat where the local conditions are suitable, otherwise the invading population very quickly dies out. Invasion success thus depends on the specific local conditions where an invasion is occurring. Global conditions, such as the number of local habitats in the metapopulation, have no influence on the success or not of an invasion.

Persistence of the resulting H-P-Q community, however, does depend on the number of patches. As for the simple one host–one parasitoid metapopulation, persistence depends on the heterogeneity of parasitism between habitats being maintained by self-organizing spatial patterns, which themselves depend on there being sufficient space available. The spatial patterns of an H-P-Q community are on a larger scale than when the hyperparasitoid is absent. For example, in one series of simulations with $\mu_P = \mu_H = 0.1$, $a_P = 0.3$, and $\lambda = 2$, the H-P community needed approximately sixty-four patches for a 50% probability of persistence, whereas with a hyperparasitoid of similar dispersal rate the H-P-Q community needed a minimum of 121 patches for a 50% probability of persistence (Wilson et al., in preparation).

Interesting trade-offs tend to occur in metapopulation persistence between dispersal rates and competitive ability measured in parasitoids by their searching efficiencies (see Tilman 1994, Tilman and Downing 1994, and Rees 1995 for comparable trade-offs in plant communities). Thus, although hyperparasitoids with high attack rates *or* dispersal rates are better able to invade a host-parasitoid metapopulation, if they have both high attack *and* dispersal rates they tend to overexploit their hosts (i.e., parasitoid *P*) and consequently fail to persist. Such a trade-off is illustrated in Figure 4.7. Hyperparasitoids that have low attack rates are too uncompetitive—they cannot even invade (region B). Hyperparasitoids that both are good dispersers and have high attack rates are able to invade, but they then overexploit their hosts, and the resultant community breaks up with either the extinction of both the hyperparasitoid and parasitoid or just the hyperparasitoid (region A), much as in the region of complex dynamic instability shown in Figure 4.6. In between the two regions A and B the balance between attack rates and dispersal is such to allow coexistence of the three species (region C).

Coexistence of competing parasitoids (H-P-P) or competing hosts (H-H-P) is greatly restricted compared to the H-P-Q system above (Hassell et al. 1994; Comins and Hassell 1996). For a single local population system, and without additional features promoting intraspecific stability, one parasitoid or one host will always drive the competing species to extinction. Competitive exclusion also occurs within the equivalent metapopulations, except in relatively small regions of parameter space. For example, with two competing parasitoids, this region of coexistence occurs where the dispersal rates of the two parasitoid species differ by an order of magnitude, and when the slower dispersing species also has a greater attack rate. Thus when coexistence does occur in metapopulations it seems to depend upon a kind of fugitive coexistence (Hutchinson 1951; Levins and Culver 1971; Horn and MacArthur 1972; Hanski and Ranta 1983; Nee and May 1992; Hanski and Zhang 1993), balancing dispersal rates and competitive ability. Interestingly, and best seen where there are clear-cut spirals, this coexistence tends to be associated with

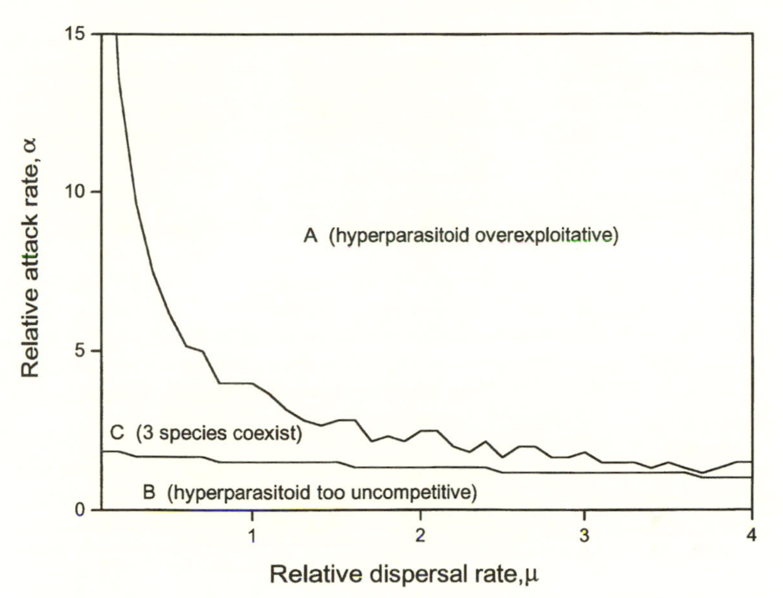

FIGURE 4.7. Persistence plot for the host-parasitoid-hyperparasitoid community. Both axes plot the dispersal and attack rates of the hyperparasitoid relative to the parasitoid, where $\alpha = a_q/a_p$, $\mu = \mu_q/\mu_p$, $a_P = 0.3$, $\mu_N = \mu_P = 0.1$, and $\lambda = 2$. The lines plot the boundaries of persistence when $n = 20$ (four hundred patches). The regions are explained in the text.

the slowly dispersing parasitoids being confined to the central foci of the spirals (Figure 4.8). Here the slow dispersers have an attack rate advantage in the areas where host density fluctuates the least, and the fast dispersers have a colonization advantage in the areas where host density is fluctuating most strongly (the areas swept by the spiral arms). Persistent spatial separation of the two competing parasitoid species is thus maintained by this mechanism. Similarly, for persistence in the H-H-P community the less dispersive host has either a larger intrinsic rate of increase or a smaller parasitoid attack rate, and, with certain parameter combinations, the less dispersive species host is restricted to spiral foci, in exactly the same way as for the less dispersive parasitoids. Since these foci are relatively static, the

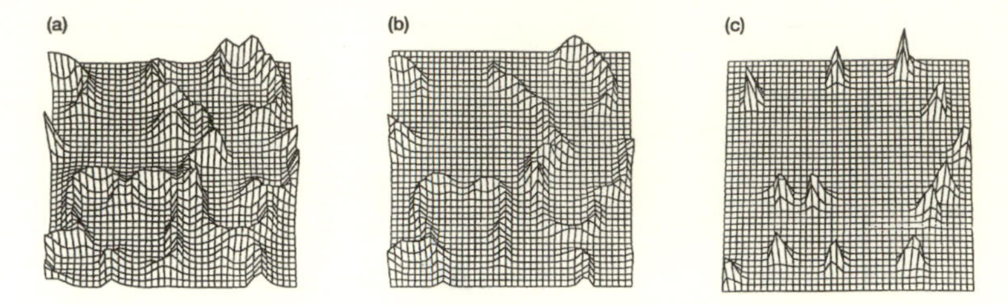

FIGURE 4.8. Maps of the spatial density distribution of hosts (a), highly dispersive parasitoids (b), and sedentary parasitoids (c), in a snapshot from the dynamics of a persistent P-H-P system with $\lambda = 2$, $\mu_N = 0.5$, $\mu_{P1} = 0.5$, $\mu_{P2} = 0.05$, and $\alpha = 1.3$. The grids must be mentally superimposed in order to perceive the relationships between the densities of the various species. Spiral foci exist at the ends of the "mountain ranges" in the leftmost figure (excluding ends at the edges of the grid). In the time evolution of the system the "mountain ranges" are the peaks of population density waves and are thus in continuous motion. The foci, by contrast, remain in almost exactly the same place, for indefinitely long times. (From Comins and Hassell 1996.)

less mobile species appears to occur only in isolated, small "islands" within the habitat, much as if these were pockets of favorable habitat. As the dispersal rates become less divergent between the species, and providing that coexistence is still feasible, the niche of the less dispersive species spreads further into the arm of the spirals.

Attempting to detect the patterns predicted from metapopulation models from real populations poses enormous logistical (and financial) problems, just from the imposed scale of the studies (but see Hanski, Chapter 2). If it emerges that metapopulation effects are a key part to understanding population dynamics, ecologists will be forced to reappraise their approach to fieldwork. Hopefully, useful tricks will emerge; for instance, it may prove possible to piece together from limited data on the population density distributions particular patterns of delayed covariance that are diagnostic of different kinds of spatial, and hence metapopulation, dynamics (Comins and Hassell 1996).

SUMMARY

This chapter has reviewed some dynamical affects of spatial structure in host-parasitoid systems at both local and metapopulation scales. Heterogeneity at the local population level is most conveniently expressed in terms of individual-to-individual variability in the risk of hosts being parasitized, which may be either dependent (HDD) or independent (HDI) of local host density. These can be identified and quantified from straightforward data on the spatial patterns of parasitism in relation to the host distribution. Despite the level of detail in many of these studies, it is encouraging that fairly sweeping criteria can be found that are designed to capture the overall effects of spatial heterogeneity within relatively simple criteria for estimating the variation in the risk of parasitism within the host population. The overall conclusion is that local populations can be stabilized by sufficiently heterogeneous distributions of parasitism, particularly if this is independent of the host distribution (HDI).

Metapopulations of hosts and parasitoids introduce another dimension to the dynamics. Although metapopulations share broadly the same conditions for local stability as the constituent populations, the two become very different when the local populations are individually unstable. Now a variety of striking spatial patterns of abundance can be observed associated with different dynamics of the metapopulation as a whole. It is these spatial patterns that maintain heterogeneity in the host risk of parasitism and so promote persistence.

Invasion and persistence are not always linked in spatially distributed host-parasitoid communities, particularly for metapopulation systems. Thus, a parasitoid or hyperparasitoid may be able to invade a community from which it was absent, but the resultant community may then break up with the extinction of one or more species. Alternatively, a parasitoid or hyperparasitoid species could have a high probability of persistence in an established community but have a low probability of invading a community from which it was absent. The particular history of invasions of a community can thus have a dramatic effect on the resulting community structure. Coexistence of competing species in these metapopulations, when it does occur, not only depends upon a balancing of dispersal rates and competitive ability but also seems to be associated with some degree of self-organizing niche separation between the competing species. This is best seen when the spatial dynamics show clear spirals, in which case the relatively immobile species tends to be confined to the central foci of the spirals and the highly dispersive species to the remainder of the "trailing arm" of the spirals.

CHAPTER FIVE

Basic Epidemiological Concepts in a Spatial Context

Elizabeth Eli Holmes

INTRODUCTION

It is obvious to any observer of epidemics that the spread of disease is unavoidably spatial. Disease moves from individual to individual following the network of contacts between individuals within a population. For many host-pathogen systems, the pattern of spread is a combination of local transmission out from a focus of infection and long-distance transmission, which establishes new foci. Yet most classical epidemiological theory glosses over the spatial dimension of disease transmission and instead assumes that every individual is equally likely to contact every other. A key question is to what extent do we lose insight or are quantitatively misled by modeling the intrinsically spatial process of disease spread with nonspatial theory. Obviously, spatial models are necessary to address spatial questions, such as the velocity at which disease spreads over a landscape or the spatial pattern of disease prevalence (see Murray 1990 and Cliff et al. 1981 for reviews). However, many of the most basic and important epidemiological questions are not spatial: Will a pathogen cause an epidemic? Can it invade and persist in a population? What fraction of the population will be infected? When will epidemics occur and reoccur?

Early epidemiological theory addressed these sorts of questions using simple nonspatial models of communicable disease such as the Kermack-McKendrick model (Kermack and

McKendrick 1927):

$$\frac{dS}{dt} = -\beta SI - bS + b$$

$$\frac{dI}{dt} = \beta SI - \mu I - bI$$

$$\frac{dR}{dt} = \mu I - bR. \tag{5.1}$$

These models divide the population into susceptible, infected, and resistant (i.e., immune) classes, S, I, and R. All individuals are equally likely to contact every other individual in the population, and there are no differences among individuals except for their infection status. Individuals recover and become immune at rate μ and die at rate b. The population is assumed to be stable so that all individuals that die are replaced by births of new susceptible individuals. Susceptible individuals become infected at a rate βI that is simply proportional to the fraction infected.

From such simple nonspatial theory come some of our most fundamental principles about the dynamics of disease within populations, and although modern epidemiological models are far more realistic and introduce a variety of complexities (see Anderson and May 1991 for a review), these basic concepts are still used to understand and think about real disease systems. The purpose of this chapter is to discuss how fundamental epidemiological concepts change, qualitatively or quantitatively, when we move from a nonspatial to a spatial model of communicable disease.

Basic Epidemiological Concepts

1. The reproductive rate of a disease. The basic reproductive rate of a disease, R_0, is the expected number of new infections caused by one infected individual in a sea of susceptible individuals (MacDonald 1957).

2. Deterministic threshold theorem. Kermack and McKendrick (1927) showed that in a simple model with no influx of susceptible individuals, there exists a threshold density of susceptibles below which an epidemic cannot occur.

3. Stochastic threshold theorem. The Kermack-McKendrick model assumes that the population is large enough to be considered infinite and that densities can be considered continuous variables. Bailey (1975) studied stochastic versions of the Kermack-McKendrick model to explore the dynamics of disease in finite communities and showed that disease cannot become established in a population unless the size of the community is above the stochastic threshold.

4. Threshold for a disease to become endemic. The previous two points consider the dynamics of an epidemic. In this case, natural mortality, b, is negligible. If we consider instead the reintroduction of susceptible individuals due to births, then disease may persist within the population and become endemic. Analogous to the epidemic threshold, there exists an endemic threshold for the disease to establish and persist.

5. Equilibrium levels of disease. Kermack and McKendrick (1932) showed that Equation 5.1 predicts stable equilibrium levels of susceptible, infected, and resistant individuals. These levels are a function of the rates μ, β, and b.

6. Periodicity of epidemics. The long-term records of infectious diseases provide some striking examples of cyclic predator-prey dynamics. The Kermack-McKendrick model offers an explanation for these cycles as an intrinsic product of the interaction between hosts and pathogens (Soper 1929; Anderson and May 1983).

SPATIAL VERSUS NONSPATIAL

How does the spatial dimension change these basic concepts? To tackle this question, I discuss what two types of spatial models predict for concepts (1) through (6) above. The focus of the discussion will be on a cellular automata version of the Kermack-McKendrick model. This model is individual based and explicitly captures the notion of a spatial network of contacts between individuals. However, discussion of only one type of model leaves one blind to the degree to which the results depend on a particular model structure, and much insight into results from one type of model is gained by comparing results with other types of models. Thus, in the concluding sections, the results for the cellular automata model are compared to a partial differential equation model that also incorporates local transmission.

Kermack-McKendrick Model

Before introducing the cellular automata model, let me briefly discuss the classical Kermack-McKendrick model (Eq. 5.1). The key feature of this model is the idea that disease transmission is described by the βIS term, which is known as the mass-action assumption. In this case, the rate at which susceptibles become infected is directly proportional to the fraction infected; doubling the number infected doubles the rate of disease transmission. For this model, the reproductive rate of the disease is

$$R_0 = \frac{\beta}{\mu + b}.$$

R_0 is analogous to the intrinsic rate of increase, r, in simple population growth models. R_0 is not generally observed because it is the maximum possible growth rate, which occurs at the very beginning of an epidemic when the population is 100% susceptible. See Bailey (1975) for a modern discussion of this model and Serfling (1952) for a historical review.

Cellular Automata Model of an Infectious Disease

In the basic lattice model, sites are distributed on a square lattice on which each site has a set of physically neighboring sites. Each site represents an individual that can have one of three states: susceptible, infected, or resistant. Disease transmission is modeled as a probabilistic process. Each infected site has an equal and independent probability, q, of infecting a susceptible neighbor. Thus the probability that a susceptible site becomes infected is

1 − (the probability of not being infected)

$$= 1 - (1 - q)^{\text{number of infected neighbors}}. \quad (5.2)$$

At each time step, a site changes state based on the probabilities:

$S \rightarrow I$	$1 - (1 - q)^{\text{number of infected neighbors}}$
$I \rightarrow R$	μ
sites die and are reborn susceptible	b.

In the simulations discussed here, three different neighborhoods are compared: the four directly adjacent sites, the eight nearest neighbors, and the twenty-four nearest neighbors. Sites are updated synchronously, meaning time is discrete. Discrete time can affect the dynamics of cellular automata (Ingerson and Buvel 1984; Nowak, Bonhoefer, and May 1994), and to minimize these effects, the simulations were run with small transition probabilities to approximate continuous time.

In this discussion, the model with local transmission is often compared to the analogous model with global transmission. In the global model, the "neighborhood" is the entire population, and every infected site is equally likely to infect any susceptible site in the entire population. The rate of disease transmission is: (the proportion susceptible) × (the probability that a susceptible site becomes infected), that is,

$$S\left[1 - (1 - q_g)^{IN}\right] \approx (q_g N) IS$$

for q_g small. The parameter q_g is the probability that an infected individual infects a susceptible neighbor for the global model. Since transmission can be described as a linear function of IS for q_g small, the cellular automata model with global transmission is the analogue to the Kermack-McKendrick model with $\beta = q_g N$.

The definition of the disease reproductive rate in the cellular automata model is analogous to the definition for the Kermack-McKendrick model. It is the mean infectious period times the rate at which new infections are caused by an infected site that is surrounded by susceptible sites. Thus,

$$R_0 = \frac{qN_n}{\mu + b}$$

where N_n is the number of neighbors (4, 8, 24, or N). When a simulation of the model with local transmission was compared to one with global transmission, R_0 was kept identical in the simulations, but the new infections were distributed either locally or globally. Specifically, and to make the R_0's equivalent,

$$q_n = q_g \frac{\text{population size}}{\text{number of neighbors}} \tag{5.3}$$

where q_n and q_g are the transmission probabilities for the local and global models.

EPIDEMIOLOGICAL PRINCIPLES IN A SPATIAL CONTEXT

Epidemic Threshold

The Kermack-McKendrick threshold for an epidemic to occur in a population is $R_0 > 1/S$. The Kermack-McKendrick threshold is the motivation behind the concept of "herd immunity." This idea says that to protect a population from a disease, it is not necessary to vaccinate the entire population; it is enough to vaccinate a proportion $(1 - 1/R_0)$. One of the basic results from the cellular automata model with local transmission is that the Kermack-McKendrick threshold is overly conservative. In the cellular automata model with local transmission, each infected individual interacts with a relatively small neighborhood. From the infected individual's perspective, its "world" quickly fills with other infected individuals, and the rate at which it causes new infections rapidly declines. To overcome this severe depletion of the local susceptible pool, the disease must have a higher R_0 in order to cause an epidemic (e.g., $R_0 = 1.29$ for four neighbors and $R_0 = 1.13$ for eight neighbors). This means that when one takes into account localized transmission, one needs to vaccinate a smaller proportion of the population. As an aside, the stochastic nature of transmission is often much more important than the local nature of transmission if only a few infected individuals are introduced into a susceptible population. In a stochastic model, there is some probability that the disease will go extinct by chance alone even though $R_0 > 1$. In a stochastic lattice model, this probability of chance extinction is $(1/R_0)^a$ where a is the initial number infected (Whittle 1955; Kendall 1965). When $a = 1$ and R_0 is small, this probability is quite high.

The traditional Kermack-McKendrick threshold described above is the threshold for the disease to increase when it enters a population. However, spatial models of disease introduce a new type of the epidemic threshold, which is the threshold reproductive rate for a pandemic. A pandemic is an epidemic that affects the entire population. The Black Plague in the 1300s and 1500s, the flu epidemic of the early 1920s, and the smallpox epidemics that decimated the Americas are real-world examples of pandemics that spread across entire continents. The cellular automata model predicts that there is a threshold R_0 for a pandemic to occur. At the reproductive rates near the threshold for a small innoculum of infection to increase in a population, the disease will begin to spread from the initial site of infection, but then die out. With a higher basic reproductive rate, the disease will spread farther and farther until, with a high enough R_0, it is able to spread throughout a very large population (Figure 5.1). The pandemic threshold for the cellular automata model with transmission to the four nearest neighbors is $4 < R_0 < 6$ (Kuulasmaa 1982). If R_0 is below this, the disease causes a small, spatially limited epidemic. If it is above, it can spread throughout even an infinite population. The probability of a pandemic also increases as R_0 increases above the pandemic threshold. From analogy with similar models (Ball 1983; Bramson, Durrett, and Swindle 1989) the

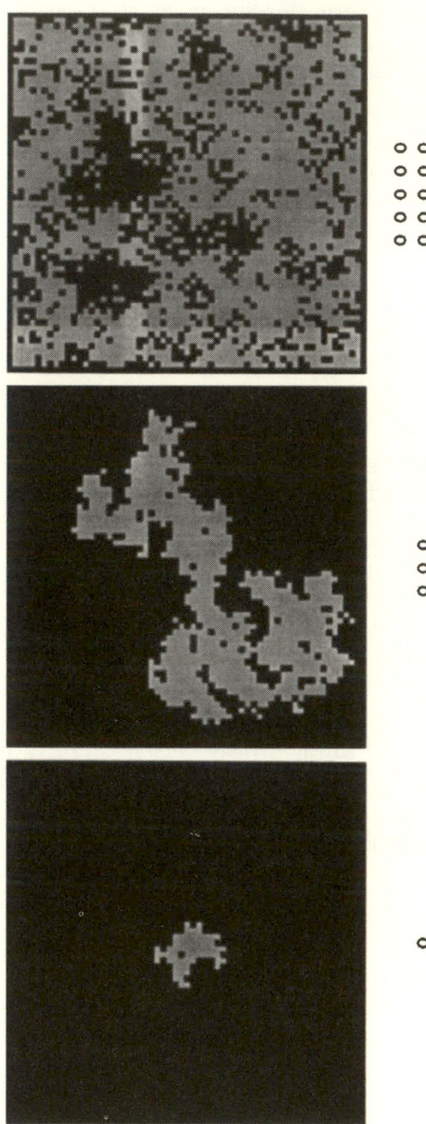

pandemic threshold depends on the size of the contact neighborhood and should tend to $R_0 = 1.0$ as the neighborhood size increases.

Threshold Community Size

The stochastic threshold theorem states that there is a threshold community size for a pandemic to occur in a population. This idea stems from work on a model equivalent to the cellular automata model with global transmission. In this model, smaller population size leads to an increased probability that the incipient epidemic will go extinct in the early stages merely by chance. If we assume that there are no births and no loss of immunity, the threshold population size is $N_t = \mu/\beta$ (Bailey 1975). Below this, the epidemic dies out with certainty, and above the threshold, pandemics occur with probability $1 - N/N_t$ where N is population size.

There is no real equivalent of this idea in the cellular automata model with local transmission. From the infected individual's perspective, the world is the same whether it is in an infinite population or in a population barely larger than its neighborhood size. Instead of a threshold community size, however, there exists a threshold neighborhood size below which the disease dies out. The existence of this threshold can be surmised by noting that the pandemic threshold is $4 < R_0 < 6$ when spread is to the nearest four neighbors, and that as the neighborhood size increases, the threshold approaches unity (see the previous section). In simulated epidemics that are started from one infected individual (Figure 5.1), the spatial size of the epidemic increases dramatically as the neighborhood size increases (note that R_0 is held constant between simulations).

FIGURE 5.1. Epidemics with increasing neighborhood size. The patterns show the distribution of susceptible (black regions) and immune individuals (gray regions) at the end of an epidemic. Each epidemic was started with one infected individual at the center of the grid. In each case, $R_0 = qN_n/\mu = 2.67$. The contact neighborhood is shown under each simulation (○ = neighbor of the x individual). $\mu = 0.15$, $b = 0$.

Equilibrium Levels of Disease

The previous discussion concerned growth and spread of a disease within a population. A disease can also persist at some basal level within a population, in which case it is said to be endemic. The Kermack-McKendrick model and the cellular automata model with global transmission predict that if a disease is able to cause an epidemic, then given sufficient time, the disease will become endemic at the equilibrium levels of

$$\hat{S} = \frac{\mu + b}{\beta} = \frac{1}{R_0}$$

$$\hat{I} = \frac{b}{\beta}(R_0 - 1)$$

$$\hat{R} = \frac{\mu}{\beta}(R_0 - 1). \qquad (5.4)$$

Simulations of the cellular automata model with global versus local disease transmission show that, in general, localized transmission reduces the equilibrium level of infection within the population. Correspondingly, the equilibrium levels of susceptibility are higher. The lower equilibrium level of infection can be explained by the rapid depletion of the local susceptible pool which effectively reduces the transmission of the disease and creates isolated pockets of susceptibility (Figure 5.2). In a simulation with global transmission, infection is uniformly distributed, and such pockets of susceptibility do not occur.

The degree to which localized transmission decreases the infection levels, however, depends greatly on the reproductive rate of the disease and the length of the infectious period relative to an individual's lifespan. In Figure 5.3, the absolute difference in the susceptible equilibrium fraction observed in simulations of the local versus global model is shown as a function of R_0 and the ratio of lifespan to infectious period (μ/b). When the disease is chronic (when the infectious period is long relative to lifespan), the model with local transmis-

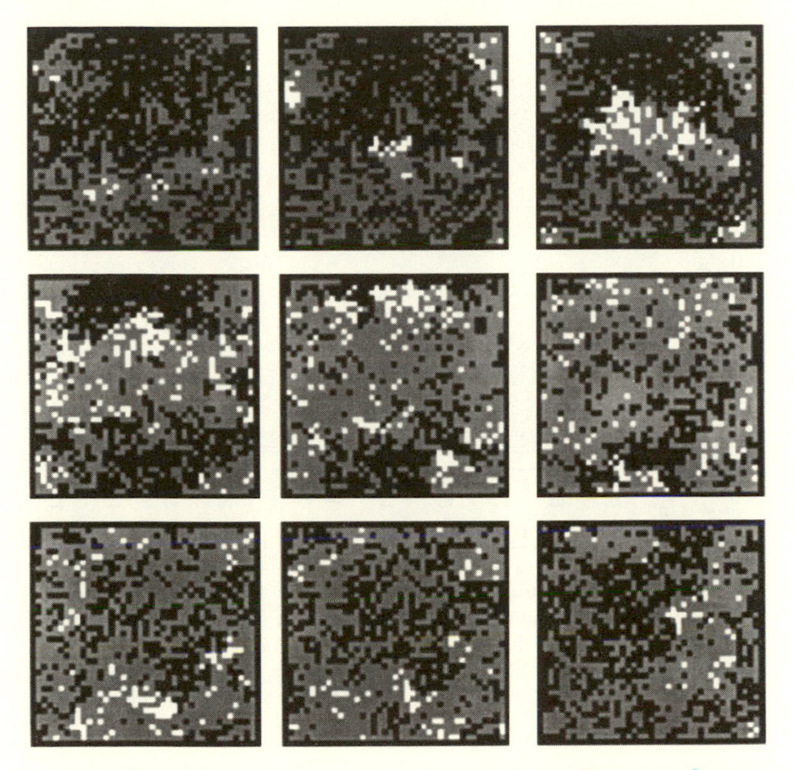

FIGURE 5.2. Endemic disease patterns. Sequential snapshots are at forty-cycle intervals. Black is susceptible; white is infected; gray is immune. The sequence begins one thousand cycles after an epidemic started with 50% infected and 50% susceptible in a random pattern. $R_0 = 4$, $\mu = 0.1$, $b = 0.01$ on a 40×40 grid with eight neighbors.

sion predicts the equilibrium levels of susceptibility are similar to those predicted by the simple nonspatial model. Figure 5.3 shows results when the neighborhood is the four nearest neighbors. The differences between the local and global models, however, shrink rapidly for larger neighborhood sizes. Localized transmission most strikingly reduces the level of infection in a population when the disease is acute, the disease reproductive rate is low, and the contact neighborhood is small.

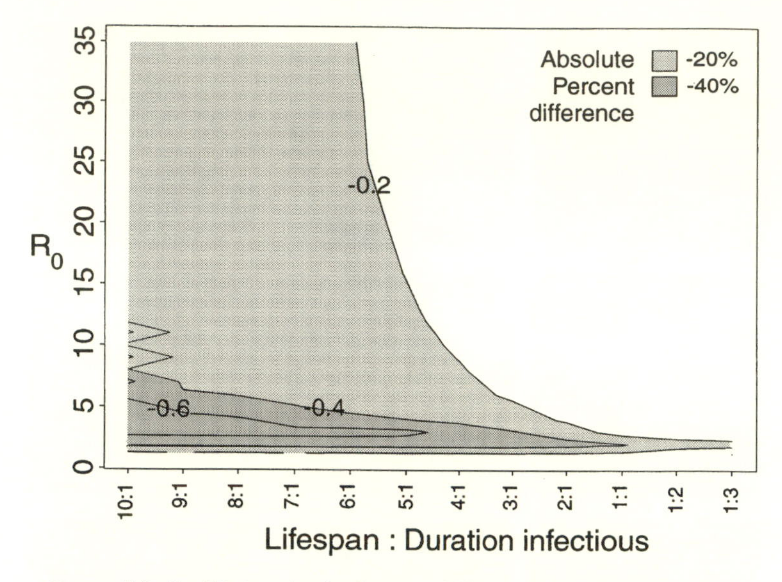

FIGURE 5.3. Equilibrium level of susceptibility in the cellular automata simulations with global versus local transmission (four neighbors). The shaded regions show the parameter space in which local transmission leads to a significantly higher equilibrium fraction of susceptible individuals. The absolute difference is calculated as the (level susceptible in the global model) – (level susceptible in the local model).

The Threshold for Endemicity

Localization transmission not only decreases the prevalence of disease in the population, but it also introduces a new type of endemic threshold. As discussed above, in a model with global transmission, the threshold for the disease to become endemic is $R_0 > 1.0$. It is the same regardless of the lifespan or the duration of the infectious period. When transmission is localized, in contrast, the ability of the disease to persist depends greatly on the ratio of the lifespan to the infectious period. If the infectious period is too short, the disease cannot persist even though R_0 is high. For example, in simulations with transmission to the nearest four neighbors on a 400×400 grid, the disease rapidly went extinct when the ratio of the lifespan to the infectious period was greater than four. For the same parameters in a simulation with global transmission, the disease persisted and stabilized at a 10% infection level. In these simulations, disease extinction could be due to chance extinction when the number of infected sites is small. The probability of stochastic extinction by time t is approximately (Renshaw 1991)

$$\left(\frac{\mu t}{1 + \mu t}\right)^{\hat{I}N}$$

where $\hat{I}$ is the equilibrium fraction infected. However, the rapid extinction at large grid sizes suggests that a threshold is being crossed in the simulations.

Analytical results from work on contact processes also imply that a new threshold exists. Loosely, one can argue as follows. If the rate at which individuals die is high compared to the rate at which they lose infection ($b \gg \mu$), then we have approximately a simple contact process in which

$S \rightarrow I$ from contact with neighboring infecteds

$I \rightarrow S$ at a constant rate, b, due to death of an infected and birth of a susceptible.

This process has been studied extensively, and the disease becomes endemic if $1.33 < R_0 < 4$ (Durrett and Levin 1994a,b). Numerically, the critical R_0 is $R_c \approx 1.64$ (Durrett 1991). If there are no births ($b = 0$), we have the simple epidemic with no recovery, and as discussed previously the disease dies out on an infinite grid if $R_0 < 4$. With this information, one can conjecture how the critical R_0 for persistence varies with the ratio of the lifespan to the infectious period, μ/b (Figure 5.4). Specifically, the critical R_0 increases as the infectious period shrinks relative to the lifespan. Thus as the lifespan increases, one can pass from a region where the disease can persist into a region where it cannot. Note that such a threshold does not exists when transmission is global; as the lifespan increases, $\hat{I}$ decreases and the probability of stochastic extinction increases, but a threshold is not crossed.

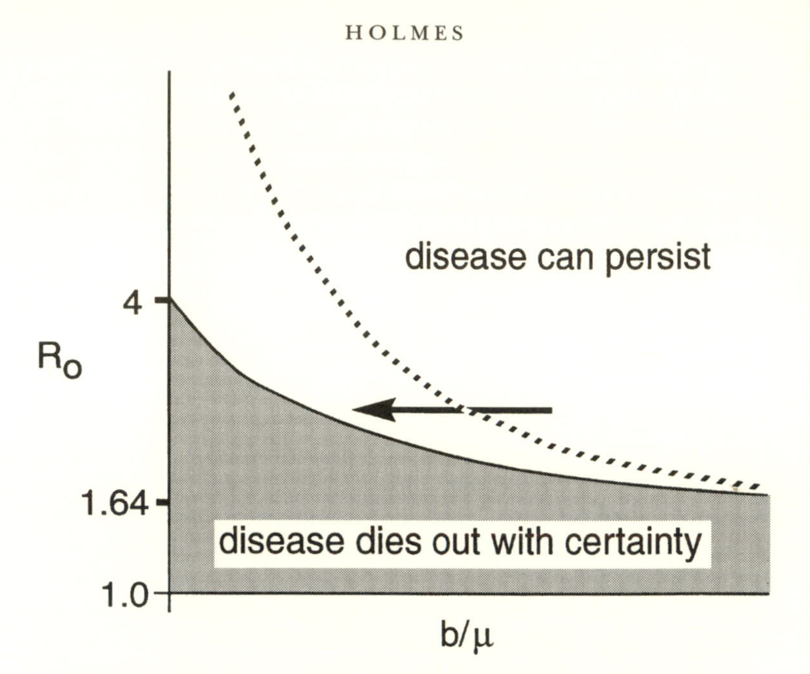

FIGURE 5.4. The critical reproductive rate for endemicity. When the ratio of the infectious period to the lifespan, b/μ, is very large, $R_c \approx 1.64$. When $b/\mu = 0$, R_c is at least greater than four. As b/μ is decreased (keeping R_0 constant), one can pass from a region where the disease can persist through a threshold b/μ to a region where the disease cannot persist. Note that the shaded area is the minimum region where the disease will die out because the critical R_0 at $b/\mu = 0$ is a lower bound. The actual line for the critical R_0 (designated by the dashed line) will lie somewhere above the shaded area.

The Periodicity of Epidemics

The seemingly regular recurrence of epidemics is certainly one of the most striking features from the historical records of directly communicable disease. For example, in large, non-immunized populations, measles epidemics tend to reoccur every one to two years; chickenpox and mumps, every two to four years; diphtheria epidemics, every four to six years; and smallpox, every five years (Anderson and May 1985a). The Kermack-McKendrick model presents a simple mechanism to explain this periodicity. Analogous to the predator-prey cycles that occur in Lotka-Volterra models, cycles of disease reflect the tendency of the disease to decimate the host and then to decline until susceptibility increases again in the host population. The Kermack-McKendrick model predicts that epidemics (Soper 1929; Anderson and May 1992) reoccur every T years where

$$T \approx 2\pi \sqrt{\frac{L(D_i - D_l)}{R_0 - 1}} . \tag{5.5}$$

L is the mean lifespan $= 1/b$, D_i is the infectious period $= 1/\mu$, and D_l is the latent period. The time between epidemics is determined by how quickly susceptibility returns, reflected by L, and how quickly the disease increases in the population, reflected by $(R_0 - 1)/(D_i + D_l)$. Simulations show that the cellular automata model with global transmission produces epidemic cycles that are analogous to those in the Kermack-McKendrick model. The cycles in both models are damped, meaning their amplitude decreases with time; however, a variety of factors, such as seasonality in transmission rates, demographic stochasticity, age structure, and time lags can cause the cycles to persist (Anderson and May 1992).

To study the effect of local transmission on epidemic cycles, I simulated the cellular automata model with local transmission using neighborhoods of eight and twenty-four (400×400 grid). Damped cycles of epidemics appear in these simulations. As in the models with global transmission, the time between epidemics is negatively related to R_0 and positively related to lifespan. However, in the local transmission model, the time between epidemics is up to three times longer (Figure 5.5).

What causes these differences? Some insight can be gained by considering a plot of new infections per time step versus the product of the fraction infected times the fraction susceptible, IS. In a model with mass-action mixing, for example, the Kermack-McKendrick model or the cellular automata model with global transmission, this plot is linear because the number of new infections per time step is βIS (or $qNIS$ in the cellular

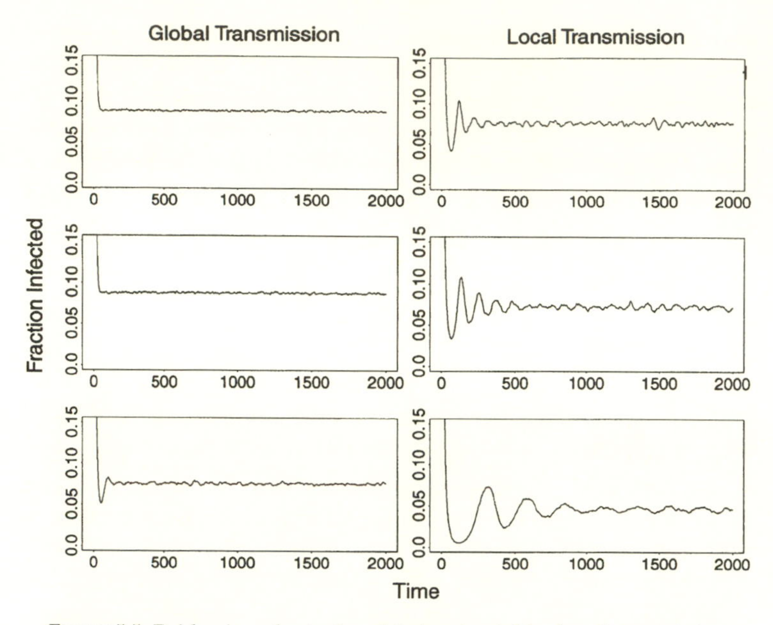

FIGURE 5.5. Epidemic cycles in the global and neighborhood transmission models. Simulations were started with 50% infected and 50% susceptible with infection randomly distributed on the lattice. Top, middle, and bottom rows are R_0 = 35, 20, and 5, respectively. $\mu = 0.1$, $b = 0.01$.

automata model). In the cellular automata model with local transmission, disease transmission is not via mass action, and this plot can be highly nonlinear. However, with time, the spatial pattern of infected, immune, and susceptible sites settles into an equilibrium, and at this equilibrium, the number of new infections per time step is approximately a linear function of *IS*. The slope of this line gives an estimate of the effective transmission rate, β_e which can be much less than the intrinsic transmission rate, $q_n N_n$. All this means is that near equilibrium the cellular automata model with local transmission can be approximated by a simple mass-action model with a new transmission parameter β_e. Indeed, the observed equilibrium fractions of infected, susceptibles, and immune in the local model are those predicted from the global model using β_e, b, and μ (Eq. 5.4). This lower effective transmission rate largely explains

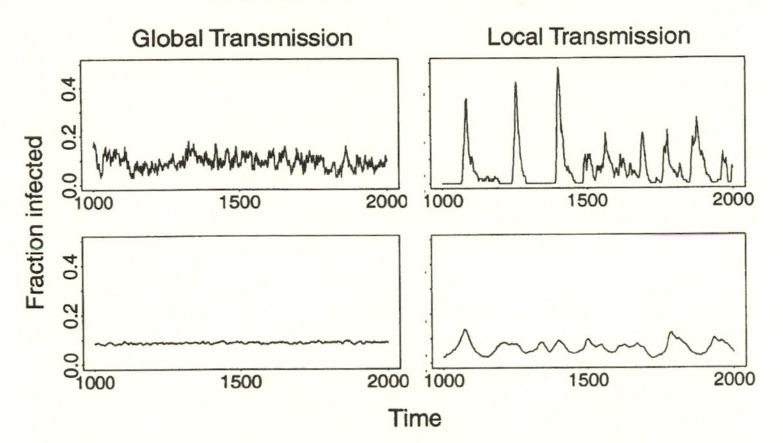

FIGURE 5.6. The infection within a subregion of the population versus within the total population. The top graphs show the fraction infected within a 10×10 subsection of the total grid. The bottom graphs show the fraction infected within the entire grid. On the left is the cellular automata model with global transmission, and on the right is the cellular automata model with a neighborhood of eight. $\mu = 0.1$, $R_0 = 25$, $b = 0.01$, 100×100 grid.

the lengthened periods of epidemic cycles. For example, replacing β by the β_e in Equation 5.5 gives a good prediction for the observed time between cycles in the local model (Figure 5.5).

At the level of the whole population, infection levels cycle smoothly. At the local level, the picture is very different. In the local transmission model, foci of infection move around in space (Figure 5.2). In any one small region, a cycle of epidemic occurs: Immunity increases, and the disease dies out, immune individuals die and are replaced by susceptible individuals, the disease reenters, and a new epidemic occurs. Cycles of epidemics sweep through the local population at irregular intervals, and the interepidemic period is a roughly negative function of R_0 (Figure 5.6). However, these cycles do not necessarily correspond to the cycles at the population level, and unlike the epidemic cycles at the population level, they do not damp out with time. Such stochastic local epidemics have been used to explain the irregularly occurring epidemics of measles in collections of small communities (Bartlett 1955).

The Character of Epidemics

The classic bell-shaped form of an epidemic is typical of diseases such as smallpox, measles, chickenpox, and rubella that have a high reproductive rate and short infectious period relative to the immune period (large μ/b). To look at the effect of local transmission on the size and duration of an epidemic, I compared epidemics in the cellular automata model using neighborhoods of four, eight, or twenty-four with those in the cellular automata model with global transmission. The reproductive rate of the disease was identical between comparisons. The models were compared over a range of R_0 and initial fraction infected. The epidemic started with one infected individual in the center of the grid, and there were no births or deaths ($b = 0$).

In general, an epidemic in the cellular automata model with local transmission sputtered slowly along compared to an epidemic in the model with global transmission (Figure 5.7), unless R_0 was well below the pandemic threshold. In this case, the epidemic in the neighborhood model died quickly. The maximum difference between epidemic durations occurred when the disease reproductive rate was just above the pandemic threshold (Figure 5.8b). At this R_0, the disease began to spread slowly through the entire population. The size of the epidemic was also generally smaller in the model with local transmission; however, the difference depended greatly on R_0 and was at a maximum between the pandemic threshold and $R_0 = 1$ (Figure 5.8a). At $R_0 = 1$, disease dies out quickly in both models, and above the pandemic threshold, the disease may spread slowly but eventually affects the entire population.

COMPARISON WITH OTHER SPATIAL MODELS

Clearly localized disease transmission changes the impact of disease on a population. However, the results discussed above emerge from a cellular automata model. This model is stochastic, individuals come in discrete units (it is not possible to have half an individual), and space is broken up into discrete patches.

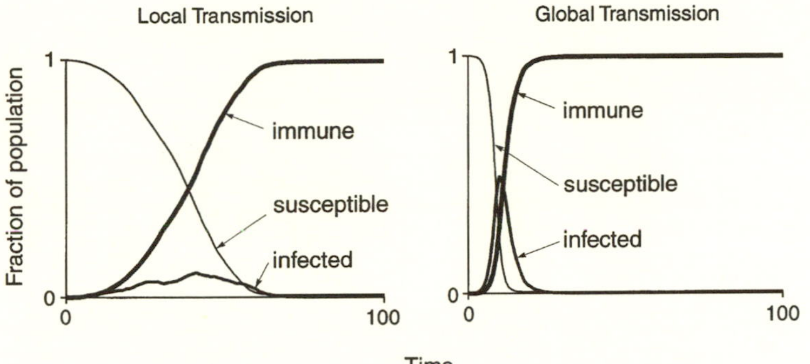

FIGURE 5.7. Epidemics in the cellular automata model with local versus global transmission.

To what extent are the results dependent on this cellular automata framework instead of localized transmission per se? Insight can be gained by looking at a partial differential equation model, the spatial Kermack-McKendrick model, which has localized transmission but a different basic framework. This model is deterministic, allows infinitesimal individuals, and evenly spreads individuals out on a plane rather than divides them among discrete patches.

Spatial Kermack-McKendrick Model

In the spatial Kermack-McKendrick model (see comments by D. G. Kendall in the paper by Bartlett 1957), susceptible individuals become infected by contact with surrounding infected individuals that are within some radius r. Within that radius, transmission occurs through mass-action mixing:

$$\frac{\partial S}{\partial t} = -\beta S\tilde{I}$$

$$\frac{\partial I}{\partial t} = \beta S\tilde{I} - \mu I$$

$$\frac{\partial R}{\partial t} = \mu I. \qquad (5.6)$$

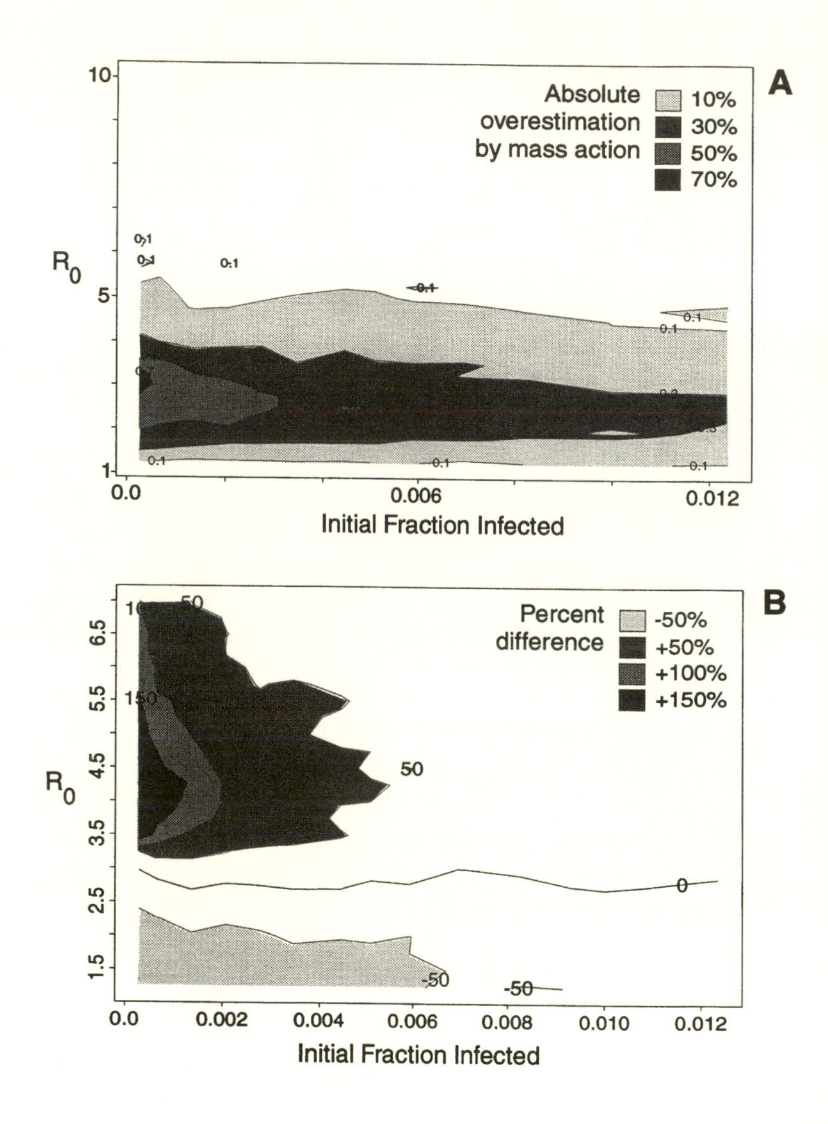
A
Absolute
overestimation
by mass action
10%
30%
50%
70%
R0
Initial Fraction Infected
B
Percent
difference
-50%
+50%
+100%
+150%
R0
Initial Fraction Infected

In this form, it is a partial differential equation model of an epidemic with no recovery; $\partial S / \partial t$ is the change in the proportion of susceptibles at location (x, y). The form of the model is quite similar to the Kermack-McKendrick model (Eq. 5.1) with b set to zero, but the contact rate is given by $\beta S\tilde{I}$ instead of βSI, where $\tilde{I}$ is a weighted function of the number of infected individuals within a radius, r, of a susceptible individual. The weighting is defined such that the total contact rate, $\beta\tilde{I}$, remains constant as the contact radius is reduced. In this way a susceptible individual is not less likely to become infected if they have a small contact radius; they simply contract the disease from a closer neighbor. The effect is to keep R_0 constant while studying the effect of shrinking the contact radius.

Epidemic, Endemic, and Neighborhood Thresholds

In the cellular automata model with local transmission, there are two types of epidemic threshold: the epidemic threshold for a small introduction of pathogen to increase locally before dying out, and the pandemic threshold for a pathogen to spread throughout the entire population. In the cellular automata model, both the epidemic and pandemic thresholds are greater than the simple epidemic threshold, $R_0 > 1$, from the nonspatial Kermack-McKendrick model. In contrast, it can be shown that both the epidemic and pandemic thresholds in the

FIGURE 5.8. Comparison of epidemic size and duration with global versus local transmission. (A) Shaded areas show the absolute difference in the size of the epidemics. Thus, an epidemic that infects 50% of the population in the global and only 20% in the local model gives an absolute difference of 30%. Local transmission causes the greatest decrease in epidemic size when R_0 is between 1.0 and the pandemic threshold. (B) The shaded areas show the parameter space where local transmission lengthens or shortens the duration of an epidemic. When R_0 is small, epidemics are small and die out quickly in the local model. As R_0 increases, the disease begins to spread throughout the population, and the epidemic duration is generally much longer than in the global model. The percentage difference is $100 \times$ (Neighborhood − global duration)/(global duration). Neighborhood equals the four nearest neighbors.

spatial Kermack-McKendrick model are the same, $R_0 > 1$, as for the nonspatial Kermack-McKendrick model. This contrasts greatly with the pandemic threshold ($4 < R_0 < 6$) in the cellular automata model with local transmission.

The difference between the pandemic thresholds in the cellular automata versus the spatial Kermack-McKendrick model is striking and emphasizes the implications of the different assumptions in the two models. In the spatial Kermack-McKendrick model, infinitesimal individuals exist; thus one can have infinitesimal infections produced in an infinitesimal time step, and these infinitesimal infections can then produce infection in the next time step. In the cellular automata model, individuals come in units of one, and transitions occur in discrete jumps. Also, in the spatial Kermack-McKendrick model, the contact rate per susceptible is always a linear function of the proportion infected (rate $= \beta\tilde{I}$). In the lattice model with local transmission, the rate saturates (rate $= 1 - [1 - q]^{IN_n}$).

The two models also differ in terms of a threshold neighborhood size. In the cellular automata model, there is a threshold neighborhood size below which the disease dies out and above which it can spread throughout the entire population. This threshold does not exist in the spatial Kermack-McKendrick model. As long as $R_0 > 1$, a pandemic can occur, regardless of the contact radius. The difference between the models emphasizes again the significance of the assumption that individuals come in discrete units.

Finally, the models differ with regard to the endemic threshold for persistence of disease in a population. In the nonspatial Kermack-McKendrick model, this is simply $R_0 > 1$; in the cellular automata model, the threshold is higher, and it depends on the ratio of the lifespan to the infectious period (Figure 5.4). The spatial Kermack-McKendrick model is again like the nonspatial model. The threshold is $R_0 > 1$ and is independent of the ratio of the lifespan to the infectious period. Furthermore, the equilibrium levels of susceptibility, infection, and immunity for the spatial Kermack-McKendrick model with recovery are given by the formulas from the nonspatial model (Eq. 5.4).

The Spatial Pattern of Disease

The cellular automata model produces fascinating, heterogeneous spatial patterns of disease that persist through time (Figure 5.2). The spatial Kermack-McKendrick model, in contrast, can produce traveling waves of infection, but at equilibrium it produces a uniform distribution of disease. For example, after introduction of the disease at one point in space, the pathogen spreads outward in a circular wave but then leaves uniform densities of susceptible, immune, and infected individuals in its wake. If there is no recovery, then the uniform densities will be $(1 - 1/R_0)$ susceptible and $1/R_0$ immune. If there is recovery, the densities will be those given in Equations 5.4.

The difference between the models stems from two basic differences in frameworks of the models: (1) in the cellular automata model, space is broken into discrete patches, and (2) in each patch, there are unstable dynamics. Point (2) comes about because neither susceptibility, immunity, nor infection are a stable endpoint for an individual. Instead individuals go through an inevitable cycle: susceptibility to infection to immunity and back to susceptibility. This combination of patches and unstable dynamics within a patch is the root of the interesting spatial patterns in a wide variety of cellular automata and coupled-lattice models.

To see how the patterns depend on unstable local dynamics, imagine instead that each patch in the cellular automata model represents a subpopulation and that in each subpopulation the dynamics are described by the Kermack-McKendrick model (Eq. 5.1). Like the cellular automata model, this is an epidemiological model in which there is a collection of discrete sites with local transmission between sites. However, now the local dynamics are stable because in an individual subpopulation the densities of susceptible, infected, and immune individuals will go toward the equilibrium values given in Equations 5.4. With patches and stable local dynamics, interesting spatial patterns of infection do not form. Instead, a uniform density at the equilibrium levels is produced (given by Eq. 5.4).

Are Cellular Automata Special?

The earlier discussion of the cellular automata model suggested that localized transmission changes the basic epidemic and endemic thresholds and creates new types of thresholds that are not seen in nonspatial models. At the end of this comparison between the cellular automata and partial differential equation models, however, it may seem that the changes in the fundamental epidemiological concepts have more to do with the specific framework of the cellular automata than with localized transmission per se. However, remember that the framework of the cellular automata is biologically intuitive. Individuals are discrete units. When the unit of a patch is an individual, then space is clearly divided into discrete patches. For the types of communicable diseases discussed here, disease dynamics are unstable at the level of the individual. The take-home message of this comparison between models is not that the results from the cellular automata model are somehow special, but that it is not localized transmission alone that is important. The combination of local transmission, discrete patches, and local unstable dynamics is key.

SUMMARY

One of the general principles to emerge from analyses of cellular automata models is that it is more difficult for a disease to persist when one considers the local nature of disease transmission instead of assuming that the population is like a well-mixed soup in which transmission is effectively global. In nonspatial models, disease persists if the reproductive rate is greater than unity. When transmission is highly localized, the reproductive rate must be much higher both for large epidemics to occur and for a disease to become endemic. Notably, this quantitative difference in the epidemic and endemic threshold does not occur in a partial differential equation model with local transmission. This points out that the combination of local transmission and the discrete individuals is important when studying the effects of localized transmission. Second, the cellular automata model with local transmission introduces two new types of thresholds. The first is a threshold neighborhood size. In the lattice model, disease dies out quickly if the contact neighborhood is small and the reproductive rate is below some critical level. As the contact neighborhood is increased, the disease will at some point persist and be able to cause a pandemic. The second is a threshold ratio of lifespan to the duration of the infectious period. In the models with global transmission, the disease can persist even if the infectious period is extremely short. In the cellular automata model with local transmission, the disease goes extinct if the infectious period is too short. Finally, in cellular automata models, local transmission greatly lengthens time between epidemics. The interepidemic time can be three to four times longer than in the analogous model with global transmission. Interestingly, the difference can be explained in large part by a lower effective transmission rate in the model with local transmission.

While it is clear that local disease transmission alters many of these basic concepts from classical epidemiological theory, the magnitudes of the changes depend in large part on the parameters of the models. The effects of local transmission can be swamped by high interaction rates between individuals, moderate neighborhood size, or high turnover at the individual level. This begs the question of whether local disease transmission changes disease dynamics in the real world, and answering this question means estimating the model parameters for real diseases. However, the basic Kermack-McKendrick model and the cellular automata models are cartoons that incorporate a few basic aspects of real diseases. Because they are so simple, it is nontrivial to make meaningful associations between model parameters and data collected on real diseases. Here is a simple example of the types of parameter estimation problems that arise: Figure 5.3 shows the change in the impact of disease on a population if one assumes local versus global transmission. The plot axes are the disease reproductive rate and the ratio of the lifespan to the infectious period. Lifespan and infectious period could be estimated for real organisms and their diseases, but in the model, organisms are assumed to die and lose infection at constant rate. If this is not the case for the real organism and disease (which will often be true), then lifespan and infectious period for the real organisms do not correspond to those parameters in the model. A more fruitful use of these simple models is to develop a general qualitative understanding of the behavior of complex spatial systems. Without such a foundation, the analysis of spatial models that include more biological realism becomes lost in a muddle of spatial and nonspatial effects.

CHAPTER SIX

Measles: Persistence and Synchronicity in Disease Dynamics

Neil M. Ferguson, Robert M. May, and Roy M. Anderson

INTRODUCTION

Directly transmitted viral and bacterial diseases of children that induce lasting immunity on the recovery of the host—such as measles—have long been the focus of attention for mathematical biologists and epidemiologists (Hamer 1906; Kermack and McKendrick 1927; Soper 1929; Bartlett 1957; Anderson and May 1991). This is in part due to their public health importance and the fact that the introduction of mass vaccination programs in many countries creates a replicated, very large-scale population perturbation experiment (Anderson 1989).

Although simple mathematical models of these systems are able to capture many features of the observed dynamics, such models have tended to exaggerate the ease with which the diseases could be eradicated by mass cohort immunization. This has triggered much recent interest in the spatial pattern of infection in highly immunized communities. The availability of a wealth of finely spatially stratified data arising from national and international infectious disease surveillance systems (especially good in North America and parts of Europe) has resulted in the development of a set of increasingly sophisticated spatially explicit models (Grenfell et al. 1995; Lloyd and May, in preparation; Rhodes and Anderson 1996).

This body of work is largely published in the epidemiology and mathematical biology journals and is hence not commonly referred to by ecologists interested in spatial dynamical problems in single-species or multispecies communities (Hassell, Comins, and May 1991). However, such epidemiological studies are of great relevance to ecologists for three reasons. First, directly transmitted virus-human host interactions are one of the simplest two-species systems in which a great deal is understood concerning the biology and typical course of infection in the host, the route of transmission between hosts, and the demography of the host population. Second is the availability of large spatio-temporally stratified datasets based on disease notification. Third, the availability of effective vaccines against many communicable childhood diseases has effectively resulted in a series of large-scale population perturbation experiments, where the reproductive success of the parasite has been perturbed by the introduction of mass immunization. The pattern of changes in the incidences of infection and disease post-immunization provide an excellent template against which to test ideas concerning the factors that determine pathogen population size and, more explicitly, the role of spatial factors in disease persistence. Study of the latter factor is often facilitated by heterogeneity in the level of immunization or host population density in different spatial locations. It is rare in ecological study that such long-term and detailed records are available of fluctuations in the abundances of two interacting species.

For these reasons, this chapter focuses on the role of spatial factors in the dynamics of a directly transmitted viral infection, specifically the measles virus, in human communities. Although at first sight the work may appear of most relevance to epidemiological study and disease control, these systems reveal many points of general interest in ecology and in particular in the assessment of the importance of spatial heterogeneity.

BASIC DATA AND DYNAMICS

The transmission cycle of measles is typically described in terms of four disease states: Susceptible, Exposed (but not infectious), Infectious, and Recovered (SEIR). An affected individual passes through each state in turn, spending around one week in each of the exposed and infectious states, before acquiring lifelong immunity. Since measles is highly infectious, the mean age at infection prior to vaccination was low—four to five years in most developed countries and one to two years in the developing world (Anderson and May 1985a,b)—with virtually everyone being infected by adulthood. Before the introduction of vaccination programs, measles used to be a major cause of childhood mortality and morbidity in the developed world, and it remains a serious public health problem in much of the rest of the world today.

Measles case notification data have been widely collected in many different communities and countries for many years (Cliff, Haggett, and Smallman-Raynor 1993), resulting in one of the most complete epidemiological datasets for a single disease yet collated. In many countries notification data were collected down to a relatively fine spatial scale; in the United Kingdom, for instance, notifications were collated separately for the fifteen hundred local boroughs across England and Wales, providing an unrivaled opportunity to compare the behavior of spatial transmission models with real data.

Except for small isolated communities (such as islands; Rhodes and Anderson 1996), measles incidence patterns are universally strongly periodic, with the majority of datasets exhibiting predominantly two-year cycles, and the remainder giving annual or triennial cycles (Cliff et al. 1993).

SIMPLE TRANSMISSION MODELS

It is relatively easy to translate the SEIR definition of infection status into a simple deterministic model of disease transmission within a population. We define X to be the number of susceptibles, H to be the number of infecteds (but not infectious), Y to be the number of infectious individuals, and Z to be the number of (immune) recovereds. The dynamics are then modeled by

$$\frac{dX}{dt} = \mu N - (\lambda + \mu)X$$

$$\frac{dH}{dt} = \lambda X - (\sigma + \mu)H$$

$$\frac{dY}{dt} = \sigma H - (\nu + \mu)Y. \tag{6.1}$$

For constant population N, Z can be calculated from $N = X + H + Y + Z$. The first equation above describes the rate at which people are born into the susceptible class (μN) and the rate at which susceptibles are infected (λX). The second equation describes the rate at which susceptible individuals move into the infected class (λX), and the rate at which infected individuals become infectious (σH). The last equation describes the rate at which individuals recover from infection (νY). All classes of individuals also experience a natural mortality with a per-capita rate of μ. Clearly, the mean latent and infectious periods are therefore given by $1/\sigma$ and $1/\nu$, respectively, while the mean lifespan is $1/\mu$.

The key parameter in this set of equations is λ, termed the *force of infection*, and defined to be the per-capita rate of infection for susceptibles. If we assume that an individual is equally likely to meet any individual in the population—the homogeneous mixing assumption—then λ is just given by

$$\lambda = \frac{\beta}{N} Y \tag{6.2}$$

where β, the transmission coefficient, represents the product of the mean number of contacts per unit time, and the probability of infection per infectious contact. Y/N is then just the fraction of all contacts who are infectious. This simplified definition of the force of infection gives rise to a nonlinear *mass-action* (Hamer 1906; Kermack and McKendrick 1927; Mollison 1984) term—$\lambda X = \beta XY/N$—in the SEIR equations above.

A stochastic version of the SEIR model can easily be constructed—the terms on the right-hand side of Equation 6.1 are merely considered to be the rates at which individuals move between disease states in a Monte Carlo simulation.

Early versions of the SEIR model, using constant β (Hamer 1906; Kermack and McKendrick 1927; Soper 1929; Bartlett 1957), exhibited simple dynamics; damped oscillations converging to a fixed point, which corresponds to the endemic equilibrium state of the disease. Since this is clearly at variance with the observed sustained cycles, annual seasonal forcing was introduced via the transmission parameter, usually as $\beta(t) = \beta_0[1 + \delta \cos(\pi t)]$ (Bailey 1975; Dietz 1976; Fine and Clarkson 1982a). This represents seasonal variation in transmission due to climate, mixing, or both patterns in children influenced by the structure of the school year. Here δ is termed the fractional forcing amplitude. The addition of seasonality dramatically changes the dynamics, giving rise to coexisting stable cycles, chaotic attractors and repellors, a complex bifurcation structure, and dynamical intermittency. This complexity, superficially resembling the observed dynamics, accounts for the popularity of the SEIR model as a description of the transmission of childhood diseases.

Stochastic SEIR models (Olsen and Schaffer 1990; Schaffer et al. 1990; Rand and Wilson 1991; Engbert 1993; Engbert and Drepper 1994; Bolker and Grenfell 1995a,b) add a further term representing the importation of infectives from an external source. This has the effect of restarting transmission after disease extinction. Typically, such a term modifies the definition of the force of infection thus:

$$\lambda = \frac{\beta}{N}(Y + Y_0) = \frac{\beta}{N}Y + \lambda_0 \qquad (6.3)$$

where Y_0 specifies the immigration of infectives per unit time or λ_0 specifies a "background" force of infection. The SEIR model with immigration represents the simplest possible spatial model (Engbert and Drepper 1994; Bolker and Grenfell 1995a,b; Ferguson, Nokes, and Anderson 1996b) of transmission, a single population coupled to an external reservoir of infection. Y_0 effectively acts as a minimum bound on Y (keeping Y above unrealistically low values of minute fractions of an individual) and for realistic levels of immigration is often sufficient to eliminate most of the deterministic chaotic behavior much discussed in previous work (Aron and Schwartz 1984; Schaffer and Kot 1985; Olsen, Truty, and Schaffer 1988; Olsen and Schaffer 1990; Bolker and Grenfell 1993, 1995a,b). This significant effect of an immigration term gives an indication of the possible importance of spatial structure. It also lends support to one of the major points we wish to stress in the ecological study of spatial phenomena—that stochastic effects are central to the interpretation of observed spatial pattern.

MIXING HETEROGENEITY AND AGE-STRUCTURED MODELS

Although the seasonally forced SEIR model certainly captures much of the gross structure of measles transmission dynamics, it suffers from a number of flaws. The level of seasonal forcing required to generate realistic time series is too large. The SEIR model (without immigration) also tends to generate epidemic troughs that are much too deep, and unrealistically high immigration levels of infectives are required to reproduce realistic critical population sizes in the stochastic SEIR model. More importantly for epidemiologists, the SEIR model only tracks total incidence and does not allow comparison with age-structured serological and incidence data. It is also insufficiently realistic to permit detailed investigations of the potential effects of different vaccination programs (Anderson and May 1983, 1985a,b, 1991; Anderson, Grenfell, and May 1984; Tudor 1985; Hethcote 1988; Babad et al. 1995). Finally, its assumption that the force of infection is constant across all ages is clearly at variance with estimates obtained from serological data (Grenfell and Anderson 1985).

These deficiencies can largely be overcome by introducing heterogeneity into the previously homogenous mixing pattern of the SEIR model. The easiest way of doing this is to introduce age into the model in the form of a discrete cohort structure (Schenzle 1984). Children are born continuously at rate B into the first cohort, but cohort members only move onto the next cohort at the beginning of each (school) year. Mortality is zero for the first twenty cohorts and is set to a constant rate μ in the last cohort. The deterministic Realistic Age Structured (RAS) model is described by the following equations:

$$\frac{dX_i}{dt} = \mu_i N - (\lambda_i + \mu_i) X_i$$

$$\frac{dH_i}{dt} = \lambda_i X - (\sigma + \mu_i) H_i$$

$$\frac{dY_i}{dt} = \sigma H_i - (\nu + \mu_i) Y_i. \tag{6.4}$$

Here the subscripts denote a cohort-specific variable, with $1 \leq i \leq 21$. An additional disease state, representing newborn hosts protected by maternal antibodies, is often added to achieve a better match to observed serological profiles (which record the prevalence of infection by age), but it makes little difference to the overall dynamics. Again, Z_i can be calculated from $N_i = X_i + H_i + Y_i$, since each cohort's total population is fixed. It is important to note that the discrete movement between cohorts at the start of each year acts as a significant seasonal forcing independent of any variation in β. Here λ_i is the force of infection experienced by susceptibles in cohort i and can be expressed thus:

$$\lambda_i = \sum_j \frac{\beta_{ij}}{N} Y_j + \lambda_0. \tag{6.5}$$

Here β_{ij} is the transmission coefficient from cohort j to cohort i, otherwise known as the WAIFW (who acquires infection from whom) mixing matrix (Anderson and May 1985b). It is this matrix that encompasses all mixing heterogeneity. Typically it is expressed in terms of broader age classes that form clear social groupings, most commonly 0–5 (preschool), 6–10 (primary school), 11–20 (secondary school/college), and 21+ (adults). It is this incorporation of heterogeneous mixing between age groups that is at the core of the RAS model's success —age structure alone (i.e., all β_{ij} identical) does not significantly improve on the realism of the SEIR model.

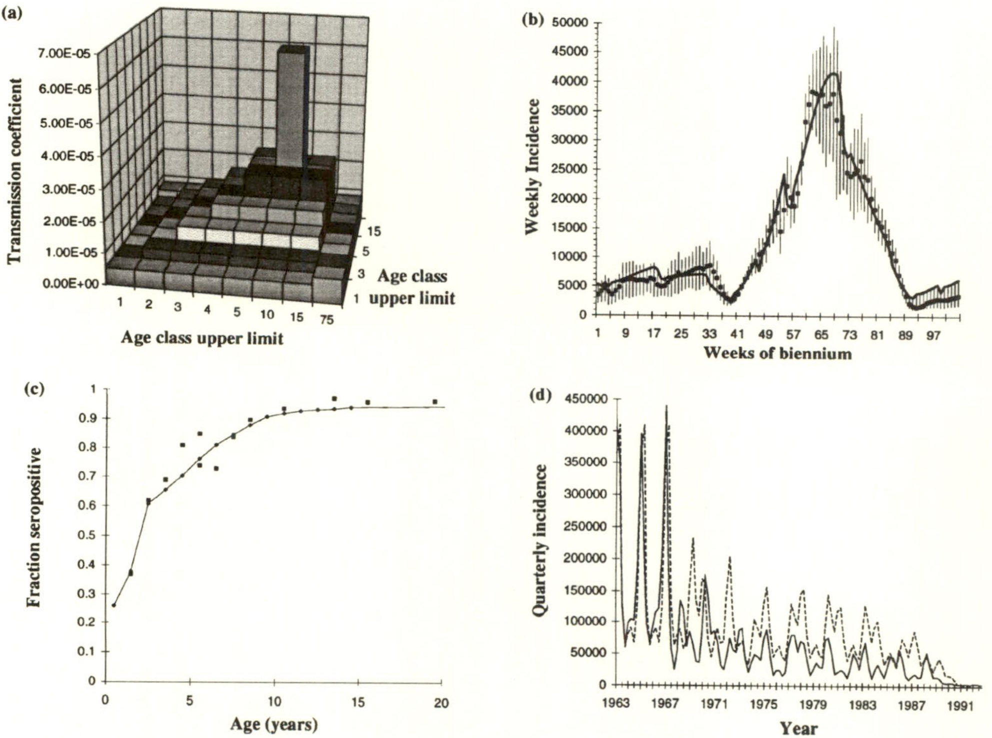

(a)
Transmission coefficient
7.00E-05
6.00E-05
5.00E-05
4.00E-05
3.00E-05
2.00E-05
1.00E-05
0.00E+00
1 2 3 4 5 10 15 75
Age class upper limit
1 3 5 15
Age class upper limit
(b)
Weekly Incidence
50000
45000
40000
35000
30000
25000
20000
15000
10000
5000
0
1 9 17 25 33 41 49 57 65 73 81 89 97
Weeks of biennium
(c)
Fraction seropositive
1
0.9
0.8
0.7
0.6
0.5
0.4
0.3
0.2
0.1
0
0 5 10 15 20
Age (years)
(d)
Quarterly incidence
450000
400000
350000
300000
250000
200000
150000
100000
50000
0
1963 1967 1971 1975 1979 1983 1987 1991
Year

The RAS model is remarkably successful in reproducing pre vaccination dynamics (Figure 6.1b) and both pre- and post-vaccination serological profiles (Figure 6.1c) and has been extensively used to explore vaccination strategies in both the deterministic and stochastic frameworks (Anderson and May 1983, 1985a,b, 1991; Schenzle 1984; Tudor 1985; Hethcote 1988; Babad et al. 1995), for both measles and other childhood diseases (Anderson and May 1991; Garnett and Grenfell 1992a,b; Ferguson, Anderson, and Garnett 1996). However, in the case of measles, the RAS model is poorer at reproducing incidence figures post-vaccination than pre-vaccination, consistently overestimating the expected numbers of cases (Babad et al. 1995; see Figure 6.1d). A key question, therefore, is whether the incorporation of space can correct this overestimation error and better describe the persistence of epidemics in populations.

EXPLICIT SPACE: METAPOPULATION MODELS

The multipatch version of the stochastic RAS (or SEIR; see Bartlett 1957) model is easy to formulate: Dynamics within patches are still described by Equation 6.4, but the force of infection experienced by age class i in patch a is now given by

$$\lambda_i^a = \sum_b \sum_j \frac{\beta_{ij}^{ab}}{N} Y_j^b + \lambda_0^a; \qquad (6.6)$$

that is, β now represents the transmission probability from age class j in patch b to age class i in patch a. For simplicity, β_{ij}^{ab}

FIGURE 6.1. RAS model dynamics. Parameters as in Babad et al. (1995). (a) WAIFW mixing matrix (transmission coefficients between age classes) for Babad et al. (1995) age class definitions. (b) Best fit of simulated time series (solid line) to mean U.K. notification data 1956–65 (square markers ± 1 S.D.). (c) Comparison of model (solid line) and observed (square markers) serological profiles for 1981 (twelve years into vaccination). (d) Simulated (dashed line) and observed (solid line) U.K. measles incidence 1963–93.

is usually factorized thus:

$$\beta_{ij}^{ab} = \mathbf{m}^{ab}\beta_{ij} \tag{6.7}$$

where β_{ij} represents the WAIFW mixing matrix between age classes used in the basic RAS model and $\mathbf{m}^{ab}$ describes mixing between patches. The off-diagonal elements of $\mathbf{m}$ typically represent the ratio of between-patch to within-patch transmission, with the on-diagonal elements usually being assigned value unity—assuming identical population densities for all patches. While we will use the factorized form of β_{ij}^{ab} here, it should be noted that by doing so we are assuming that age structure of between-patch mixing is the same as that of within-patch mixing. This is likely to be far from accurate, since older children and adults tend to be more mobile than younger children. The matrix $\mathbf{m}^{ab}$ specifies all spatial heterogeneity and structure in this model. We use two forms of $\mathbf{m}^{ab}$: all patches coupled to each other ("uniform coupling"), and patches lying on a square grid, each with a maximum of four nearest neighbors.

Persistence and Space

Neither the RAS or SEIR models successfully predict the correct critical population size required for persistence of measles infection without the inclusion of a substantial level of immigration of infection from external sources. Figure 6.2a shows the relationship between monthly fade-out fraction (the proportion of months with no infection) and community size as calculated from a wide range of empirical datasets (Bolker and Grenfell 1995a,b). It should be noted that the "island" data exhibit significantly higher fade-out fraction than the data from U.K. or U.S. cities of the same population. This is indicative that cities experience much stronger epidemiological coupling to neighboring areas than more isolated communities.

By adjusting the rate of immigration/importation of infectives—represented by λ_0, the external force of infection —it is possible to generate remarkably accurate fits to these

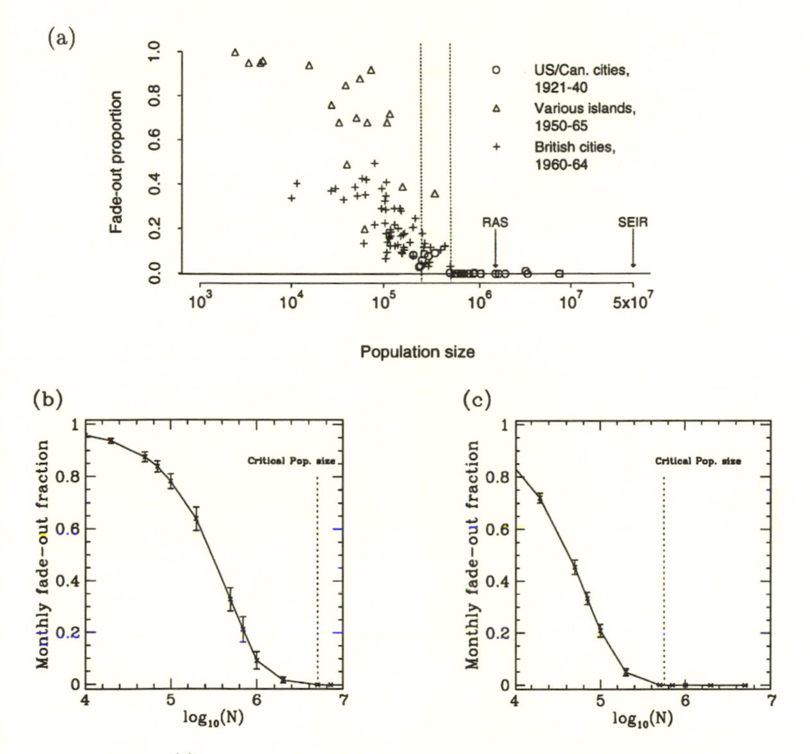

FIGURE 6.2. (a) Fade-out fractions for measles calculated from a range of empirical datasets (Bartlett 1960; Black 1966; Shaw 1990). Dotted lines indicate estimated critical community size of 250,000–500,000 (Bartlett 1957, 1960), and arrows indicate predictions of RAS and SEIR models with low immigration. Graph reproduced with permission from Bolker and Grenfell (1995a). Mean (of thirty runs) fade-out fraction generated by stochastic RAS model with (b) $\lambda_0 = 10^{-4}$, and (c) $\lambda_0 = 10^{-3}$. Parameters as in Ferguson et al. (1996b).

two types of fade-out fraction behavior. Figure 6.2b shows the fade-out fraction measured from multiple realizations of the stochastic RAS model with $\lambda_0 = 10^{-4}$ (approximately 6.2 infectives per million per annum, using the parameters given in Ferguson et al. 1996b). Figure 6.4c shows the results obtained with $\lambda_0 = 10^{-3}$. While neither plot represents a best fit to the island or city data in Figure 6.2a, it is clear that Figure 6.2b closely resembles the island-type behavior, while Figure 6.2c resembles the city type. However, in both cases the tail of the curves is too shallow, giving critical population sizes that are considerably too large.

In the case of the "island" simulations, λ_0 truly represents random immigration of infection from an external source (e.g., holiday visitors), while the magnitude of λ_0 required to match the city data is too large to be anything but a crude measure of epidemiological coupling to neighboring areas. The relative magnitude of λ_0 in the latter case—up to 1% of the overall force of infection—indicates that accurate modeling of measles transmission between cities warrants the use of metapopulation models.

Moving to use of the metapopulation model, Figure 6.3a shows the average fade-out fraction of a patch within a twelve-patch model with 0.1% coupling between patches and a value for λ_0 representative of true importation of cases. The critical population size is approximately 250,000 in the completely coupled case (i.e., all patches nearest neighbors), in agreement with the empirically measured estimates from city incidence data. The simulation performed using a square lattice, rather than uniform coupling, gives a higher fade-out fraction due to the lower number of nearest neighbors (4) and thus a lower overall mean coupling between patches.

An alternative way of viewing the metapopulation model is to ask how patchiness *within* a city- or island-sized population affects persistence. In this case we measure the fade-out fraction for all patches combined as a function of the combined population of all patches. This is typically what has been examined in most previous work (Bolker and Grenfell 1995a,b), due to the computational intensiveness of stochastic simulations of large populations. Figure 6.3b shows the results of such simulations using the same parameters as before. The curves for both patch network topologies achieve a reduction in the critical population size to a more realistic level (for isolated island populations) of around two million. It is interesting that the fade-out fraction for the square lattice patch topology is lower than for the completely coupled case (for $N < 600{,}000$) —the opposite trend to that seen when measuring the fade-out fraction for individual patches. This is because in the square

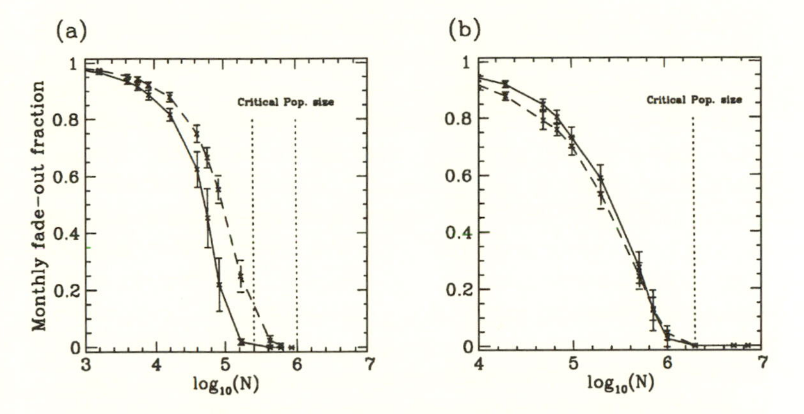

FIGURE 6.3. (a) Mean (of ten runs) fade-out fraction of individual patch from twelve-patch stochastic RAS model with 0.1% interpatch coupling and $\lambda_0 = 10^{-4}$. (b) Mean fade-out fraction for all patches from same ten runs.

lattice model the weaker mean coupling between patches means that patches on opposite sides of the lattice are more likely to be desynchronized than in the uniform coupling case, lowering the probability of simultaneous fade-out in all patches.

Synchronicity and Vaccination

We have already referred to the role that dynamical synchronicity between different spatial regions plays in measles epidemic patterns, concentrating on its role in affecting extinction behavior. In essence, as coupling between patches decreases so too does the level of synchronization, thereby enhancing persistence by making rescue effects in epidemic troughs more likely. However, decreased synchronicity also has a major effect on the periodicity and amplitude of epidemic cycles, making low-amplitude annual cycles the norm for extremely weakly coupled systems with large numbers of patches. The reason is clear: For very weak coupling, patches essentially behave independently, so the behavior of the total population is an average over multiple biennial epidemics (that are intrinsic to the dynamics of the two-species interaction), half of which have their maxima in even years, and half in odd.

One of the best ways of characterizing this behavior is by using the mean spatio-temporal correlation function between patches. This is generated by calculating the mean cross-correlation function (Chatfield 1989)—for a range of time lags—between all pairs of patches a particular distance from each other. In this context, "distance" is just the minimum number of links in the patch network between the pair of patches; for example, nearest neighbors are distance 1 apart, next nearest neighbors distance 2, etc. The distance 0 function is just the mean auto-correlation function across all patches. The one-year lag value of the cross-correlation function is particularly useful for measles, since its value typically increases as synchronicity falls, due to increasing numbers of patches having their epidemic maxima falling on different years of the biennium. Another summary statistic frequency used (Bolker and Grenfell 1995a,b; Grenfell et al. 1995) is the mean cross-correlation coefficient averaged over all pairs of patches—which is just the zero lag value of the cross-correlation function (see, e.g., Chatfield 1989 for a discussion of time-series analysis methods).

As an example, Figure 6.4a shows the spatio-temporal correlation function calculated from normalized measles notification data of the twelve Inner London boroughs (excluding the City of London, and using 1965 boundaries) from 1950–68 (prior to vaccination being introduced). Strong peaks are seen at zero and two-year lags. The effect of vaccination on synchronicity in this system is startling (Bolker and Grenfell 1995b). Figure 6.4b shows the difference between the spatio-temporal correlation function calculated from the London data in the period 1950–68 (Figure 6.4a) and that calculated from the period 1969–87 (post-vaccination). A drop of around 0.3 is seen in the zero-lag correlation, with a corresponding increase of around 0.15 in the one-year lag value—consistent with an effective drop in epidemiological coupling. There is also a smaller fall in the cross-correlation at two years lag.

An important point to note is the lack of significant distance-dependent effects in the London dataset—especially when coupling is strong. Thus epidemic spread in London is remarkable in its rapidity and homogeneity, especially consid-

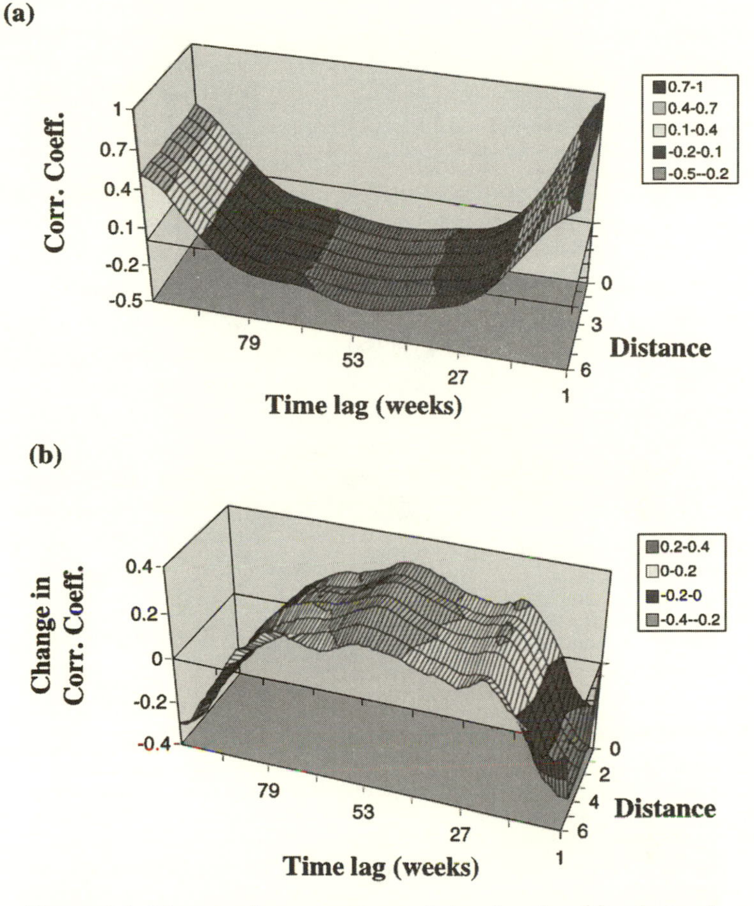

FIGURE 6.4. Mean spatio-temporal correlation function (a) calculated from notification data for the twelve Inner London boroughs (1950–68). (b) Change in spatio-temporal correlation function after vaccination for Inner London boroughs (from 1969–87 data).

ering the widely differing social conditions (density, age distributions) in different boroughs. Similar results are seen when one analyzes data from different English cities.

One can use the metapopulation RAS model to simulate these systems in an attempt to understand the mechanism underlying this decorrelation. Figure 6.5a shows the

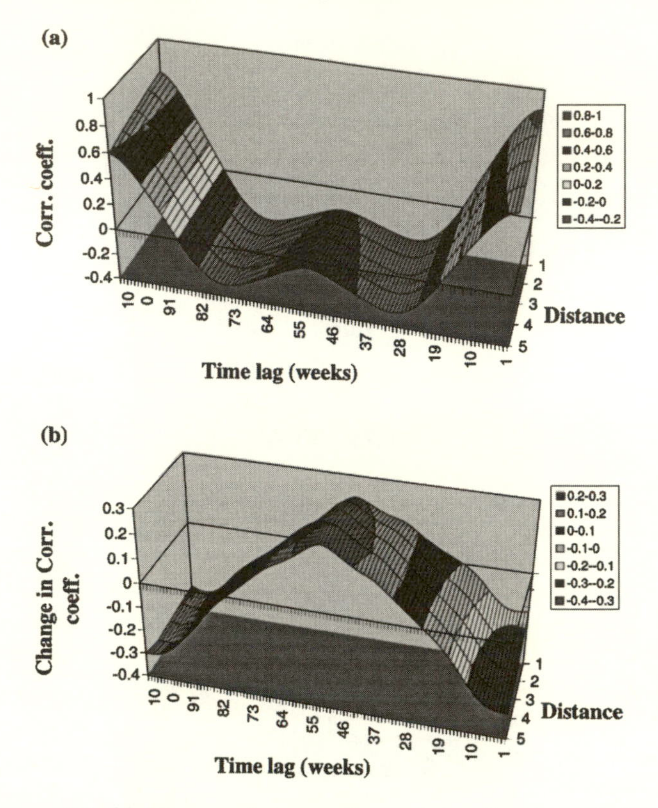

FIGURE 6.5. (a) Mean spatio-temporal correlation function calculated from twelve-patch RAS model with 3% interpatch coupling, 275,000 per patch (average of ten runs). (b) Change in spatio-temporal correlation function after vaccination for twelve-patch RAS model (parameters as for [a]).

pre-vaccination spatio-temporal correlation function from a set of simulations of the London boroughs (using a twelve-patch RAS model with 3% interpatch coupling), and Figure 6.5b shows the change in this function after the introduction of vaccination (30% coverage—though this makes little difference). Although the behavior seen is qualitatively similar to the data ((Figure 6.4), the model only produces a drop of 0.17 ± 0.02 in the cross-correlation coefficient and overestimates the increase in the correlation function at one year lag. Note that, as a whole, cross-correlation is weaker between English cities than between London boroughs by approximately 0.1—both before and after vaccination; that is, in both cases a drop of around 0.3 in correlation is seen. This independence of the magnitude of decorrelation following vaccination from the pre-vaccination correlation (and thereby the interpatch coupling) is also not seen in simulations, where the drop in correlation decreases markedly with initial correlation and coupling.

Hierarchical Spatial Structure

Deeper understanding of the role of space in the persistence of measles and the patterns of fluctuations in incidence over time is likely to be aided by further data analysis based on fine stratification by spatial location, and at a range of spatial scales, combined with the study of models that incorporate a hierarchy of spatial scales (i.e., mixing within a family unity within a school, within a community, and between towns and cities). At present it is unclear which spatial scale dominates the observed pattern, although evidence is accumulating that both micro (within family) and macro (between cities) are of great importance (Bolker and Grenfell 1995a,b; Rhodes and Anderson 1996). The introduction of mass vaccination provides some vital clues to the respective significance of events occurring at differential spatial (and social) scales, since it may have induced a phase transition in the transmission dynamics through its importance at all scales. One approach is to examine the predictions of models that incorporate a hierarchy of different scales (a computationally intensive approach) to see if better fits to observed trends can be obtained. A simple beginning is illustrated in Figure 6.6a, which shows the mean between boroughs (100,000 people each) and between cities (500,000 people each) cross-correlation functions generated from a model with two spatial scales—five cities (with a 0.2% coupling between them), and five boroughs in each city (with a 2% coupling between each borough in a city). Figure 6.6b shows the change in this correlation function following the

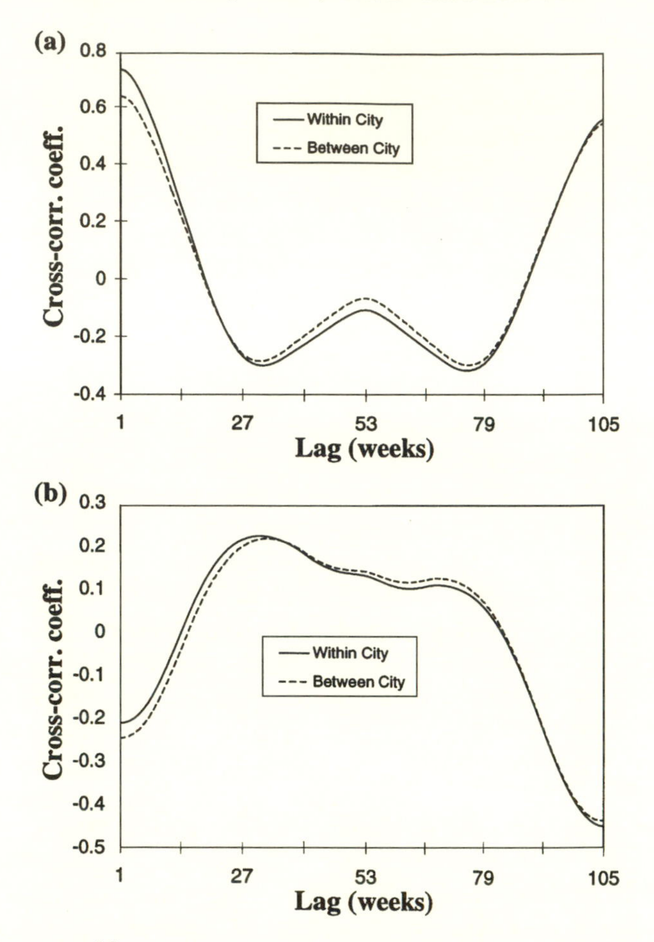

FIGURE 6.6. (a) Mean pairwise cross-correlation function calculated from twenty-five-patch RAS model with two scales (2% coupling between boroughs, 0.2% between cities, 100,000 per patch, average of ten runs). (b) Change in mean pairwise cross-correlation function after vaccination (parameters as for [a], using U.K. vaccination coverage data).

introduction of vaccination. Comparison with Figure 6.4b shows a dramatic improvement in the realism of the model in capturing the observed pattern from the London boroughs. However, the pre-vaccination cross-correlation function still fails to fully reproduce the observed behavior.

It is likely that use of explicitly spatial models will help to remove the existing discrepancies between incidence time-series predictions of the simple RAS model and the notification data record (Babad et al. 1995). Although the RAS model produces a remarkably good fit to the pre-vaccination time series, agreement is much poorer post-vaccination. In particular, the nonspatial RAS model predicts biennial epidemics with maxima much higher than actually observed. However, one effect of the observed 40% drop in correlation would be to significantly reduce epidemic peak heights—an effect completely ignored in the nonspatial model.

It is as yet unclear whether the use of metapopulation structure substantially improves model *predictivity*. In general, time series of the incidence of measles become less predictable (in terms of predicting epidemic peaks) after the dynamics are perturbed by mass vaccination (Fine and Clarkson 1982b). Metapopulation models better reproduce this drop in predictability (probably associated with decorrelation), but this does not necessarily mean that they are better than nonspatial models as tools to predict any particular future incidence time series, merely that they provide a better description of the ensemble behavior of the measles transmission system. However, such models do offer the potential to test more accurately the potential outcomes of different control programs, based on mass immunization.

SUMMARY

We have demonstrated how building increasingly detailed population models of the transmission of a simple viral infection agent within its host population allows us to better describe the observed pattern of infection. The progress in this field of population ecology is tightly linked to the availability of very detailed longitudinal and spatial records of the abundance of the pathogen and the demography and behavior of its human host. The datasets are replicated by country and further include the results of very large-scale population experiments induced by mass vaccination. Programs for mass vaccination differ between countries, both in their detailed implementation (e.g., age at vaccination) and scale (level of vaccination). As such, much is likely to be gained in the future by comparative studies between countries with good records of disease incidence. These databases represent a unique collection of information whose value is not widely appreciated by ecologists. This is unfortunate because many basic concepts in population ecology are well illustrated with the theory surrounding the study of simple direct lifestyle pathogens and the analyses of the extensive epidemiological data: for example, the roles of space in promoting parasite persistence and seasonality in perpetuating long-term cycles, and the importance of stochasticity for most spatially structured systems.

The addition of metapopulation structure to simpler transmission models improves persistence and at least partially captures the decorrelating effect of vaccination. Early results using multilayered metapopulation structure—modeling patchiness both within and between cities—suggest that the coupling of different scales amplifies the magnitude of decorrelation. Additional insight may be provided by complementary model designs that attempt to capture the behavior of the host at different spatial locations, using lattice (Rhodes and Anderson 1996), dyad, or neighborhood (Keeling and Rand in preparation) structures.

Although spatial models provide valuable insight into disease dynamics, their practical application is hindered by the difficulty of parameter estimation—a problem familiar to many ecologists studying competition, predator-prey, or host-parasite systems. Theoretically, epidemiological coupling between cities might be quantified by measuring human mobility patterns, but as yet such work is in its infancy. At a smaller scale, although much more is known about the social networks of children and adolescents in the sociological context, this information has not yet been quantified in an epidemiologically meaningful way. Similar problems are met when considering the potential effects of environmental and demographic heterogeneity. The exclusion of these effects from the models presented here is likely to account for some of the existing discrepancy between models and data.

Finally, we hope we have convinced the reader that infectious disease transmission within human communities represents a test-bed system for many ecological theories of space, persistence, and synchronicity. In particular, the measles virus transmission system offers ecologists the potential to test models and ideas against the template provided by some remarkably detailed spatially structured longitudinal datasets.

ACKNOWLEDGMENTS

We thank Bryan Grenfell, Ben Bolker, David Rand, Matt Keeling, Alun Lloyd, Chris Rhodes, Geoff Garnett, and Jonathan Swinton for useful conversations on the work presented here, and the Wellcome Trust for providing fellowship (N.M.F.) and research grant (R.M.A.) support. R.M.M. thanks the Royal Society for research support.

CHAPTER SEVEN

Genetics and the Spatial Ecology of Species Interactions: The *Silene-Ustilago* System

Janis Antonovics, Peter H. Thrall, and Andrew M. Jarosz

INTRODUCTION

To population geneticists, the idea that limited dispersal may influence subsequent evolutionary processes not only has been part of the fabric of the discipline but has also been a driving motivation for a large number of theoretical and empirical studies designed to clarify, elaborate, and illustrate (some may even say, deify) the early, seminal ideas of Sewall Wright (1931, 1940, 1943). He argued that evolution should be more rapid in subdivided populations than in a panmictic population because chance effects in small semi-isolated populations permit characters to move from one adaptive state to another through an intermediate state of lower fitness. This idea has received much attention from evolutionary biologists (Wade 1992; Barton 1992; Whitlock 1995), and there is now a particularly large literature on the influence of limited dispersal (''isolation by distance'') on spatial genetic substructuring (Epperson 1993). However, the relationship of much of this work to ideas in ecology has remained tenuous. One of the main reasons is that studies on genetic substructuring have focused primarily on neutral genes. Yet neutral genes are of little direct interest to the ecologist concerned with the causal processes determining numerical abundance. Apart from studies on the interaction of gene flow and selection (Endler 1977), consideration of how spatial substructuring of individuals influences selection response is rare in the population genetics literature. Some time ago, Levin and Kerster (1975) showed that if individuals occupy positions on a spatial array, then the dynamics of selection can be greatly affected by seed- and pollen-dispersal distributions. Yet there have been few subsequent attempts to comprehensively "test" the robustness of population genetic theory in spatially extended populations. That this theory may be drastically changed is now being suggested by ecologically motivated studies. For example, Molofsky et al. (1997, in preparation) and Durrett and Levin (1997) have shown that some forms of positive frequency-dependent selection can maintain polymorphism in stochastic spatially distributed systems. This is not possible in single, unstructured populations. Similarly, the degree of spatial aggregation of individuals can greatly affect the conditions for coexistence under competition (Pacala 1986a; Kreitman, Shorrocks, and Dytham 1992). Therefore the outcomes of even quite simple types of interactions among species (qua genotypes) may be qualitatively different when considered in a spatial context.

It is tempting for an ecologist to argue that "genetics is only in the details" and to subsume genetic variation as just another form of heterogeneity among the many inescapable and therefore perhaps ignorable complications of an already complex discipline. For example, at a recent symposium on disease in natural populations (Grenfell and Dobson 1995), there was a discussion section focused on the issue of "Is genetics just another heterogeneity?" In coevolutionary systems, genetic composition is likely to interact strongly with numerical abundance and vice versa. In host-pathogen systems, for example, the presence of genetic variation can greatly influence numerical dynamics and coexistence (May and Anderson 1983; Antonovics 1994). Conversely, purely genetic models that omit numerical dynamics fail to capture conditions for maintenance of resistance polymorphisms (Antonovics and Thrall 1994) and fail to predict how pathogen virulence might evolve (Lipschitz and Nowak 1994).

In the context of spatially structured populations and metapopulations, there are important reasons why it is productive to include a genetic perspective. First and foremost, viewing connectedness simply as a property of colonization ignores the connectedness that comes from gene exchange. Gene-flow distances can be very different from colonization distances; this is particularly obvious in plants, where pollen is dispersed quite differently from the seeds. Thus genetic rescue (e.g., input of resistance genes into a diseased population) may be as important as ecological rescue (input of propagules).

Second, colonization events (and extinction events if they are gradual) are almost invariably accompanied by genetic drift ("founder effects"); genetic stochasticity accompanies demographic stochasticity. This may create large differences in the genetic composition of founding populations with these differences declining through time as a result of gene flow among populations (McCauley 1993). In the metapopulation that we have been studying, newly founded populations of the plant *S. alba* are more differentiated with regard to allozyme and chloroplast DNA markers than are long-established populations (McCauley, Raveill, and Antonovics 1995). In host-pathogen systems (and analogous coevolutionary interactions) such chance effects may result in a severe dislocation of any local correspondence between host resistance genes and pathogen virulence genes (Jarosz and Burdon 1991; Frank 1997). The fractionation processes will become even more severe as the numbers of genes and alleles involved in the interactions increase and may lead to locally unstable and unpredictable dynamics ("We are ready to see that host-parasite genetics is like the weather"; Frank 1997).

Third, because colonization and extinction result in increased variation among populations, it becomes important to explore the possibility that group selection may be an effective force. The potential consequences of group selection can be enormous (Gilpin 1975; Boerlijst, Lamers, and Hogeweg 1993; Kelly 1994). However, evaluation of group selection in nature requires knowledge of metapopulation dynamics at both a genetic and an ecological level.

Fourth, conflicts between selection within populations and selection for colonization or population persistence (e.g., allocation to dispersal *vs.* competitive success) may limit character evolution (Roff 1994; Olivieri, Michalakis, and Gouyon 1995), and this in turn may limit evolutionary response to extinction (Meagher, Antonovics, and Primack 1978) or to range extension at species boundaries (Carter and Prince 1981). Metapopulation genetics may therefore play a crucial role in explaining limits to species distributions and predicting evolutionary responses to environmental change.

Therefore, while population geneticists can point with pride to the achievements of their discipline in taking spatial processes into account, there remain many unexplored areas, and there is room for new ideas and insights. There is also a need for many of the ideas to enter the mainstream of population genetics. For example, although all individuals live in spatially explicit situations, very few population genetics texts consider dispersal as a primary fitness component; at best, migration is introduced as a complication of the Hardy-Weinberg law and then promptly forgotten.

In this chapter our primary goal is to use our studies on the *Silene-Ustilago* host-pathogen system to illustrate how the interaction of genetic variation with population dynamics is critically important for host-pathogen dynamics on a broad regional scale. Our secondary goal is to illustrate some methodological principles regarding the study of spatially extended populations. In particular, we hope to convince the reader that the simultaneous study of multiple populations is not necessarily a dauntingly impossible task; indeed, it may actually be easier and more informative than the detailed study of a few target populations.

THE *SILENE-USTILAGO* METAPOPULATION

We have been studying populations of the short-lived perennial plant *Silene alba* (white campion) and its associated fungal pathogen *Ustilago violacea* (anther smut). The disease has an intriguing biology. Infection results in anthers that produce fungal spores rather than pollen (Figure 7.1). The disease is pollinator transmitted, and diseased plants are sterilized. This system has added interest because the transmission properties of the disease have much in common with other sexually

(a)
(b)

transmitted diseases (Thrall, Antonovics, and Hall 1993; Lockhart, Thrall, and Antonovics 1996; Roy 1994; Kalz and Schmid 1995).

In our study area, in the Allegheny Mountains of western Virginia, the host plant is almost entirely restricted to roadsides, and, moreover, the pathogen is restricted to this one species (Antonovics et al. 1995b). Because the plant is distributed in patches of differing sizes and spacings, which may coalesce or separate due to colonization and extinction events, we do not define a population in terms of the patches themselves but count numbers of diseased and healthy individuals within contiguous forty-meter segments of roadsides (Antonovics et al. 1994). Local landmarks (unusual trees, driveways, telephone poles, etc.) are used to demarcate each segment. The scale of forty meters includes perhaps one or two, but not many, genetic neighborhoods (as estimated from spore-, pollen-, and seed-dispersal distances). Moreover, by pooling field data from adjacent segments, we have found that the patterns of disease incidence are remarkably robust over several scales (Figure 7.2). The census includes several thousand segments spanning 150 kilometers of roadsides, of which about four to five hundred are occupied by *S. alba* in any one year. For the past nine years, we have counted the number of diseased and healthy individuals within each segment, followed by a recensus later in the season to check extinctions of the host or pathogen. We generally make no attempt to map individuals within a segment to a precise location (except to help relocate, say, rare diseased individuals or new colonists). Our census is therefore simple and rapid, and fieldwork can be completed by three crews of two to three people in one week.

FIGURE 7.1. (a) A diseased plant of *Silene alba* from Virginia showing flowers with the conspicuous black centers that result from the production of spores by smutted anthers. (b) Sections of healthy and diseased flowers of *Silene alba*. Top panel: male flowers. Bottom panel: female flowers. Note that in females the smut fungus induces production of stamens with anthers that carry smut spores, and the female gynoecium is rudimentary and sterile.

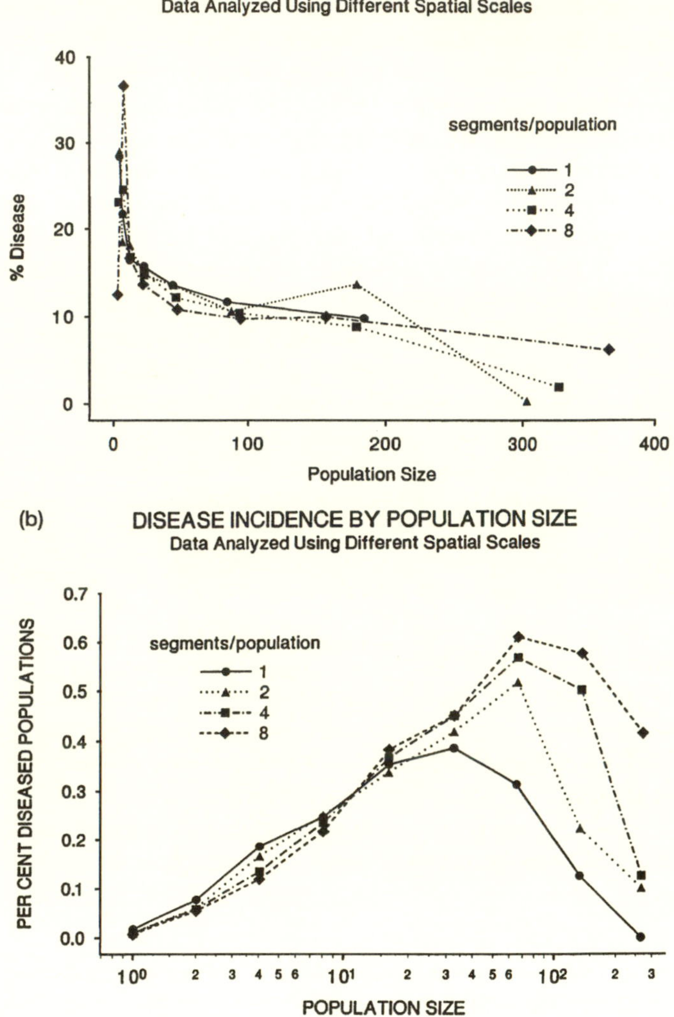
(a)
DISEASE PREVALENCE IN DISEASED POPULATIONS
Data Analyzed Using Different Spatial Scales
segments/population
1
2
4
8
% Disease
40
30
20
10
0
0
100
200
300
400
Population Size
(b)
DISEASE INCIDENCE BY POPULATION SIZE
Data Analyzed Using Different Spatial Scales
segments/population
1
2
4
8
PER CENT DISEASED POPULATIONS
0.7
0.6
0.5
0.4
0.3
0.2
0.1
0.0
10⁰
2
3
4 5 6
10¹
2
3
4 5 6
10²
2
3
POPULATION SIZE

Colonization, extinction, and population interconnectedness play an important role in the dynamics of this pathosystem (Antonovics et al. 1994; Thrall and Antonovics 1995). Between 1989 and 1993 the survey included from 412 to 494 occupied segments (which we call populations) per year. Of these, from 16%–19% were diseased, with an average disease frequency of 24%–42%. The populations have a high turnover rate: Extinction rates of healthy populations have been 14%–22%, and the disease has been lost from host populations at a rate of 19%–36%. The relative constancy of the system is maintained by correspondingly high colonization rates: Over this period, colonization rates were 15%–29% for healthy populations and 23%–45% for the disease. Extinction rates are higher for small populations, and colonizations are decreasing functions of distance from preexisting populations (Thrall and Antonovics 1995). Growth rate of healthy populations is density dependent, with the disease having a marked impact on population growth. In particular, high levels of disease shift population growth rates from positive to negative values (Figure 7.3). However, there is no significant difference in the extinction rate of diseased and healthy populations when corrected for population size. The impact of the disease on population extinction is therefore gradual; the disease results in a declining population growth rate, and small population size in turn presages an increased probability of extinction. The overall

FIGURE 7.2. (a) Relationship between the size of diseased populations and disease prevalence (percent infected individuals) for different sampling scales of the metapopulation. The different sampling scales are generated by pooling adjacent roadside segments into successively larger groups (as shown in the key). Data are from the metapopulation census (1988–93). To account for ascertainment bias, percent infection is calculated as $100 \times (D - 1)/(T - 1)$, where D and T are number diseased and total number, respectively. (b) Relationship between the probability that a population is diseased and the size of the population for different sampling scales of the metapopulation. The different sampling scales are generated by pooling adjacent roadside segments into successively larger groups (as shown in the key). Data are from the metapopulation census (1988–93). The points on the x-axis are means of logarithmically increasing size classes.

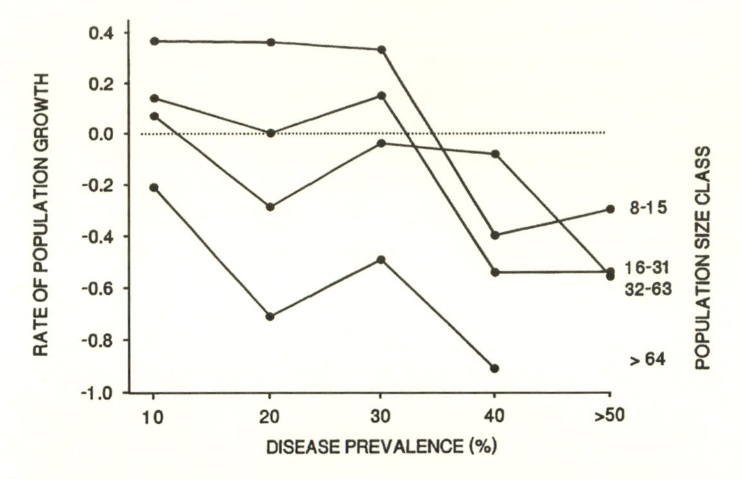

FIGURE 7.3. Growth rate of diseased populations of different sizes as a function of disease prevalence. Growth rate is measured as the log (numbers at time $t + 1$/numbers at time t); prevalence is the frequency of diseased individuals at time t. Data are for successive censuses of the metapopulation for the period 1988–93. Populations in the smallest size class (< 8) showed no significant relationship of growth rate with disease frequency and are not shown.

effects of the pathogen on host abundance are therefore difficult to infer directly, but using the simulation described below, we have shown that the presence of the pathogen can more than halve the number of occupied segments in the metapopulation as a whole. Such long-term regional consequences would be imperceptible from a simple, one-time descriptive study of disease incidence.

Superimposed on this metapopulation structure is the "complication" that there is substantial genetic variation for disease resistance in the host plant. Some genotypes are almost completely resistant, yet others are very susceptible; moreover these differences are highly heritable (Alexander 1989; Alexander, Antonovics, and Kelly 1993; Biere and Antonovics 1995). However, the precise genetics underlying the resistance is not known. Additionally, there are large fitness costs associated with resistance in the absence of the disease; more resistant plants flower later in the season and produce fewer flowers

(Alexander 1989; Biere and Antonovics 1995). Unexpectedly, the fungus appears to be relatively uniform with regard to its virulence, and therefore this host-pathogen system does not follow the classical gene-for-gene scenario (Burdon 1987; Jarosz and Burdon 1991).

THE *SILENE-USTILAGO* SIMULATION

In any large-scale study, experiments that manipulate the entire system are almost impossible. No doubt the metapopulation experimentalist dreams of the day when military-style spending (perhaps accompanied by military-style invasion and coercion) are part of ecology, but in the interim the only recourse is to develop simulation models of the system and to study it "experimentally" by manipulating parameters and conditions on a computer.

Because our goal has been understanding, not management, we have not tried to develop a totally "realistic" model, as might be the case if we wished to make precise predictions about the future fate of *Ustilago* and *Silene* in Giles County, Virginia. Instead, our strategy has been to begin with simple and general heuristic models of pollinator (or sexually) transmitted diseases and to add minimal complexity in a stepwise fashion so as to capture particular features of a real-world metapopulation. By having an understanding of the simpler single-population models, we can then assess the importance of the added features of spatial structure.

Within Population Dynamics

We assume that resistance is determined by a single locus with two alleles, but that the pathogen is genetically uniform. We have no knowledge of the number of genes involved in resistance and make the one-locus assumption for simplicity. We assume that plant reproduction occurs early in the season and is followed by infection, overwintering death, or both. The latent period between spore deposition and the appearance of infected flowers averages over six weeks (Alexander et al. 1993), and plants that become diseased in the first season still show a substantial reproductive success in that year (Alexander 1990; Biere and Antonovics 1995). We also assume that there is no host recovery and that the death rates of healthy and diseased individuals are the same. Recovery can occur, but it is usually from late-season infections that are unlikely to impact greatly on the overall dynamics. Death rates of diseased plants can be greater than those of healthy plants in some years (Alexander and Antonovics 1995; Thrall and Jarosz 1994b), although the overall averages are usually not significantly different. If we let X_t, Y_t, and N_t represent, respectively, the numbers of healthy hosts, infected hosts, and total host population size at time t, then the within-population dynamics can be represented by the following equations (for brevity, we only present the equation for the ith host genotype):

$$X_{i,t+1} = X_{i,t}[b_i + (1 - P_i)(1 - d)] \tag{7.1}$$

$$Y_{t+1} = \left(Y_t + \sum_i P_i X_{i,t}\right)(1 - d) \tag{7.2}$$

where d is the death rate and b_i is the recruitment rate (number of seeds reaching adulthood); a cost of resistance is included by assuming that b_i is greater for the less resistant genotype. P_i is the probability that a healthy plant becomes diseased. Under nonlinear frequency-dependent disease transmission, this is given by

$$P_i = 1 - \exp\left(-\beta_i \frac{Y_i}{N_t}\right). \tag{7.3}$$

The parameter β represents the effectiveness of disease transmission and can be expanded to take into account both number of contacts (i.e., pollinator visits) and per-contact infection rates (Thrall, Biere, and Uyenoyama 1995; Antonovics et al. 1995b). Seeds of each genotype are produced according to Mendelian expectations based on the frequency of genotypes in the pollen pool (including immigrant pollen) and the fecundity and genotype of each female parent.

We assume the recruitment rate b_i declines hyperbolically as population density (N_t) increases such that per-capita reproduction is given by

$$b_i = \frac{\lambda_i}{\gamma N_t + 1} \tag{7.4}$$

where λ_i is the maximum reproductive rate of the ith host and γ is the strength of density dependence. Hyperbolic functions are good representations of density-dependent growth in plant populations (Harper 1977; Thrall, Pacala, and Silander 1989). In the simulation, we assign carrying capacities to the component populations by varying γ. We assume that the carrying capacity of the populations is variable, and we infer the distribution of carrying capacities from the average size of healthy populations that have persisted for the whole census period.

Among Population Dynamics

We assume that the dynamics within each segment are deterministic, whereas the dispersal and extinction/colonization phases are stochastic. This allows us to distinguish chance processes associated with spatial structure from other chance effects occurring within small populations. It also allows a clearer comparison of the simulation outcomes with results from single-population models.

Seed, pollen, and spores are dispersed using dispersal curves that approximate the empirical data for new colonizations of the plant and the pathogen (Antonovics et al. 1994). Following dispersal, new numbers of hosts and pathogens are calculated for each patch, and the dynamics are repeated. We consider initial colonists and subsequent migrants as part of the within-population dynamics. At the end of each time interval, for both disease and host, a probability of extinction is calculated for each occupied segment using the relationship between extinction probability and populations size as determined from the census data (Thrall and Antonovics 1995).

Unless otherwise indicated, the following estimates of the disease transmission parameter (β_i), the birth rate (λ_i), and the death rate (d) were used in the simulation. For susceptible and resistant hosts, respectively, values of β_i were 5.8 and 0.4, and values of λ_i were 2.0 and 1.5. The death rate for healthy and diseased individuals was assumed to be 0.5. These values were obtained by pooling data from several field experiments (Thrall and Jarosz 1994a,b; Alexander and Antonovics 1995). An example of the graphical output of a typical simulation run is shown in Figure 7.4a. It can be seen that as the disease spreads in an area it reduces the number of occupied sites, and often the disease goes locally extinct as a result of both the increased spacing among occupied sites and the local increase in the resistance gene (not shown).

IN VITRO METAPOPULATION EXPERIMENTS

In this section we use the simulation of the *Silene-Ustilago* system to carry out "experiments" that explore issues pertaining to the interaction of numerical and gene frequency

FIGURE 7.4. (a) Example of graphical output of the metapopulation simulation. The horizontal axis represents a linear array of six hundred roadside segments, and the vertical axis represents a time period of three hundred years starting at the top. Each screen pixel is therefore one roadside segment at one time interval. White (background) represents unoccupied segments. Gray represents healthy populations, dark gray represents diseased populations with less than 50% diseased individuals, and black represents heavily diseased populations (more than 50% diseased). (b) Graphical output of the metapopulation simulation illustrating the operation of Wright's shifting balance theory (Wright 1931). The form is as in (a) except that the shading represents the genotypic frequencies of *AA*, *Aa*, and *aa* in each roadside segment. Gray represents populations fixed (frequency > 99%) for *a*, dark gray represents polymorphic populations, and black represents populations fixed for *A*. There is heterozygote disadvantage (relative fitnesses are *AA* = 1, *Aa* = 0.5, and *aa* = 0.67), but *A* is initially at a lower frequency (10% in all populations). Single-population deterministic models predict that *A* should be eliminated; instead, it spreads because it attains a high frequency in local sites due to stochastic founding events, as predicted by the shifting balance theory.

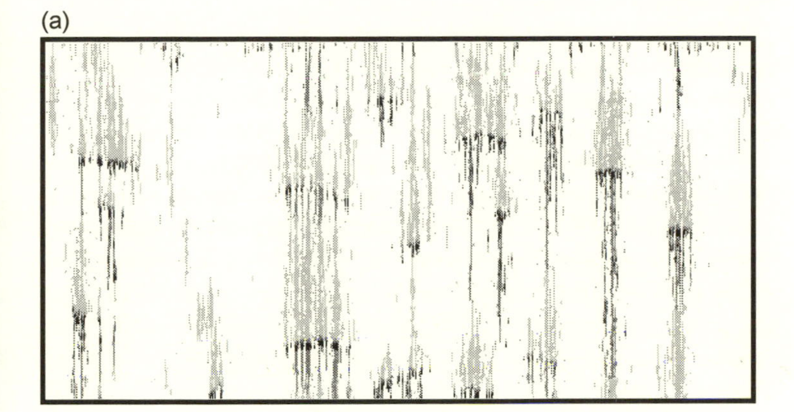

dynamics in spatially extended populations. We first investigate how the expectations from single-population dynamics are changed by metapopulation structure, in populations where the host is either genetically uniform or where it shows genetic variation for resistance.

In our second experiment, we "pretend" (using our simulation) that we are ecologists who have made field observations of the disease, but that we have done so without knowledge of the underlying genetics. To do this we run our simulation assuming the transmission parameters and reproductive outputs are weighted averages of the component genotypes (as would happen if we made measurements without regard to the underlying genetic heterogeneity). We then ask whether assuming no genetic variation makes a difference to the predictions about subsequent coexistence and abundance of the host and pathogen. Given that extra effort is needed to characterize genetic parameters in a field study, it is important to know whether genetics is simply "another form of heterogeneity" that has little effect on the average trajectory of the system, or whether it can have substantial effect on the dynamics.

In the final experiment we illustrate how the outcome of selection in single populations can be quite different from the outcome of selection in a spatially explicit and more realistic ecological context. When we first simulated selection on alleles at a single locus in an earlier *Silene-Ustilago* model, we found that the rate of loss of a deleterious allele was far slower in the metapopulation than would have been predicted in a single large population (Thrall and Antonovics 1995). Here we use the example of selection at a single locus when there is heterozygote disadvantage to illustrate the potential operation of Wright's shifting balance theory in the *Silene* metapopulation that we have been studying.

Experiment 1: *The effect of metapopulation structure on coexistence in genetically uniform versus genetically variable host-pathogen systems*

We begin by considering how host-pathogen coexistence is affected by different levels of host resistance in a single genetically uniform population. If the hosts are uniformly as resistant as the most resistant genotypes that we find in the field, then the pathogen cannot persist in a single population (Table 7.1a; lowest value of β). On the other hand, if the hosts are uniformly less resistant (i.e., as susceptible or even more so than the most susceptible genotypes found in the field), then hosts and pathogens coexist (Table 7.1a; high values of β). Because we assume reproduction in any given season always occurs prior to disease transmission (Eqs. 7.1–7.4), it is not possible for the pathogen to cause the extinction of the host. However, where disease transmission occurs prior to reproduction, the pathogen can cause the extinction of the host (Antonovics 1992; Thrall et al. 1995).

We next compare the predicted single-population dynamics with the outcomes in the metapopulation simulation. The results (Table 7.1a) show that in genetically uniform populations it is harder for the disease to persist in the metapopulation as a whole than in a single population. This is because in small populations the disease has a high extinction probability, and all newly founded populations are initially healthy because the disease is not seed transmitted. Only when the host is quite susceptible (β is 2 or greater) does the disease persist consistently (90% of the time or more) in the metapopulation (Table 7.1a). Even when the disease persists, a substantial fraction of the populations are healthy because they are newly established and have not yet acquired the disease.

We now introduce genetic variation for resistance and susceptibility, and an associated cost of resistance (as we have found in our real-world populations), and again examine single versus metapopulation dynamics. For brevity, we term the less resistant genotype "susceptible" and the more resistant genotype "resistant." We use these terms to evoke contrasting properties, acknowledging that susceptibility is really the inverse of resistance. In a single population, genetic polymorphism is more likely when the genotypes differ greatly in their resistance and susceptibility (Figure 7.5); this happens over quite a broad range of resistance costs (as in other models of this type; Antonovics and Thrall 1994). When the genotypes differ less in their resistances, either the resistant or the susceptible allele goes to fixation (depending on the cost of

TABLE 7.1. Effect of varying plant resistance to the anther smut disease on disease frequency, in either a single population or in the metapopulation. The single-population values are from deterministic models, and the metapopulation values are means of ten runs of the *Silene-Ustilago* metapopulation (standard errors are not shown but were generally 5–10% of the means). Values for % persistence refer to persistence for three hundred generations out of one hundred runs; values for % disease are averages of diseased populations only. Models are described in the text.

(a) Disease frequency in genetically uniform hosts	Resistance (β)				
	0.41	1.25	1.50	2.00	5.86
Single Population					
% individuals diseased	0	13.2	21.8	32.2	48.5
Metapopulation					
% disease persistence	0	9.0	38.0	90.0	91.0
% sites diseased	—	6.6	22.2	55.5	43.0
% disease	—	42.1	36.0	19.5	23.2

(b) Disease frequency and resistance in genetically variable hosts	Resistance (β) of *RR*			
	0	0.41	1.50	2.00
Single Population				
% diseased	2.9	3.25	21.9	48.5
Metapopulation				
% disease persistence	14.0	51.0	79.0	80.0
% sites diseased	16.2	22.5	42.0	50.8
% disease	34.5	33.5	41.1	44.3
Frequency of Resistant Allele (%)				
Single Population	81.5	87.8	100.0	0.0
Metapopulation	33.8	31.7	2.3	0.0

Note: Resistance of the susceptible genotype is set to $\beta = 5.86$.

resistance). The effects of decreasing the resistance of the more resistant genotype are somewhat surprising (the row of X's in Figure 7.5; Table 7.1b): as its resistance is decreased (keeping costs the same), the resistant allele actually reaches a higher frequency, and there is a region over which it becomes fixed. The reason is that as the resistance of the most resistant genotype decreases, disease frequency increases, and this fur-

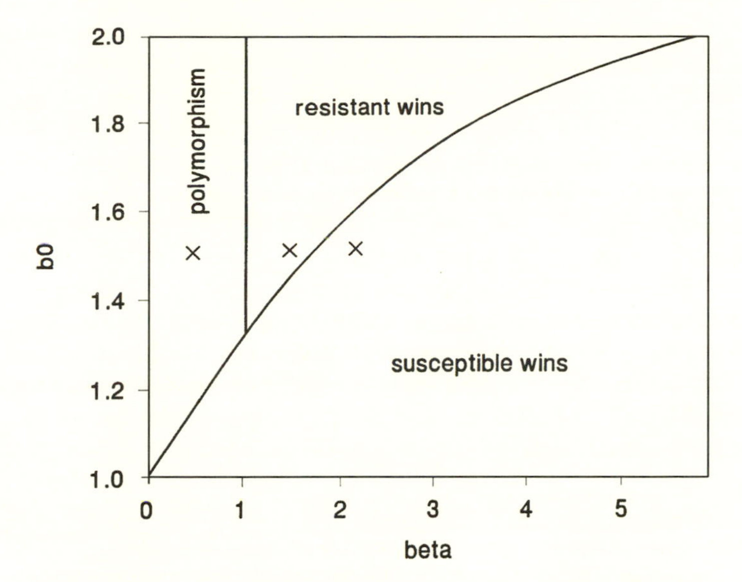

FIGURE 7.5. Phase diagram illustrating regions of polymorphism and monomorphism when resistance is determined by one locus with two alleles, *R* (resistant) and *r* (susceptible), in a single-population deterministic model. See the text for the model description. It is assumed that *rr* has a transmission coefficient of 2 and a reproductive output of 2. The diagram shows the equilibrium regions for increasing values of the transmission coefficient and increasing values of the reproductive output of the more resistant *RR* genotype (we assume heterozygotes are intermediate). Resistance of this genotype decreases from left to right, and the cost of the resistance decreases from top to bottom.

ther increases the selection on this allele. This continues until the selective advantage due to resistance becomes too small relative to the cost, at which point the susceptible allele now goes to fixation. If we consider resistances and costs typical of the *Silene-Ustilago* system, the resistant allele is predicted to increase to a high frequency (87.8%) and disease levels should be very low (3.2%) (Table 7.1b). Given such low equilibrium values, one might expect that the disease would almost never persist in a metapopulation context.

However, in the metapopulation simulation the disease persists 56.7% (n = 20) of the time. Moreover, when it persists, the percentage of populations that are diseased is substantial (22.5%) as is the prevalence of the disease in these populations (33.5%). There are several reasons for the high level of persistence of the disease in the metapopulation. First and foremost, many newly founded populations contain only susceptible individuals: In these populations the disease can establish rapidly and reach a high frequency. Second, the initial disease frequency in most populations where the disease has just arrived will be greater than the low equilibrium frequency expected in a single isolated population. Third, these two effects will interact—the high incidence of disease in sites with susceptible hosts will result in higher disease colonization rates of all populations, that is, there is an increase in the effective pathogen growth rate over the metapopulation as a whole.

Whereas in a single population, decreasing the resistance of the most resistant genotype results in an increase in its frequency (see above), in the metapopulation the reverse effect is seen. Indeed, at values where the resistance allele would go to fixation in a single population, it persists only at a low frequency in the metapopulation ($\beta = 1.5$; Table 7.1b). The reason is that when the resistance allele is less extreme, it reaches a higher final equilibrium frequency, but it takes much longer to attain that frequency. For example, if the resistant allele is started at a frequency of 10%, after twenty-five generations it reaches 42.6% in the case of $\beta = 0.41$, but only 12.5% when $\beta = 1.50$. Continual extinctions and recolonizations therefore maintain the resistance allele at a low frequency because within most populations equilibrium is not reached before there is extinction of either the host or the pathogen.

Experiment 2: The effect of ignoring genetics on the future dynamics of a metapopulation

In this experiment, we ran the simulation for a substantial period of time (three hundred generations) and then assumed we were ecologists estimating population parameters without regard to the underlying genetics. To do this, the disease transmission coefficient and host reproductive output were calculated from the weighted averages of the component genotypes (as would happen if we made measurements on individuals randomly sampled from the metapopulation). Because we assumed an exponential model of transmission, the disease transmission coefficient was calculated as follows: We first calculated the weighted average of the probability that the genotypes would become diseased at the extant disease prevalence and then calculated the transmission coefficient that would give this same probability in a uniform population with the same disease prevalence. Disease prevalence was calculated as the overall frequency of disease in diseased populations. Only runs where the disease had persisted for three hundred generations were included.

When the simulation was continued assuming the populations were genetically variable, the disease only persisted for a further three hundred generations in 40% of the runs ($n = 40$). However, when we assumed the populations were genetically uniform, the disease nearly always persisted (85%, $n = 40$). In the genetically variable populations, the periodic spread of the highly resistant genotypes resulted in large fluctuations in the frequency of disease and a high probability of global extinction. In the uniform population, the moderate but relatively constant level of host susceptibility resulted in much more stable dynamics and longer-term persistence.

Experiment 3: *Selection against heterozygotes*

In 1931 Wright suggested that evolution might be more rapid in a group of smaller populations connected by occasional migration than in a continuous panmictic population because chance effects in small populations might permit the evolution of traits whose intermediate condition might be disadvantageous. The simplest single-locus scenario of this is when the alleles underlying the trait in question show heterozygote disadvantage. In a single population with heterozygote disadvantage, the intermediate allele frequency equilibrium is unstable, and a rare allele cannot invade even if it is superior in fitness in its homozygous condition to the alternative, more common, allele.

In our simulation, we excluded the disease completely and reset the fitness values (as given by the birth and death rates) of the host genotypes such that heterozygotes were at a disadvantage. We illustrate the case where the relative fitnesses of *AA*, *Aa*, and *aa* were set to 1, 0.5, and 0.67, respectively. In a single infinite population, there is an unstable equilibrium at $A = 0.25$, and below this frequency, *a* always wins (even though *aa* has a lower fitness than *AA*). In the metapopulation simulations we introduced *A* at an initial frequency of 0.1 in all populations.

Contrary to single-population expectations, in 20% ($n = 25$) of the runs, the *A* allele spread to fixation (Figure 7.4b), although often it took several thousand generations to do so. There were long periods when the population was characterized by spatially separated patches that were close to fixation for either *A* or *a*. As expected, either decreasing the degree of heterozygote disadvantage or increasing the initial allele frequency of *A* increased its likelihood of spreading in the metapopulation. These results show that Wright's conjecture that metapopulations may be theaters for evolutionary processes that are impossible in large single populations has cogency in this particular simulated real-world metapopulation. Moreover, it is likely that our results underestimate the likelihood of such effects because in our simulated metapopulation the numerical and allele frequency dynamics within each population were deterministic.

SUMMARY

In this chapter we have used a combination of field, theoretical, and simulation studies to illustrate several principles relating to the impact of genetics on metapopulation dynamics. It is clear that genetics is not "just another" heterogeneity; averages of the genetic values fail to accurately capture the dynamics of the system. Spatially explicit processes create situations that amplify the impact of genetic variation. In population genetics, it has long been recognized that limited dispersal will result in the spatial redistribution of genetic variation such that variation increases among subpopulations but decreases within patches. As emphasized by Frank (1997), chance effects associated with colonization and extinction processes in metapopulations can result in the dynamics of coevolutionary systems becoming highly unpredictable from single-population theory. Our study shows that the redistribution of genetic variance during the colonization phase can have profound effects even in the simplest of cases (we assumed genetic variation only in the host).

For a biologist, the final arbiter of the importance of a theoretical construct is whether it generates explanatory power that can lead to a greater understanding of the natural world. Theory can provide powerful insights into what is plausible, but it needs to be integrated with empirical studies to assess whether particular processes are actually likely in nature. Nowhere is this currently more true than in the area of population substructuring and its relationship to group selection (Wilson 1983). Therefore we need more empirical data on spatially distributed populations in the real world, with an emphasis on studies that gather information about genetic and ecological connectedness, and extinction and colonization dynamics. Ironically, our own study was initiated by an interest in finding populations as targets for detailed individual study, until we realized the feasibility of extending the sampling to a regional scale.

At first sight, it would seem that there are many deterrents to the study of populations on a regional scale. More effort may be necessary to study many versus a few populations, and colonization and extinction may be rare events and hard to document. If populations form a patchy continuum, it is hard to know at what scale a study should be done or how populations that are components of the larger metapopulation should be defined. In addition, experiments are likely to require large spatial and time scales. Our study was greatly facilitated by both the accessibility and linear arrangement of the roadside populations. Of particular importance was our decision to study the system as a "connected lattice," using a grid superimposed on a patchwork of populations. This enabled us to obtain crude census data on each segment of the grid quickly (obviating the mapping and detailed study of the fate of individuals) and to focus more on events at a regional scale. After nine years of experience with this study we have become convinced not only that the study of metapopulations in nature is very feasible (especially in plants) but that major misunderstandings can emerge from studying only one or a few populations as a guide to long-term dynamical outcomes. Thus, in the study of single populations, it is unclear what criteria should be used to choose a "representative" population, how population boundaries should be defined, how to test causal processes when replication is limited, and when to discount the "unusual disaster"; there is also the mundane problem of how to avoid damaging plots in the process of intensive measurement and sampling. Whereas experimental field studies on metapopulations are likely to be labor intensive, computer simulations developed interactively with both descriptive field data and smaller manipulative studies can become a valuable surrogate for such experiments.

PART III

Competition in a Spatial World

Because all organisms require and consume a range of resources, many of which are in short supply, interspecific competition has long been viewed as one of the major factors controlling both the distribution and abundance of species and multispecies coexistence. Interspecific competition repeatedly has been demonstrated in laboratory, greenhouse, and field experiments. Such competition experiments have reinforced Hutchinson's paradox of diversity because they have frequently shown that a single species emerges as the winner in competition, whereas numerous species coexist in natural communities. For example, common garden competition experiments at Cedar Creek have revealed a simple competitive hierarchy on low-nitrogen soils, with the best nitrogen competitor displacing five inferior competitors within three to six years. Nonetheless, these same five grass species seem to persist for decades, and presumably will do so for millennia, in nearby grasslands. Space provides a major solution to this paradox of diversity. The consideration of space has several ramifications.

First, Lehman and Tilman (Chapter 8) consider a multispecies competition version of an implicitly spatial model, Levins's "metapopulation" model. Although this model does not explicitly include spatial location, its underlying rules of site occupancy and dispersal abstract some of the essence of space. They summarize the derivation in Tilman (1994) of multispecies coexistence on a single limiting resource in a spatially homogeneous habitat and then extend this by considering both the sympatric evolution of high-diversity communities in homogeneous habitats and the robustness of this multispecies coexistence to more realistic models that include the actual mechanisms of resource competition.

Next, Pacala and Levin (Chapter 9) explore plant competition in greater depth and with a more detailed consideration of spatial structure. They use the approach of "moment closure" to derive intriguing results that explain why it is that species that compete strongly in controlled laboratory or common garden experiments often apparently compete so weakly in nature. Building on the classical concept of Lotka-Volterra competition coefficients ($\alpha_{i,j}$), they demonstrate that the explicit inclusion of space modifies competition coefficients, reducing the intensity of competition, in spatial habitats. Space has these effects because neighborhood interactions and local dispersal cause species to be clumped and spatially segregated. The magnitude of such effects depends in intriguing ways on the intensity of competition and the scales over which neighborhood competition and dispersal occur.

Tilman and Lehman (Chapter 10) then ask what effects habitat destruction should have on the composition and diversity of communities. This model predicts that habital loss should cause the biased extinctions of abundant species if they are poor dispersers and good competitors. Such an "extinction debt" is shown to be robust with respect to many more realistic complications that were not included in earlier formulations. However, as might be expected, the inclusion of explicitly spatial patterns of habitat destruction can greatly modify the number of species threatened with extinction following a given degree of habitat destruction. In total, the consideration of explicit space illustrates the limitations of implicitly spatial models, such as metapopulation models, as quantitative predictors of extinctions.

The final two chapters of this part deal with space in more qualitative ways and leverage this simplicity to address a series of broad, whole-community "scaling laws" (*sensu* Levin and Pacala, Chapter 13). Cornell and Karlson (Chapter 11) address a fundamental question—is it local processes, such as interspecific competition, or regional processes, such as the size of the species pool and migration, that control local diversity? They propose a method for distinguishing between local and regional processes and then review, evaluate, and synthesize eighteen relevant field studies. Their analyses, which suggest that both regional and local processes control local species richness, thus support and extend the concepts developed in the first two chapters in this part.

In total, these four chapters demonstrate the rich patterns that may emerge when the threads of space are interwoven with the processes of competition. They illustrate that consideration of space can provide insights on many scales, and with respect to a variety of issues. Perhaps most importantly, when processes are viewed spatially, several surprising questions and paradoxes emerge, none of which would have drawn our attention in a nonspatial world.

CHAPTER EIGHT

Competition in Spatial Habitats

Clarence L. Lehman and David Tilman

Early models describing local competition among species were interpreted as showing that two or more competing species could not coexist on a single resource (Lotka 1925; Volterra 1928). This was called the competitive exclusion principle, which was later generalized to the statement that n species could not coexist on fewer than n resources or limiting factors (e.g., MacArthur and Levins 1964; Levins 1968). When a single resource was explicitly included in such a model, the species formed a competitive hierarchy, with poorer competitors displaced as the resource was depleted by superior competitors (Tilman 1982).

The competitive exclusion principle immediately raised a paradox of diversity, however (Hutchinson 1961). A given habitat, such as a prairie or a lake, contains hundreds of species, but the number of limiting resources—nutrients, light, space, and so forth—is relatively small, likely fewer than a dozen. Resolving the paradox became a central issue in theoretical ecology. Early erroneous theoretical predictions about diversity arose from limited assumptions in classical competition models —including the assumption that the environment is homogeneous and perfectly well mixed. The early equations of competition had no explicit spatial coordinates (x, y, z) describing locations of individual organisms. Hutchinson conjectured that, among other things, relaxing this assumption would help resolve the paradox of diversity. The question addressed in this chapter is: How do predictions from ecological models of competition change when spatial structure is included?

ADDING SPATIAL STRUCTURE

Implicitly Spatial Formulations

As outlined in Chapter 1, space is implicitly included in metapopulation-like models (Levins 1969), with a dynamical variable $p(t)$ representing the portion of habitat sites occupied by an individual of the species at time t. A species is characterized by its mortality rate m and its colonization rate c:

$$\frac{dp}{dt} = cp(1 - p) - mp. \tag{8.1}$$

This basic formulation can be expanded into an abstract model for interspecific competition among individual plants (Tilman 1994):

$$\frac{dp_i}{dt} = c_i p_i \left(1 - \sum_{j=1}^{i} p_j\right) - m_i p_i - \sum_{j=1}^{i-1} c_j p_i p_j. \tag{8.2}$$

Here p, c, and m have the same meaning as in Equation 8.1, except now they are indexed by species number i. In Equation 8.2, species are ranked by their competitive ability, as in a model of competition for a single resource, with species 1 being the best competitor. For historical reasons, it is convenient to call Equation 8.2 a metapopulation model, but it is important to keep in mind that in Equation 8.2 and throughout this chapter, we model individual organisms occupying individual sites, not groups of populations.

Colonization and mortality are explicit (through parameters c_i and m_i respectively), but the competitive hierarchy itself is implicit. The hierarchy results because the sums in Equation 8.2 include terms for superior competitors (those of lower index i) but exclude terms for inferior competitors (higher index i). In nonspatial models, only the best competitor, species 1, survives. Does the addition of implicit space change this exclusive survival of the best competitor and thus potentially resolve Hutchinson's paradox?

Tilman (1994) showed that under Equation 8.2, any number of species can stably coexist in an implicitly spatial homogeneous environment, even though the best competitor immediately displaces all others locally. Such coexistence requires both an interspecific trade-off in competitive ability versus dispersal ability and a limit to similarity of these traits. Although multispecies coexistence might seem paradoxical under such strict local displacement, it occurs because neither the best competitor nor any group of competitors can occupy all sites. Rather, at equilibrium, some proportion of sites will be empty, and there will be constant turnover in occupancy of all sites. Empty sites can be available "homes" for poor competitors if they are sufficiently good dispersers. In other words, coexistence occurs because local displacement by the best competitor is never permanent. No individual lives forever, and when an individual dies, its local site is free for colonization. If better competitors do not occupy all the available space —and according to Equation 8.2, they never can (Tilman 1994)—then there is always room for sufficiently mobile fugitives (e.g., Horn and MacArthur 1972; Armstrong 1976; Crawley 1990; Tilman 1994).

Thus arises a fundamental trade-off among competing species—the ability to hold a site (being a good competitor) versus the ability to get to a site (being a good colonizer), and this trade-off has been explored in several forms (MacArthur and Wilson 1967; Levin and Paine 1974; Werner and Platt 1976; Platt and Weis 1977; Tilman 1982, 1990, 1994). The trade-off can be as simple as energy allocated to roots traded for energy allocated to seeds. More energy to roots can mean better nutrient acquisition, and that makes a better competitor at a nutrient-limited site. More energy to seeds can mean more seeds, or more mobile seeds, or longer-lived seeds, and that makes a better colonizer. Trade-offs like this are both logically compelling and empirically observed (Tilman 1990).

Hence the mere existence of spatial structure, acting through individual mortality and colonization, alters conclusions about the ultimate outcome of competition—local exclusion but regional coexistence—even when spatial structure is only implicit. It is important to realize, however, that the early nonspatial models of competition are not strictly *incorrect*—they are simply *local*. They strictly apply only to a small local site, or to a perfectly well-mixed site, or to a site in which there is no limitation to colonization. Implicitly spatial formulations of competition take the conclusions of nonspatial models as premises, apply those premises to each point of space, then ask what conclusions follow for the entire region.

Explicitly Spatial Formulations

Explicitly spatial formulations give each organism a position demarked by some coordinates, and as a result, different points may have different characteristics—say, different nutrient levels or different light penetration (e.g., Pacala and Tilman 1994; Pickett and Cadenasso 1995). To apply the model of Equation 8.2 in explicitly spatial conditions, we use a spatial simulator with myriad sites for individual organisms, as described in Chapter 1. We have simulated numerous explicitly spatial cases, often involving altered habitat (Tilman, Lehman, and Yin 1997), and compared the results to analytical results derived for Equation 8.2. Our experience reveals two rather different results, depending on parameter values. On the one hand, many simulations of the explicitly spatial model readily approach the analytical implicitly spatial model as the size of the local neighborhood increases. This is true not just for qualitative behavior, but also for quantitative relationships among colonization rates, mortality rates, and species abundances (see Chapter 1). On the other hand, certain simulations diverge radically from predictions of the analytical model. The reason for the divergence sheds some light on the composition of communities. The single-species Equation 8.1, which applies to the best competitor, is a deterministic equation that assumes either an infinite habitat or infinitely subdivisible organisms. The explicitly spatial simulator is stochastic, both in colonization and mortality, and it uses of course a finite habitat. It thus should be a closer approximation to nature. We seek circumstances under which finitude and stochasticity cause the explicitly spatial formulation to behave differently from the implicitly spatial Equation 8.2.

Transient Behavior

Finite time is one point of departure of explicit simulations from analytical models. Models without spatial structure may reach an equilibrium relatively quickly, but spatial structure can greatly slow the timescale in competitive systems (Shmida and Ellner 1984; Hubbell and Foster 1986; Tilman 1994; Hurtt and Pacala 1995). For example, Equation 8.2, perturbed from equilibrium, can take thousands of simulated years to settle down (see figure 4 in Tilman 1994). During this equilibration, the composition of the community fluctuates widely; moreover, natural communities may be perturbed repeatedly and never settle down. This makes consideration of nonequilibrium behavior important in ecological systems (Byers, Hansell, and Madras 1992) and provides a good reason to examine the transient behavior of simulations.

Demographic Stochasticity

Finite space and a corresponding finite number of organisms is another point of departure of explicit simulations from analytical models. At its equilibrium of $\hat{p} = 1 - m/c$, Equation 8.1 has eigenvalue $\lambda = m - c$ (Tilman 1994), whose magnitude determines the strength of the force restoring the population to equilibrium after perturbation. The more negative the eigenvalue, the stronger is the restoring force; the closer the eigenvalue is to zero, the more nearly neutral is the system. Notice that the equilibrial abundance $\hat{p}$ depends only on the *ratio* of m and c, whereas the restoring force λ depends on the *difference* between m and c. Thus a small difference between m and c may allow stochastic tendencies to knock the system away from equilibrium, overwhelming deterministic tendencies to draw it back.

When the difference between m and c is small, fluctuations in the population can be dramatic, often causing the population to drift to extinction in surprisingly short times. Consider two cases with the same equilibrium value but different eigenvalues. In the first case (Figure 8.1A), the habitat contains 198 sites, and the colonization rate is twice the mortality rate

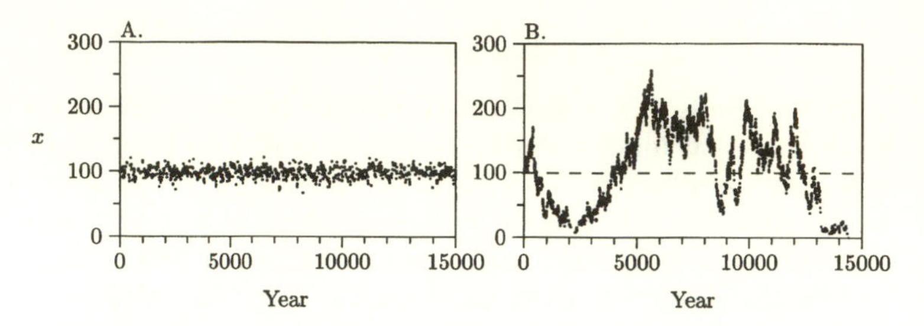

FIGURE 8.1. Effects of stochasticity on the best competitor. (A) Small oscillations in population with $m = 0.05$, $c = 0.1$, and $w = 198$ available habitat sites. The vertical axis is the number of individuals in the population; the horizontal axis is time. With these parameter values, the population remains near its equilibrium value of $(1 - m/c)w = 99$. (B) Wide swings in population with $m = 0.05$, $c = 0.050505$, and $w = 9{,}900$ available habitat sites. The vertical axis is the number of individuals in the population; the horizontal axis is time. The dashed line marks the equilibrium value $(1 - m/c)w = 99$.

($m = 0.05$, $c = 0.10$). That makes $\hat{p} = 1 - m/c = \frac{1}{2}$, so the equilibrium is $\frac{1}{2} \times 198 = 99$ sites occupied. In the second case (Figure 8.1B), the habitat contains 9,900 sites, and the colonization rate is only slightly greater than the mortality rate ($m = 0.05$, $c = 0.050505\ldots$). That makes $\hat{p} = 1 - m/c = \frac{1}{100}$, so the equilibrium is $\frac{1}{100} \times 9{,}900 = 99$ sites. The two cases have the same equilibrium site occupancy, but have markedly different dynamical behavior (Figure 8.1A,B), owing to the difference in eigenvalues. Trajectories in the first case (Figure 8.1A) appear to be locked in a narrow band around the equilibrium. In contrast, trajectories in the second case (Figure 8.1B) visit values far from equilibrium and consequently often drift to extinction. The time to extinction varies enormously—in many cases it is less than five hundred years, in rarer cases more than thirty thousand years.

Effect on the Competitive Hierarchy

Coexistence in these models occurs because individual mortality leaves open sites and limited random dispersal can never consistently fill all open sites. The best competitor sees a relatively simple environment; it can colonize any site not already occupied by a member of its own species. In these models, the only stochasticity in its dynamics arises from mortality and colonization of its own species, as just described. The second competitor is affected not only by stochasticity in its own mortality and its own colonization, but it can also be displaced by stochastic colonization of the best competitor, and it may find sites opened by stochastic mortality of the best competitor. Thus stochastic effects are greater for the second competitor, and these effects compound throughout the competitive hierarchy (Figure 8.2B).

This compounding of stochasticity, together with stochasticity accompanying small differences between colonization and mortality, opens the question of whether simple competitive interactions within a local site will lead to predictable community structures at the regional level, as they always do in the implicitly spatial model, or whether ecological drift can overwhelm competitive interactions and lead to communities that seem to depend on random forces for their structure. Hurtt and Pacala (1995) show that in a model of spatial heterogeneity with strong local competition and strong recruitment limitation, but without a strict competitive hierarchy, drift can overwhelm, and they stress that this possibility helps reconcile the apparently contradictory observations of strong interactions in competitive experiments (e.g., Tilman 1982, 1988) and communities that are difficult to distinguish from chance assemblages of species (Hubbell and Foster 1986). Our work suggests that such stochasticity has its greatest effects on inferior competitors, especially those with low equilibrial abundances, but that even superior competitors of low abundances are not immune.

Let us illustrate this idea with the models under discussion, which have a strict hierarchy of competitive exclusion locally and a uniform homogeneous environment. Suppose the implicitly spatial model 8.2 is infused with many species of specified parameters and initial abundances. Then given those parameters and abundances, the ultimate community is predetermined, for this implicitly spatial model has no stochasticity. It may take a long time, and abundances of various species may

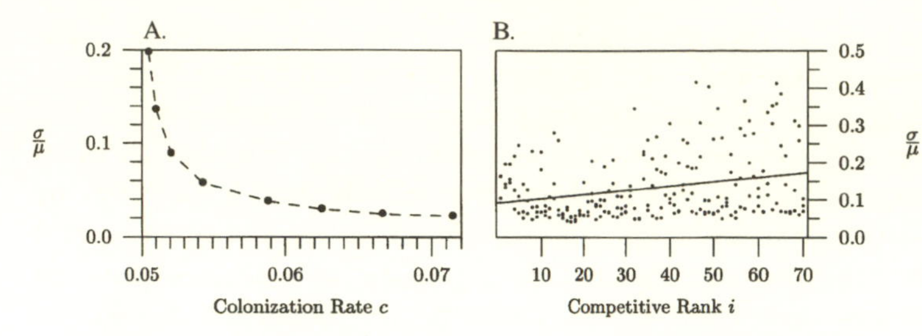

FIGURE 8.2. Variation in abundance from year to year. (A) Results for the best competitor in a stochastic version of Equation 8.2. Mortality is constant at $m = 0.05$, and the habitat contains 9,900 habitable sites. Each point was derived by starting a simulation at the corresponding equilibrium $(1 - m/c)$, running the simulation for one hundred years, and repeating the process for one hundred simulations. The population values at the beginning of each year were then used to compute the mean population and its standard deviation, and thence the coefficient of variation, which appears on the vertical axis. (B) Increase in stochasticity through the competitive hierarchy. Each point is the coefficient of variation through time derived from one of several explicitly spatial simulations with seventy species competing. Parameters are the same as those of Figure 8.3, except these results simulate continuous time.

rise and fall in transitory successional patterns, but the same final community will always result (Tilman 1994).

On the other hand, suppose identical parameters and initial abundances are applied to the corresponding explicitly spatial model with hexagonal cells described earlier. Will species abundances match the analytical predictions? The answer depends on parameters. With a community of a few abundant species (Figures 8.3A–C), species abundances may differ somewhat from the analytical predictions, but they will be roughly the same from trial to trial. In contrast, with a community of many species, none abundant (Figures 8.3D–F), species may coexist, but with little quantitative correspondence to the analytical model. In each trial a different community can arise, its composition depending on initial distributions of organisms and on chance mortality and colonization as the community develops. Thus implicit spatial structure can be sufficient to explain regional coexistence of species that cannot coexist locally, but explicit spatial structure helps to resolve the appar-

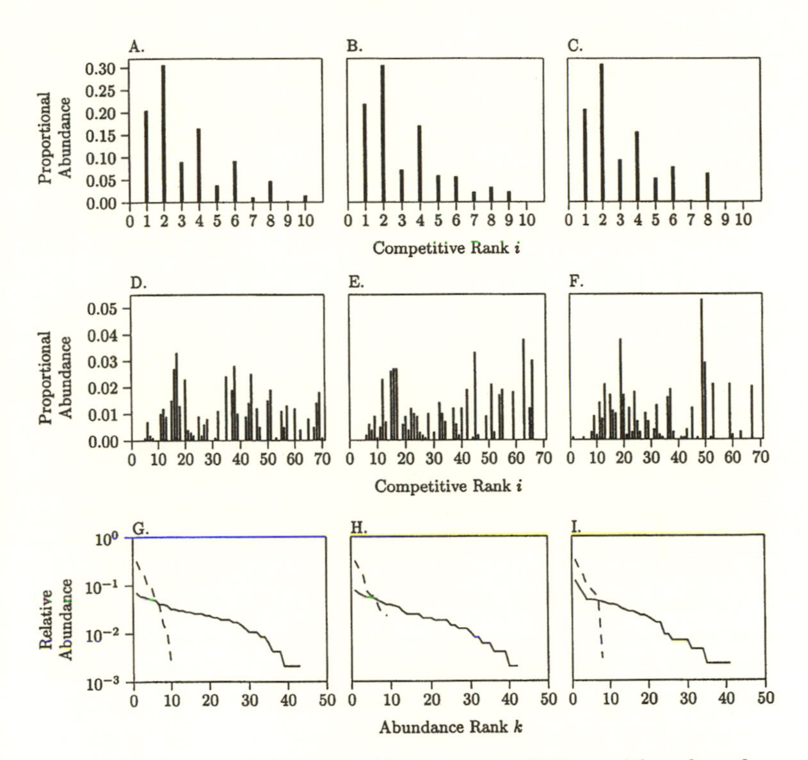

FIGURE 8.3. Community composition in an explicitly spatial analog of Equation 8.2. In A–F, the horizontal axis represents species, ranked in competitive order from the best competitor ($i = 1$) to the poorest ($i = 10$ or $i = 70$). The vertical axis represents Pi after a thousand years. Species competed in a 100×99 hexagonal array with absorbing boundaries, with dispersal once per year to the sixty hexagonal cells in four adjacent rings. Mortality was $m_i = 0.05$ for all species, and colonization was $c_i = m_i/(1 - z)^{2i-1}$, with $z = 0.3$ for parts A–C, and $z = 0.01$ for parts D–F. (A, B, C) Three typical cases with ten species competing and with the best competitors in high abundance. The composition of the communities is roughly the same from case to case for these parameter values. (D, E, F) Three typical cases with seventy species competing and with all competitors in low abundance. Of seventy species initially, approximately forty-three survived a thousand years—a different set in each case. (G, H, I) Dominance-diversity curves for communities A–F. The horizontal axis represents surviving species, ranked in order of abundance. The vertical axis is abundance. Dashed lines represent the communities directly above in A–C; solid lines represent those in D–F. The different communities in A–C have similar curves, as do those in D–F, and these curves are not unlike those of some natural communities (Hubbell 1997).

ent contradiction between local determinism versus regional stochasticity. Community composition can be strongly influenced by history and chance despite strong local deterministic dynamics (Hubbell 1979; Shmida and Ellner 1984; Hurtt and Pacala 1995), especially for highly diverse communities in which the dominant competitors are rare.

Patterns

With local dispersal and local interactions, spatial patterns form. Local competition and local colonization cause populations to become aggregated with the passage of time (Vance 1984; Pacala 1986a). Consider the best competitor. No other species displaces it from the sites it occupies, so its distribution is influenced only by its own mortality and its own patterns of colonization. Because a propagule is more likely to colonize a nearby site, local concentrations and local bare spots arise in the distribution of the population over space (Figure 8.4).

The same processes of local colonization apply to competitors of all rank, but for poorer competitors there is an additional effect. The second competitor has its own intrinsic tendency to aggregate, due to local colonization, but in addition, the environment it "perceives" is already aggregated. In other words, because the population of the best competitor is aggregated, it follows that the sites unoccupied by the best competitor are also aggregated, or at least not random. These unoccupied sites are the only places available to the second competitor for colonization. Thus the second competitor tends to aggregate in an area that is already nonrandom and hence becomes more aggregated than the best competitor. This tendency toward aggregation increases down the competitive hierarchy, rapidly reaching an asymptotic value (Figure 8.5).

More Realistic Population Dynamics

A final question concerns whether the conclusions depend upon the greatly abstracted local dynamics, or whether they continue to hold as local dynamics are made more realistic. For example, the metapopulation models discussed assume that local growth and competitive displacement are instanta-

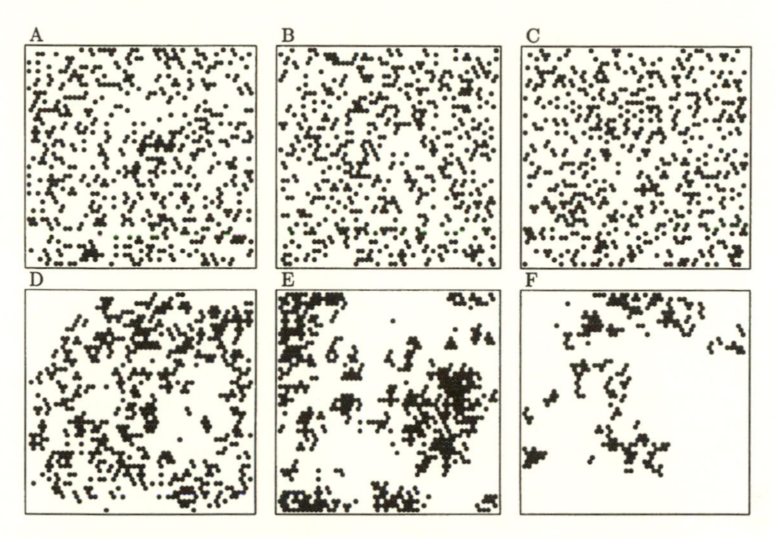

FIGURE 8.4. Aggregation developing with time. (A, B, C) Random distributions at $t = 0$ for the best three competitors in an explicitly spatial metapopulation model with local dispersal. The habitat contains 2,950 individual sites arranged in a hexagonal lattice. Dots show sites occupied by individual plants. (D, E, F) Distribution at $t = 5{,}000$. The best competitor (D) is clustered compared to the initial random distribution. The second competitor (E) and the third competitor (F) are more clustered still, because they are restricted to the already-clustered sites not occupied by better competitors. Level of clustering approaches an asymptote.

neous, but reproduction is continuous. However, for a vast array of plant and animal species, seed production or birth is an episodic event often synchronized with the seasons, whereas growth and competitive displacement occur gradually throughout the year. Thus, in reality, reproduction is better modeled as instantaneous and discrete, while growth and competitive displacement are better modeled as continuous.

We can install discrete reproduction with gradual growth and competition by considering a continuous-time differential equation at each site. Individuals increase or decrease in biomass according to the amount of limiting resource available, and the resource increases or decreases according to the amount consumed and the amount resupplied. To make this

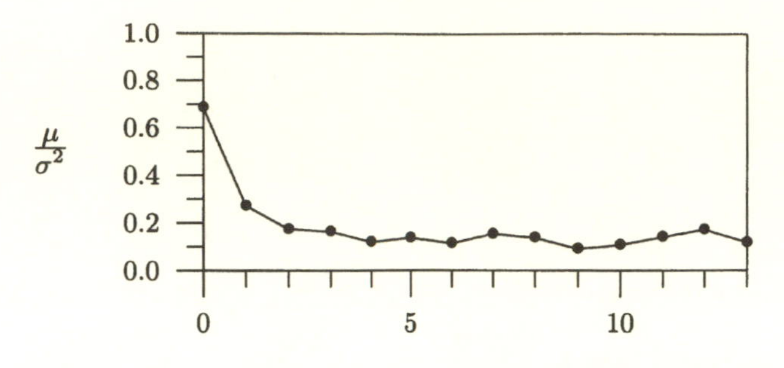

FIGURE 8.5. Random spatial distribution versus competitive rank. Spatial distribution of competing species in an explicitly spatial metapopulation model with local dispersal in a habitat of 100 × 99 hexagonal sites. The horizontal axis shows the position on the competitive hierarchy, with the best competitor numbered 1. Number 0 represents uninhabited sites. The vertical axis is a simple index of randomness (μ/σ^2) averaged over all spatial scales from 2 × 2 through 33 × 33. Deviations from unity indicate nonrandom spatial distributions. (Note that values of unity in this index do not imply randomness; Hurlbert 1990.) Randomness decreases for the first few competitors, then reaches a noisy asympotote (see Figure 8.4).

precise, we apply a specific model of resource competition (Appendix 8.1) to each hexagonal site, keeping mortality and colonization as before.

The competitors at a site interact only through the resource. The equations establish an implicit competitive hierarchy through resource values. Each species i has a characteristic level of resource called R_i^* below which its population cannot survive, but above which its population grows. The species with the lowest R^*_i displaces all others by driving the nutrient level at the site below the level at which other species can survive or invade (Tilman 1982).

Equation 8.2 entails two key assumptions: (1) A species is displaced immediately when propagules from a better competitor arrive, and (2) a species is unable to survive or invade a given site if a species higher in the competitive hierarchy is already present. In contrast, for a more realistic model, the effect of invasion by a better competitor is delayed. A better competitor can invade immediately, for the resource level maintained at the site by the poorer competitor is sufficient for the better competitor to grow. However, the better competitor displaces the poorer competitor only by reducing the resource, following Equation 8.4. This takes time. Thus, increased local dynamical realism means a delay in competitive displacement.

Second, when a better competitor dies, the resource level can recover and allow poorer competitors to invade. But it recovers as a dynamical variable, following Equation 8.3. This also takes time. Thus, increased local dynamical realism also means a delay in recolonization, at least for species low on the competitive hierarchy (i.e., the poorer competitors).

The net effect of the first of these alterations is to favor poorer competitors, while the net effect of the second is to favor better competitors. How the two balance depends on the recovery rate of the resource, the resource supply point, and the difference among species in minimal resource requirements. Figure 8.6 shows a numerical result for an environment approximated by discrete cells in a hexagonal grid. The two species coexist, but species 2 is more widespread than it would be in the corresponding nonresource model.

EVOLUTION AND A LONGER TIMESCALE

The slowing of the timescale in competitive systems with spatial structure may make it important to consider evolutionary effects. Vanderlaan and Hogeweg (1995), for example, found that certain theoretical predator and prey species can persist indefinitely when allowed to evolve, but become extinct if evolution is blocked. In all the competitive models described thus far, evolution has been blocked—characteristics of each species are fixed parameters. What can be expected at long timescales when parameters are allowed to evolve? To address this question, we examine a simple formulation of phenotypic evolution. Our phenotypic trait of interest is the position of each species along an R^*–dispersal axis, which is defined by assuming a positive correlation between R^* and dispersal ability (and a trade-off between dispersal and competitive ability,

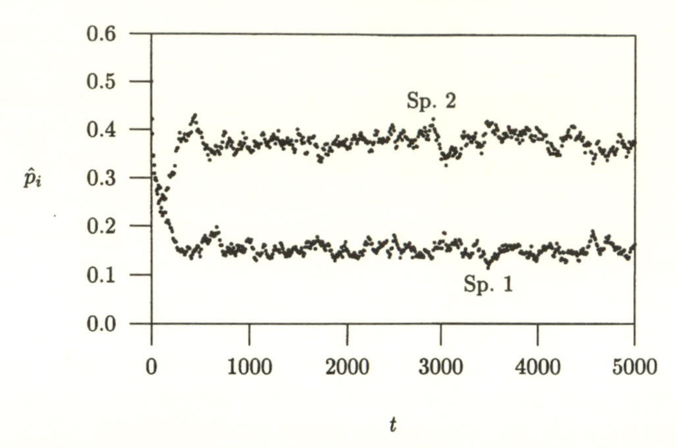

FIGURE 8.6. Coexistence of two species growing continuously and reproducing discretely while explicitly competing for a single resource. In each hexagonal cell, species compete for a single resource according to Equation 8.3. The horizontal axis is time; the vertical axis is portion of sites occupied. Colonization and mortality are such that, in the explicitly spatial analog of Equation 8.2, species 1 would occupy more sites than species 2. However, because of increased dynamical realism, including delayed displacement of species 2 by species 1, species 2 is more prevalent. Without space, species 2 would be extinct.

since low R^* corresponds to competitive superiority). This composite trait can be described by the parameter x_i. We assume that a large proportion $\alpha \approx 1$ of offspring possess the parental parameter value x_i, but a small proportion $(1 - \alpha)/2$ possess parameter value $x_i + \epsilon$, and the same proportion $(1 - \alpha)/2$ possess parameter value $x_i - \epsilon$, where ϵ is a small increment. In the infinitesimal limit, this kind of phenotypic evolution becomes simply diffusion in the parameter space (Appendix 8.2).

Solutions through Time

Figure 8.7 shows a particular numerical solution when parameters are able to evolve. The parameter x that evolves represents R^*, which establishes the competitive hierarchy—the lower the R^*, the better the competitor. Colonization increases with x, which corresponds to a competition–colonization trade-off. Mortality is the same for all phenotypes.

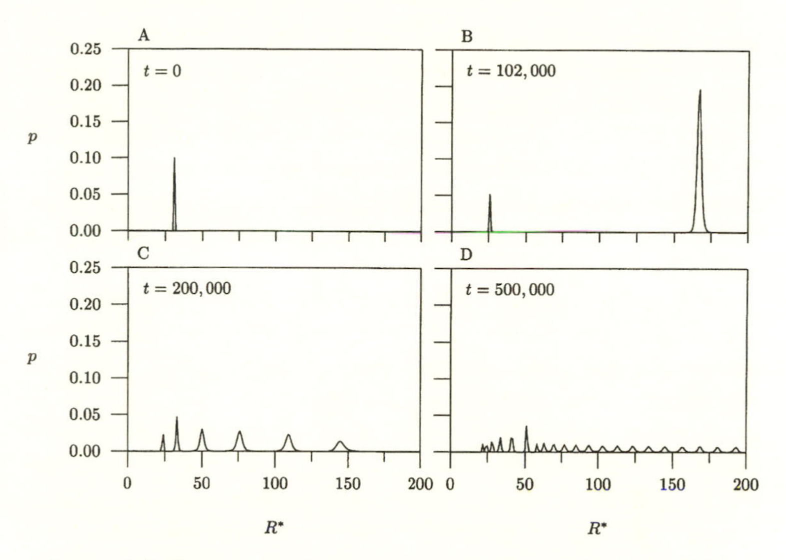

FIGURE 8.7. Patterns in phenotype space at intermediate times. The horizontal axis is the phenotypic variable x, which represents R^*, or competitive rank, and which is positively correlated with dispersal ability. The vertical axis represents abundance of the corresponding phenotype in the region. (A) At $t = 0$, a single phenotype is seeded. (B) By $t = 102{,}000$, original phenotype has evolved to lower R^*, its abundance has decreased, and a weedy species has risen to high levels. (C) At $t = 200{,}000$, six distinct peaks appear along the phenotype axis, representing six different coexisting species. (D) By $t = 500{,}000$, species have become more tightly packed, with about twenty species coexisting. This process continues until species are packed to the limits allowed by the dynamics of the system.

The system starts with a single late-successional phenotype of low, but not minimal, R^*. Phenotypic diffusion generates all phenotypes at exceedingly low population levels. Individuals of the initial phenotype spawn offspring of higher R^* and higher colonization rate, which tend to be outcompeted by their parents, and also offspring of lower R^* and lower colonization rate, which tend to displace their parents. Thus the colonization rate tends to decrease. After sufficient time, the initial species approaches its physiological limits, and its evolution slows. There is a peak in the phenotype space at the lower limit of the colonization rate (Figure 8.7A). Now a weedy species with high R^* and high colonization rate arises sympatrically (Figure 8.7B) and, with passing time, evolves to be less weedy by the same mechanism that reduced the colonization rate of the initial species. This process continues, with more and more species arising, each weedy at first but each growing less so with time.

Early in the process, even though any phenotype is possible, the phenotypes present are separated by large gaps in the phenotype space. Continuous phenotypic variation has been organized under the force of competition into clusters of phenotypes. With the passage of time, the clusters become more and more closely packed. Yet for much of the time, the clusters remain unsaturated, as real communities may be (e.g., Cornell and Lawton 1992).

What exactly is the force that forms the phenotypic peaks in this evolving competitive system? A clue lies in the limiting similarity already known in competitive systems (May 1981; Pacala and Tilman 1994; Hurtt and Pacala 1995), and specifically known for the immutable version of this system (Tilman 1994). Assuming mortality does not vary among species, then to coexist, the second competitor must have a higher colonization rate than the best competitor. But an arbitrarily higher colonization rate will not do. The colonization rate must be higher by a minimum amount, that amount determined by characteristics of the best competitor (Tilman 1994). When parameters are immutable, limiting similarity excludes certain species from the region. But when parameters can evolve, limiting similarity leads to the origin of species, that is, to sympatric speciation of hundreds of species.

We have explored evolution only in implicitly spatial competition models, and only under the simple mechanism of phenotypic inheritance with a direct trade-off between competitive ability and dispersal. The results raise many questions, and additional work is needed to determine their generality and robustness, especially the effect of random walks to extinction of rare phenotypes, of explicit genetics, and of explicit space.

SUMMARY

Implicit spatial structure, coupled with trade-offs in species dispersal and competitive traits, promotes coexistence among competing species that would otherwise be subject to competitive exclusion. Coexistence continues as spatial structure is made more explicit and as local population dynamics are made more realistic. Added realism leads to new phenomena, such as spatial patterning and increasing stochasticity down the competitive hierarchy. Explicit spatial structure coupled with stochasticity helps explain the apparent discrepancy between observations of strong competition at the fine scale and random assemblages of species at the community scale. Implicit spatial structure, interacting with phenotypic mutation and competitive dynamics, can organize a continuous array of phenotypes into discrete species-like sympatric clusters. In all these cases, the dynamics of competition over a spatial habitat can be dramatically different from the corresponding dynamics in each local site.

APPENDIXES

Appendix 8.1: Resource Competition

The following single-resource model is from Tilman (1990):

$$\frac{dR}{dt} = \alpha(R_0 - R) - \sum_{i=1}^{n} Q_i f_i(R) B_i$$

$$\frac{dB_i}{dt} = [f_i(R) - m_i] B_i$$

$$f_i(R) = r_i \frac{R}{R + K_i}. \qquad (8.3)$$

Here $R = R(t)$ is the amount of resource in the environment at time t, $B_i = B_i(t)$ is the biomass of species i at time t, Q_i is a resource-to-biomass conversion coefficient, $f_i(R)$ is the growth function for species i, r_i is the maximum growth rate of species i with unlimited resource, K_i is a half-saturation constant of the growth function, R_0 is the supply point of the resource, and α is the supply rate of the resource. In this model, if species i is alone in occupying the site, the resource will be drawn down to a level

$$R_i^* = \frac{K_i}{r_i/m_i - 1}. \tag{8.4}$$

Appendix 8.2: Phenotypic Diffusion

Instead of species with fixed parameters $x_i, x_2, x_3, \ldots, x_n$, we have in the infinite limit an unlimited number of potential "species" in a continuous parameter space $x \in \mathbf{R}^n$. Using implicit spatial structure, we replace the discrete species indices $i = 1, 2, 3, \ldots$ of Equation 8.2 with a continuous variable x, where x is simply the R^* value of the represented phenotype, and where x varies continuously across all positive values representing all possible species:

$$\frac{\partial p}{\partial t} = c(x)p(x,t)\left[1 - \int_a^x p(u,t)\,du\right]$$

$$- m(x)p(x,t) - p(x,t)\int_a^x c(u)p(u,t)\,du + \mu\frac{\partial^2 p}{\partial x^2}. \tag{8.5}$$

The point $x = a$ represents the lowest R^* of any species physiologically possible. The colonization function $c(x)$ and the mortality function $m(x)$ are algebraic and typically increase with x; μ is a constant.

This is an integro-differential reaction-diffusion equation. The first three terms make up the reaction term, which models population growth and competition over the entire spatial region for species with $R^* = x$. The fourth term $\mu\partial^2 p/\partial x^2$ is the diffusion term, which models inheritance, abstractly representing heredity, mutation, and recombination without addressing a precise genetic system. Patterns that develop in x are clusters of phenotypes, not spatial patterns.

ACKNOWLEDGMENTS

We thank Kendall Thomson, Chris Klausmeier, Diane Byers, Evan Siemann, Elena Lichman, and Charles Mitchell for cheerful and helpful discussions during the formulation of this chapter.

CHAPTER NINE

Biologically Generated Spatial Pattern and the Coexistence of Competing Species

Stephen W. Pacala and Simon A. Levin

INTRODUCTION

Simple ecological models typically are designed from the outset either primarily to expose ideas or primarily to describe particular systems. Models in the former category include assumptions that enhance mathematical tractability but make the models more difficult to test. For example, a large class of models in plant ecology relies on the assumption that the habitat is divided into a network of discrete cells with precisely one individual per cell (Skellam 1951; Horn and MacArthur 1972; Shmida and Ellner 1984; Durrett and Levin 1994a,b; Tilman 1994; Pacala and Tilman 1994). The fact that nature lacks this convenient subdivision complicates empirical measurement of quantities in the model and thus limits the model's capacity to scale up practical measurements.

In contrast, models intended to capture particular field systems are easier to estimate but tend to be considerably less tractable than are simple models of ideas. For example, over the past ten years, we and colleagues have developed, estimated, and tested models of plant communities (Pacala and Silander 1985, 1990; Cain et al. 1995; Pacala, Canham, and Silander 1993; Pacala et al. 1996). These models were defined at the level of the single individual simply because short-term data on individual performance are practical to obtain. They include every plant in a large community (10^3–10^6 individuals), each plant has a unique location, and data-defined submodels predict the fate of each plant throughout its life (birth, dispersal, survivorship, fecundity, and death). Ecological interactions included in the submodels range from the simple spatially local density dependence in neighborhood models (Pacala and Silander 1985, 1990) to the mechanistic (resource-based), and size-structured functions in the SORTIE model of forest dynamics (Pacala et al. 1993, 1996).

The most interesting general result from this body of work is that fine-scale spatial structure is sometimes caused by biological interactions among individuals and that this biologically generated heterogeneity has dramatic effects on dynamics. For each field community studied, it was possible to construct a "mean-field" (see Levin and Pacala, Chapter 12) model with spatially global interactions in place of the local interactions in the calibrated and tested models, thereby eliminating effects of biotic spatial heterogeneity. In cases without significant biotic heterogeneity (an artificially depauperate community of weeds on tilled land), the mean-field model was insignificantly different from the spatial models; both models predicted actual community dynamics and structure in the field (Pacala and Silander 1990). However, in two cases with substantial biotically generated spatial structure, the predictions of the mean-field models deviated sharply both from the spatial models and from nature. In particular, Pacala and Deutschman (1996) showed that the fine-scale spatial structure caused by interactions among individual trees is essential for maintaining both successional diversity and over one-half of the forest's living biomass.

Nonuniform spatial pattern develops in these systems because of relatively short dispersal and short-range interindividual interactions. Long-distance dispersal breaks up nonuniform spatial pattern, and long-range interindividual interactions average over it, causing dynamics to converge to a mean-field limit which can be analyzed and understood (as in the annual community studied by Pacala and Silander 1990).

However, if interactions occur over short distances, then the vital rates of plants may respond dramatically to changes in the local community caused by simple demographic stochasticity (e.g., the response of understory saplings to the death of a single nearby adult). If dispersal is also short, then the altered local abundances and vital rates will be remembered in the local spatial distribution as further changes in local community composition. In this way, small and unavoidable random fluctuations in the local spatial distribution may become first amplified and then reinserted as larger disruptions of the pattern. Repeated cycles of the process can cause inevitable fine-scale demographic stochasticity to become manifest in large-scale pattern.

Models that include stochastically seeded interplay between spatial structure and dynamics have been so intractable as to preclude even writing down the equations governing average abundances. These nonlinear, spatial stochastic processes have been the province primarily of interesting but difficult and limited theorems (Durrett and Levin 1994a). Paradoxically, our most striking result from a decade of research using data-defined models—the significance of biotic heterogeneity—has been itself the cause of our inability to understand these models fully.

To avoid the problem, most published simple models of ideas either include assumptions that prevent the phenomenon from happening or rely on computer simulations. Some common assumptions that prevent the phenomenon are that dispersal is infinitely large (e.g., Skellam 1951; Horn and MacArthur 1972; Shmida and Ellner 1984; Pacala 1986a,b; Tilman 1994), that the spatial pattern is fixed by some factor outside the model (e.g., the patch models assuming a fixed distribution of individuals per patch such as Atkinson and Shorrocks 1981; Hassell and May 1988; Ives and May 1985), and that the scale of competition is large enough to prevent significant fluctuations in local densities caused by demographic stochasticity (a tacit assumption of mean-field models including virtually all published reaction-diffusion models; see Chapter 12). The limitations of published analytically tractable spatial models are widely acknowledged, however, as evidenced by the large number of recent studies relying on computer simulations of interacting particle systems (including stochastic cellular automata and stochastic point processes, reviewed in Durrett and Levin 1994a,b).

Moment closure methods now offer a way to analyze nonlinear spatial stochastic processes to understand the causes and consequences of biotically generated spatial structure. The method relies on a non-mean-field spatial approximation to write equations governing average abundances of each species (the means or *first moments* of the stochastic process) and equations governing spatial variances and covariances (the second moments). It may be applied to virtually any spatial and stochastic ecological model, including stochastic cellular automata (Durrett and Levin 1994a,b) and point processes such as neighborhood models with finite dispersal (Bolker and Pacala 1997; Pacala 1997). We have even used it to derive equations for the moments of the complex and mechanistic forest simulation model SORTIE (Pacala and Deutschman 1996).

In this chapter we use the moment closure method to analyze some simple models of competition between two species. Our purpose is to introduce some results about the causes and consequences of biotic heterogeneity, to increase the accessibility of the methods, and to illustrate the incompleteness of current theory.

A SIMPLE MODEL OF NEIGHBORHOOD COMPETITION

Consider a community composed of two perennial plant species occupying, for simplicity, a one-dimensional habitat. Mortality is density independent at rate μ_i for species i, and an individual's fecundity, F_i, is a function of the local densities of the two species:

$$F_i(L_i, L_j) = f_i - \beta_i L_i - \alpha_{ij} \beta_i L_j \tag{9.1}$$

where L_i is the local density of conspecifics, L_j is the local density of heterospecifics, f_i is the maximum fecundity rate for species i, β_i governs the effect of conspecific neighbors, and α_{ij} governs the relative strength of inter- and intraspecific interference.

Each individual is located at a point on the line and disperses to its location from its mother with dispersal distance governed by the Laplacian density (back-to back exponentials):

$$D(x - x_m) = \frac{1}{2M_{Di}} e^{-(|x-x_m|/M_{Di})} \qquad (9.2)$$

where x is the position of an offspring, x_m is the position of its mother, and M_{Di} is the mean dispersal distance for species i. Effects of neighbors on fecundity in Equation 9.1 are assumed to decrease exponentially with distance. To determine the local density of species k ($k = i$ or j) around a species i plant located at position x, one sums the following distance weights over all species k individuals:

$$U_{ik}(x - x_n) = \frac{1}{2M_{Cik}} e^{-(|x-x_n|/M_{Cik})} \qquad (9.3)$$

where x_n is the location of a species k neighbor and M_{Cik} is the spatial scale for the competitive effects of species k neighbors on species i focal plants (analogous to the mean dispersal distance in Equation 9.2).

It is a simple matter to write a computer program that will simulate the dynamics of our two plant species. Beginning with an initial number and spatial distribution and taking Δt to be a very small time interval, one cycles repeatedly through four steps: (1) kill each individual with probability $\mu_i \Delta t$, (2) for each individual, calculate the con- and heterospecific local densities by summing the distance weights (Eq. 9.3), (3) have each individual give birth with probability $\Delta t F_i$, and (4) disperse each new offspring to its new location by drawing a pseudorandom number from the exponential probability density (Eq. 9.2). By cycling repeatedly through these four steps, one can predict the number and spatial distribution of the species at any time in the future.

With the addition of two spatial dimensions (which complicates the discussion below only a little), this model is a serviceable description of the competitive process that does apply to some field systems and can be estimated using simple field experiments (see Pacala 1986b; Pacala and Silander 1990).

However, precisely because of its fidelity to the finite scales of real communities, it is stochastic, nonlinear, and spatial and thus mathematically difficult.

In the limit as either the M_D's or M_C's tend to infinity, the model becomes mathematically identical to the Lotka-Volterra competition equations. With infinite M_C's, the model is a mean-field model because each individual interacts equally with every other individual. The parameters of the Lotka-Volterra equations are then $r_i = f_i - \mu_i$, $r_i/K_i = \beta_i$, with competition coefficients given by the α_{ij} in Equation 9.1.

With infinite M_D's but finite M_C's, we get a Poisson distribution of individuals. The macroscopic population-level model is then found by calculating the expectation of fecundity times survivorship for a randomly chosen individual. This gives a slightly different set of Lotka-Volterra equations with $r_i = f_i - \mu_i - \beta_i$, and other parameters as before. The difference between the two Lotka-Volterra limits is that in the case of infinite dispersal and finite M_C's, neighborhoods are biased by the presence of the focal individual, causing extra within-species density dependence (the extra term of minus β_i in the expression for r_i). The long-dispersal limit is the basis for most previously published analytically tractable neighborhood models (i.e., Pacala and Silander 1985; Pacala 1986a,b), the hawks and doves model with migration on a fast timescale in Chapter 12 (Eq. 9.9; see also Durrett and Levin 1994b), and many discrete-cell models (Skellam 1951; Hastings 1980; Chesson 1983; Shmida and Ellner 1984; Crawley and May 1987; Tilman 1994; Pacala and Tilman 1994).

The long-dispersal limit (with finite scales of competition) yields an average of the right-hand side of the corresponding classical mean-field limit, with the average taken over the Poisson variation in local crowding. It is important to understand that, although the Poisson limit of our simple neighborhood model differs little from the classical mean-field limit, the long-dispersal limit may differ strikingly from mean-field limits if density dependence is nonlinear (e.g., replace Equation 9.1 with an exponential decay; see Pacala and Silander 1985). Hawks and doves coexist in the Poisson limit (Chapter 12), but not in the classical mean-field limit.

Returning to the simple neighborhood model with finite scales, we begin by studying the system experimentally. Typically, the first experiment that one would perform in the field is a removal experiment to measure the strength of interspecific competition. Here we measure competition as it has been measured in literally hundreds of field experiments (see Gurevitch 1992), by removing a small number of individuals of each species in separate plots, and then measuring the resulting changes in population size. Interspecific competition is quantified as the per-capita change following heterospecific removal divided by the per-capita change following conspecific removal. We label this quantity α_{ij}^{est}. It is easy to show that the community-level strength of competition, α_{ij}^{est}, is equal to the individual-level strength of competition, α_{ij}, in either the mean-field or Poisson limit.

Figure 9.1 shows the results of one set of removal experiments, first reported in Pacala (1997). This example is for the symmetric case in which $f_1 = f_2 = f$, $\mu_1 = \mu_2 = \mu$, $\beta_1 = \beta_2 = \beta$, and $\alpha_{12} = \alpha_{21} = \alpha$, and in which all scales (the M_C's and M_D's) are equal to M. In each removal experiment, the model was first iterated a hundred time units to approximate equilibrium (on a toroidal habitat one thousand units long). Twenty percent of individuals of one species were then removed at random, and the population size changes were measured over 0.2 further time units. All experiments were replicated nine times.

Note that as expected, α^{est} approximately equals α if the scales are large (to produce the triangles shown, every individual interacted equally with every other and dispersal was Poisson). However, with finite scales (circles) α^{est} is a *humped function* of α. For α sufficiently close to one, interspecific competition at the community level (α^{est}) actually becomes weaker as interspecific competition at the individual level (α) becomes stronger. The parameter values used to produce the circles in Figure 9.1 yield an average of approximately ten to twenty-five important neighbors per-capita (ten to twenty-five individuals within the central 95% of the dispersal and competition functions). Figure 9.2 shows that the effects of finite scales are substantial even with 50–125 important neighbors

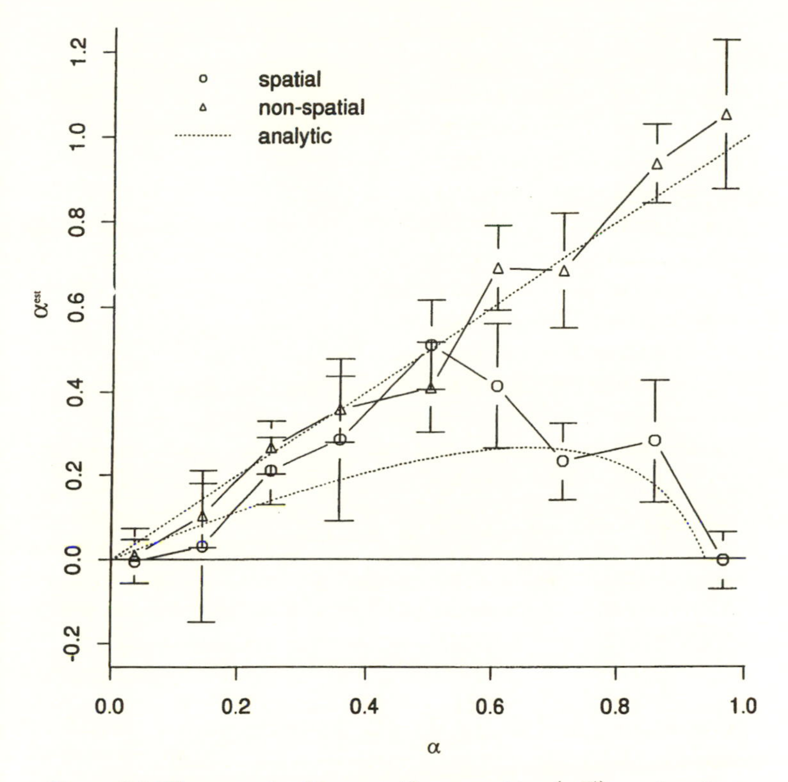

FIGURE 9.1. The strength of interspecific competition (α^{est}) as a function of the strength of interindividual interference (α) as determined using removal experiments within the point process. The mean of each of the nine replicates is shown $\pm$ one standard deviation. Circles depict runs with $M = 0.2$, and triangles depict runs with spatially global competition and infinite dispersal (Poisson scatter of offspring). The smooth curve is a plot of Equation 9.13. Parameter values were $\Delta t = 0.1$, $f = 3.2$, $\beta = 0.28$, $\mu = 0.4$, $M = 0.2$, and the habitat was a thousand units long.

per plant ($M = 1$) and that the strength of competition decreases as M decreases. Moreover, the α–α^{est} relationship appears to be humped for all finite values of M. For example, a run with $M = 5$ and $\alpha = 0.99$ (approximately three hundred important neighbors per plant on average) produced an actual strength of competition of only 0.04.

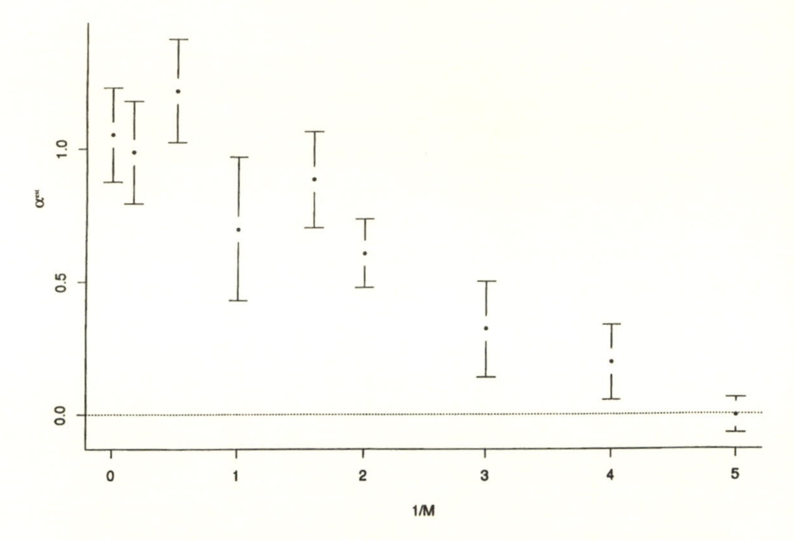

FIGURE 9.2. The strength of interspecific competition (α^{est}) as a function of the scales of competition and dispersal (M). The horizontal axis gives the reciprocal of M. Parameter values were as in Figure 9.1, but with $\alpha = 0.96$.

The results in Figures 9.1–9.2 leave us with a series of questions. Why are these counterintuitive results obtained? What eliminates competition as the species become identical (as α approaches 1)? Are these phenomena important in real ecosystems? Why are the mean-field and Poisson limits that dominate the theoretical literature so misleading and incomplete? The answers lie in the interplay between stochastically seeded spatial structure and dynamics.

MOMENT CLOSURE

A Pseudospatial Model

Before turning to the analysis of the point process itself, we first consider a simpler system that shares its locally stochastic dynamics. We make two changes in the point process. First, the environment is divided into a gridwork of identical cells, and the local densities that affect fecundity (the L's in Equation 9.1) are simply the within-cell densities. Second, when an offspring of species i is produced, it stays within its mother's cell with probability m_i and moves to a randomly chosen cell with probability $1 - m_i$ (as in many discrete-cell models such as Shmida and Ellner 1984 and Pacala and Tilman 1994). These changes to the point process impose both a patch structure that obviates information about within-patch locations and a dispersal pattern that obviates information about the spatial arrangement of patches.

Let $P(n_1, n_2)$ be the probability at time t that a randomly chosen patch contains n_1 individuals of species 1 and n_2 individuals of species 2. The probability that a patch containing n_i individuals of species i loses an individual during the small time interval Δt is simply the probability that one of the n_i individuals dies: $\mu_i n_i \, \Delta t$. Similarly, the probability that the patch gains an individual is the probability that an individual is born within the patch and does not disperse: $(f_i - \beta_i n_i - \alpha_{ij} \beta_i n_j) n_i \, \Delta t(1 - m_i) = F_i(n_i n_j) \Delta t(1 - m_i)$, plus the probability that a recruit disperses into the patch from outside it:

$$\overline{n_i F_i} \, \Delta t \, m_i \equiv \sum_{n_1=0}^{\infty} \sum_{n_2=0}^{\infty} P(n_1, n_2) n_i F_i(n_1, n_2) \, \Delta t \, m_i. \quad (9.4)$$

The above expression is simply the mean production of new species i across all patches, times the probability of dispersal.

With these definitions, we may write an expression for temporal changes in the probability distribution $P(n_1, n_2)$:

$$\begin{aligned}\frac{dP(n_1, n_2)}{dt} = & -P(n_1, n_2)\big[n_1 F_1(n_1, n_2)(1 - m_1) \\ & + n_2 F_2(n_1, n_2)(1 - m_2) \\ & + \mu_1 n_1 + \mu_2 n_2 + \overline{n_1 F_1}\, m_1 + \overline{n_2 F_2}\, m_2\big] \\ & + P(n_1 - 1, n_2)\big[(n_1 - 1) F_1(n_1 - 1, n_2)\end{aligned}$$

$$\times(1 - m_1) + \overline{n_1 F_1} m_1\Big]$$

$$+ P(n_1, n_2 - 1)\Big[(n_2 - 1)F_2(n_1, n_2 - 1)$$

$$\times(1 - m_2) + \overline{n_2 F_2} m_2\Big]$$

$$+ P(n_1 + 1, n_2)(n_1 + 1)\mu_1$$

$$+ P(n_1, n_2 + 1)(n_2 + 1)\mu_2. \qquad (9.5)$$

The problem with the system Equation 9.5 is that it describes an infinite number of equations, one for every possible combination of n_1 and n_2. We seek a simpler system that includes only some of the information in Equation 9.5.

Let N_i be the average abundance of species i across the habitat (mean number per patch). The first equation in Equation 9.6 governs the dynamics of the first moment, N_i (see the Appendix for its derivation):

$$\frac{dN_i}{dt} = N_i\left[f_i - \mu_i - \beta_i\left(N_i + \frac{\sigma_i^2}{N_i}\right) - \alpha_{ij}\beta_i\left(N_j + \frac{C}{N_i}\right)\right]$$

$$\frac{d\sigma_i^2}{dt} = \frac{dN_i}{dt} - 2\mu_i(\sigma_i^2 - N_i)$$

$$+ 2\sigma_i^2(1 - m_i)[f_i - 2\beta_i N_i - \alpha_{ij}\beta_i N_j]$$

$$- 2\alpha_{ij}\beta_i(1 - m_i)N_j C \quad (i, j) = (1, 2), (2, 1)$$

$$\frac{dC}{dt} = C[-(\mu_1 + \mu_2)$$

$$+ (1 - m_1)F_1(N_1, N_2) + (1 - m_2)F_2(N_2, N_1)]$$

$$- \alpha_{21}\beta_2(1 - m_2)N_2\sigma_1^2 - \alpha_{12}\beta_1(1 - m_1)N_1\sigma_2^2. \qquad (9.6)$$

The problem with this equation is that it depends on second moments, σ_i^2 and C, as well as on the first moments N_1 and

N_2. These second moments are σ_i^2, the variance from patch to patch in the abundance of species i, and C, the corresponding covariance between the abundances of the two species:

$$C = \sum_{n_1=0}^{\infty} \sum_{n_2=0}^{\infty} (n_1 - N_1)(n_2 - N_2)P(n_1, n_2)$$

$$= \sum_{n_1=0}^{\infty} \sum_{n_2=0}^{\infty} n_1 n_2 P(n_1, n_2) - N_1 N_2 .$$

Third moments are defined as

$$\sum_{n_1=0}^{\infty} \sum_{n_2=0}^{\infty} (n_i - N_i)(n_j - N_j)(n_k - N_k)P(n_1, n_2)$$

where i, j, and k may be equal to one or two in any combination (i.e., $i = 1$, $j = 1$, $k = 2$), and so on to fourth and higher moments. Although in this case the equations for the means depend only on the first two moments, in other cases higher moments will also be involved. The reason that third and higher moments do not arise in the equations for the means in Equation 9.6 is that the density-dependent functions (the F_i's) are linear; higher moments would result with nonlinear functions.

The second moments in the first equation in Equation 9.6 account for the effects of biologically generated spatial structure. The term $N_i + \sigma_i^2/N_i$ is the mean local density of conspecifics in the patch of a randomly chosen individual of species i (rather than in a randomly chosen patch). The mean local density grows with the between-patch variance, σ_i^2, because increasing numbers of individuals are located in clusters as σ_i^2 increases. Similarly, $N_j + C/N_j$ is the mean local density of heterospecifics in the patch of a randomly chosen species i. This density is elevated above the global mean, N_j, if the two species are spatially aggregated ($C > 0$) and depressed beneath it if the species are spatially segregated ($C < 0$). Note that if both the variance and covariance are equal to zero, then the first equation in Equations 9.6 is identical to the classical mean-field limits, whereas if the spatial distribution is Poisson ($\sigma_i^2 = N_i$ and $C = 0$), the first equation is identical to the long-dispersal limit.

The equations for the means, N_1 and N_2, do not constitute a closed system of equations, because the variances and covariances are themselves full state variables that will change through time because of local interactions and finite dispersal. To close the system, we require the equations governing the dynamics of σ_1^2, σ_2^2, and C. This illustrates the difference between Equations 9.5 and 9.6 and the kind of spatial models that dominate the theoretical literature. Simply adding diffusion terms to equations governing mean densities is not sufficient to describe stochastic processes with local interactions and finite dispersal.

In the Appendix we derive the equations governing the variances and covariances. These equations contain third moments. Because the third moments are again state variables, the equations governing the first two moments do not constitute a closed system. This illustrates the crux of the problem: Any system of equations governing the first n moments contains yet higher moments. What is needed is some closure rule that either states that some moments are negligible or expresses the higher moments in terms of lower moments.

Here we adopt the simplest closure rule. In the Appendix we show that the terms containing the third moments in the equations for the variances and covariances are negligible if both mean numbers of individuals per patch and movement rates are not too small. By omitting these negligible terms, we arrive at the approximate and closed system in Equations 9.6.

The accuracy of the approximation 9.6 is assessed simply by comparing the moments predicted by Equations 9.5 and 9.6 (Figures 9.3 and 9.4). For example, consider the symmetric case used to produce Figures 9.1–9.2 (with equal f's, μ's, β's, α's, and m's). The relationship between the community- and individual-level strengths of competition predicted by approximation 9.6 is a close approximation of the relationship predicted by the actual system (Eq. 9.5) in most cases (Figure 9.3).

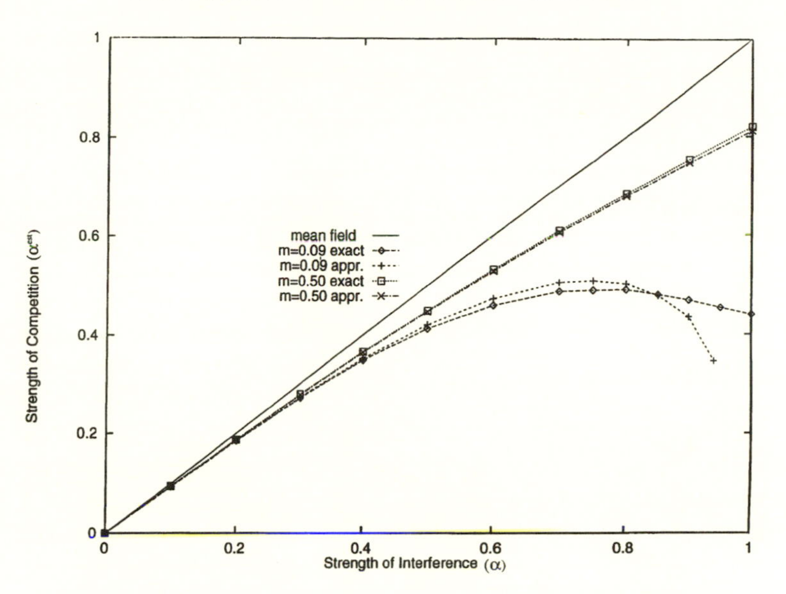

FIGURE 9.3. Equation 9.7 as predicted by the pseudospatial model (Eq. 9.5) (labeled exact) and the moment Equations 9.6 (labeled appr.). The value from the moment equations with $\alpha = 1.0$ and $m = 0.09$ is missing because the equations failed to converge for this set of parameter values. Other parameters were $f = 2.2$, $\mu = 0.88$, and $\beta = 0.066$.

The approximation fails with small m and α close to one because stochastic drift of the relative abundances within each patch is then large, and this leads to large third moments (because most patches contain primarily either species 1 or species 2).

Note that the α–α^{est} relationship given by Equations 9.5 and 9.6 may be humped, but only if the movement rate m is small. In contrast, our numerical work suggests that the relationship is humped in the point process for any finite value of M (as in Figure 9.1).

Because of the analytically tractable approximation (9.6), we are now in a position to explain the occurrence of an inverse relationship between α and α^{est}. Using the first equation in Equation 9.6, it is easy to show for a small number of

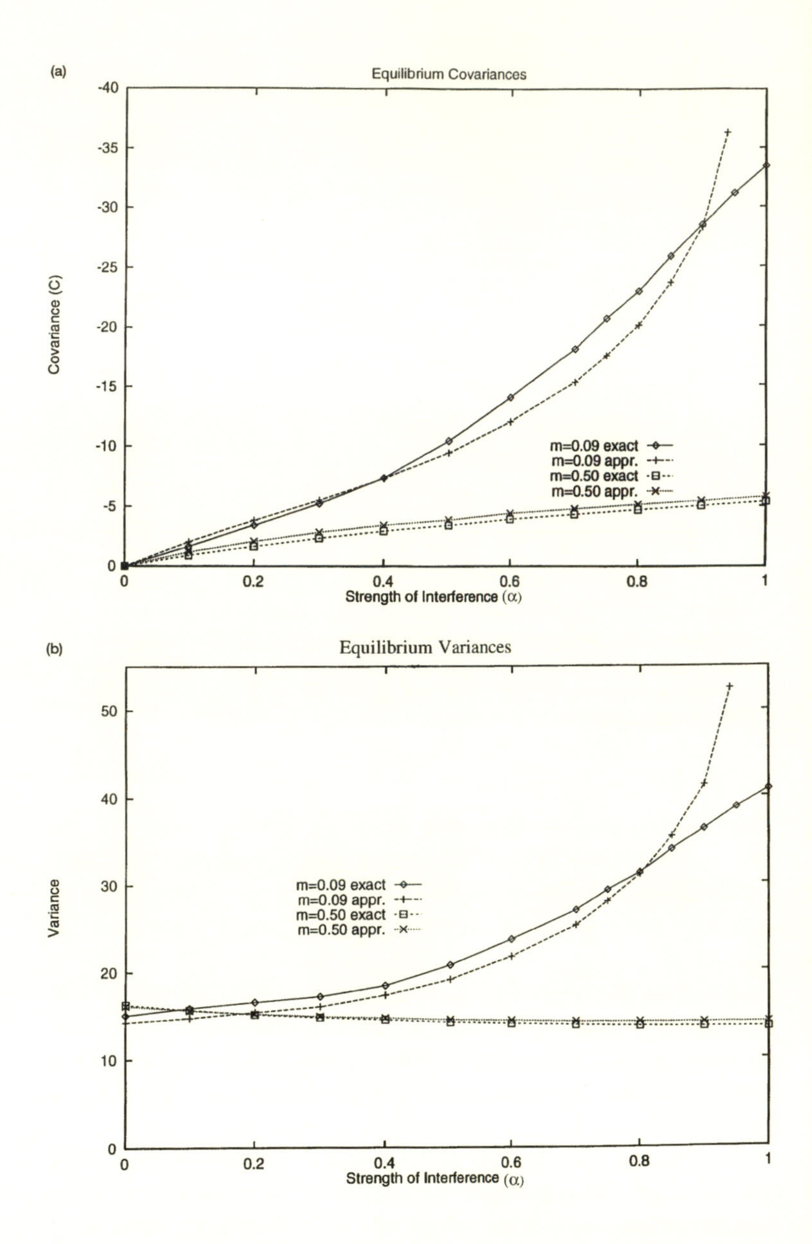
(a)
Equilibrium Covariances
Covariance (C)
-40
-35
-30
-25
-20
-15
-10
-5
0
0
0.2
0.4
0.6
0.8
1
Strength of Interference (α)
m=0.09 exact
m=0.09 appr.
m=0.50 exact
m=0.50 appr.
(b)
Equilibrium Variances
Variance
50
40
30
20
10
0
0
0.2
0.4
0.6
0.8
1
Strength of Interference (α)
m=0.09 exact
m=0.09 appr.
m=0.50 exact
m=0.50 appr.

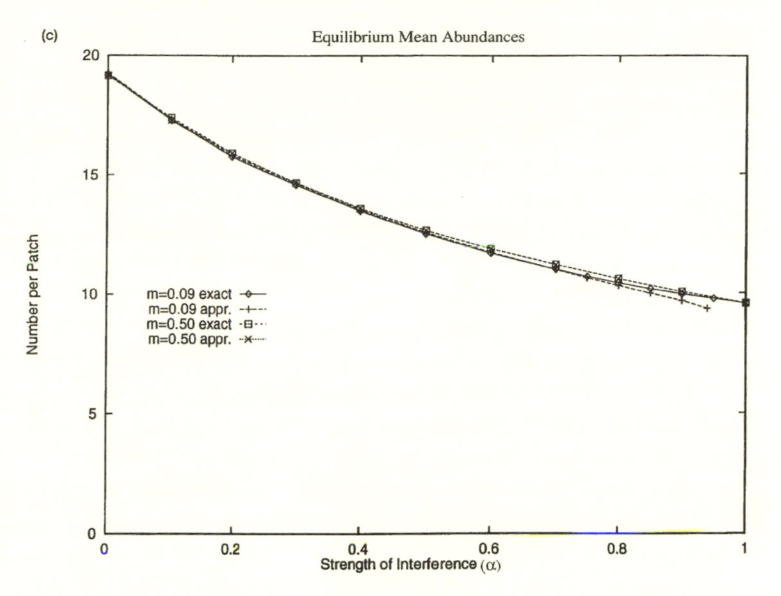

FIGURE 9.4. The moments at equilibrium predicted by the pseudospatial model (Eq. 9.5) (labeled exact) and the moment Equations 9.6 (labeled appr.). Parameter values were as in Figure 9.4. (a) Equilibrium interspecific covariance. (b) Equilibrium intraspecific variance. (c) Equilibrium means.

individuals removed in the removal experiment that

$$\alpha^{\text{est}} \approx \alpha\left[\frac{1 + \hat{C}/\hat{N}^2}{1 + \hat{\sigma}^2/\hat{N}^2}\right] \tag{9.7}$$

where the hats above the moments signify equilibrium; α^{est} is depressed beneath α because the spatial covariance is negative at equilibrium. Negative covariance implies interspecific spatial segregation. If covariance is sufficiently negative, then each patch contains primarily only one of the two species. The cause of the spontaneous spatial segregation in the model is a process directly analogous to genetic drift, which removes genetic polymorphism in small populations. Local populations in the models are kept small by density dependence, allowing relatively large random fluctuations in local relative abundance (demographic stochasticity). The random "drift" of either species toward high local relative abundance is reinforced by local dispersal, which biases the community composition of new recruits in favor of the locally most abundant species.

Note that the degree of the spatial segregation predicted by the pseudospatial model at equilibrium (the covariance in Eq. 9.7) increases as dispersive coupling decreases (Figure 9.4a). This explains the corresponding result in Figure 9.2 from the point process. Again, local dispersal facilitates local "ecological drift." The segregation also increases as interspecific competition at the individual level strengthens (Figure 9.4a) because the deterministic forces opposing drift toward local monodominance decrease in strength as α approaches one. A hump in the relationship between α and α^{est} occurs when the increasing spatial segregation overwhelms the increasing strength of individual-level competition.

Changes in m and α also affect α^{est} through changes in the variances (see Eq. 9.7). These changes generally complement those described above; within-species spatial aggregation (variance to mean ratio greater than one) develops to accompany the between-species segregation (Figure 9.4b). However, the effect of the variance is complicated by the fact that with small α, increased coupling may either increase clustering ($\sigma^2 > N$) or overdispersion ($\sigma^2 < N$) depending on the values of f, β, and μ.

It is possible to derive from Equation 9.6 useful simple formulae for the spatial moments at equilibrium. For example, if equilibrium population sizes are large and m is close to one then the covariance at equilibrium is approximately

$$\hat{C} \approx -\hat{N}_1 \hat{N}_2 (1 - m) \frac{\alpha\beta}{\mu} \tag{9.8}$$

and α^{est} is approximately

$$\alpha^{\text{est}} \approx \alpha \left[1 - (1 - m) \alpha \frac{\beta}{\mu} \right]. \tag{9.9}$$

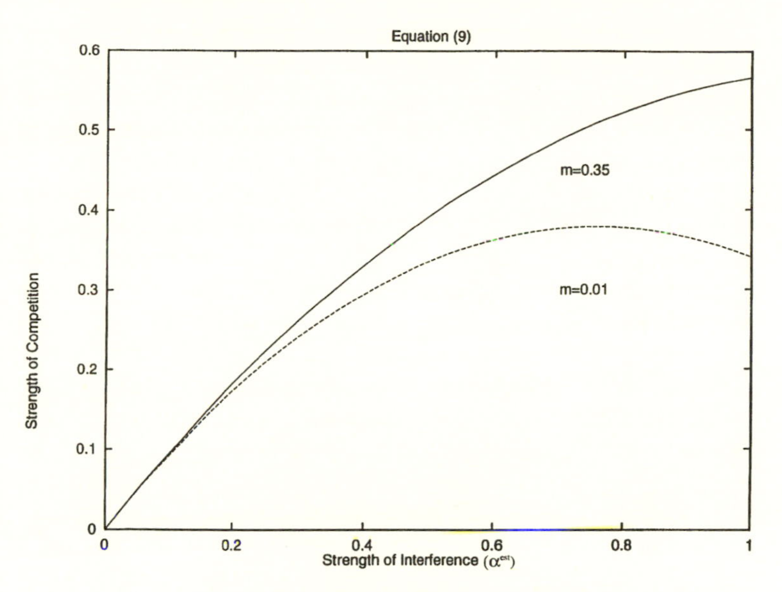

FIGURE 9.5. Equation 9.9 with $\beta/\mu = 0.67$.

Note that α^{est} is a humped function of α, but that the hump can occur for α between zero and one only if m is sufficiently small (and $\beta/\mu > 0.5$, Figure 9.5).

Finally, although we do not have the space to treat this subject fully, the model 9.5–9.6 also predicts new and unexplored mechanisms of coexistence. For example, Figure 9.6 shows an interesting case in which the two species are unable to coexist in either the mean-field or Poisson limits because $\alpha_{12} = \alpha_{21} = 1$. In either of these limits, species 2 is the superior competitor, but species 1 has a higher population growth rate in uncrowded conditions. Species 1 is weedy, with early successional vital rates, whereas species 2 has late-successional vital rates. It is easy to construct a simple submodel of resource competition to show that species 2 has the lowest R^* (*sensu* Tilman 1982). Note that the two species coexist if we give species 1 the more rapid dispersal to complement its weedy vital rates (Figure 9.6, $m_1 = 0.9$ and $m_2 = 0.5$). However, there is more happening here than a simple competition-colonization trade-off. Note that if we increase m_1 still further

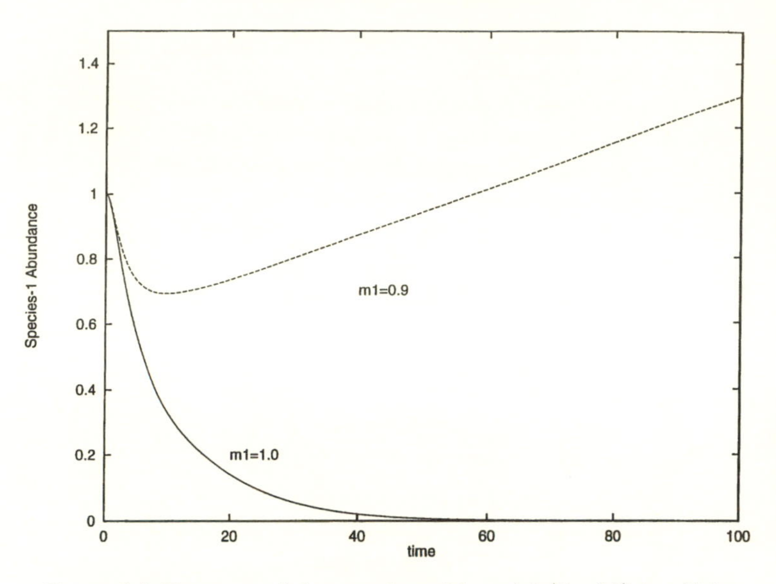

FIGURE 9.6. Two runs of the pseudospatial model (Eq. 9.5) in which species 1 competed against an equilibrium monoculture of species 2. Invasion succeeded if $m_1 = 0.9$ but failed if $m_1 = 1.0$. Other parameter values include $f_1 = 2.2$, $\mu_1 = 0.20$, $\beta_i = 0.2$, $f_2 = 1.05$, $\mu_2 = 0.8$, $\beta_2 = 0.02$, $m_2 = 0.2$, and $\alpha_{12} = \alpha_{21} = 1$. With these parameter values, species 2 had a larger equilibrium in monoculture than species 1 with either $m_1 = 1.0$ or $m_1 = 0.9$.

to 1.0, then species 1 goes extinct (Figure 9.6). If m_1 is too large, then species 1 cannot develop the degree of spatial segregation it requires to coexist with species 2.

For simplicity, we now restrict our attention to the simplest case, involving the smallest possible perturbation of the long-dispersal (Poisson) limit. We assume that $m_2 = 1$ so that species 2 always has a Poisson distribution and that m_1 is only slightly less than one. We further stack the deck against interesting behavior by assuming that $\alpha_{12} = \alpha_{21} = 1$. This eliminates the coexistence and founder control in both the Poisson and mean-field limits.

The assumption that $m_2 = 1$ and m_1 is near one ensures that spatial aggregation and segregation in the model will be small. Thus, to observe the effects of space, we must correspondingly weaken the nonspatial forces in the model. Let λ_{pi} be the eigenvalue governing the invasion of species i in an equilibrium monoculture of species j in the Poisson case ($m_1 = m_2 = 1$). It is easy to show that $\lambda_{pi} = f_i - \mu_i - \beta_i(f_j - \mu_j)/\beta_j$. We assume that both λ_{p1} and λ_{p2} are near zero, indicating near-competitive equivalence in the Poisson limit (e.g., $(f_1 - \mu_1)/\beta_1$ is close to $(f_2 - \mu_2)/\beta_2$).

Returning to the case in which $m_1 < 1$, species 1 can invade an equilibrium monoculture of species 2 (first condition below), and species 2 can invade an equilibrium monoculture of species 1 (second condition) if

$$0 < \lambda_{p1} + (1 - m_1)\beta_1(S - A)$$

$$0 < \lambda_{p2} + (1 - m_1)\beta_2\left(A - S\frac{\mu_2}{\mu_1}\right) \qquad (9.10)$$

where $S = \beta_1((f_2 - \mu_2)/\beta_2 - 1)/(\mu_1 + \mu_2)$ and $A = 1 + \beta_1/\mu_1$. Condition 9.10 is derived using a standard local stability analysis, assuming that $\alpha_{12} = \alpha_{21} = m_2 = 1$, and omitting all terms of order $(1 - m_1)^2$, λ_{pj}^2, λ_{pi}^2, $\lambda_{pi}\lambda_{pj}$, $\lambda_{pi}(1 - m_1)$, $\lambda_{pj}(1 - m_1)$ or higher.

The quantities S and A are easily interpretable as spatial effects. Quantity S describes the buildup of spatial segregation during invasion. To see this, replace dC/dt in the last equation in Equations 9.6 by zero and solve for the equilibrium covariance $C(N_1, N_2)^*$. Then divide $C(N_1, N_2)^*$ by N_1, and set N_2 equal to the monoculture equilibrium for species 2 ($N_2^* = (f_2 - \mu_2)/\beta_2 - 1$):

$$\lim_{N_1 \to 0} \frac{C(N_1, N_2^*)}{N_1} = -S(1 - m_1). \qquad (9.11)$$

Thus, S quantifies how spatial segregation grows relative to the mean during invasion.

Similarly, A describes the buildup of intraspecific aggregation by species 1 during invasion (species 2 cannot aggregate because $m_2 = 1$). Replacing $d\sigma_1^2/dt$ with zero in Equations 9.6 and solving for the equilibrium variance $\sigma_1^2(N_1, N_2)^*$, we find that

$$\lim_{N_1 \to 0} = \left[\frac{\sigma_1^2(N_1, N_2^*)}{N_1} - 1 \right] = A(1 - m_1).$$

If A is positive, then species 1 clusters as it invades because σ_1^2 grows faster than N_1. In contrast, if A is negative, then species 1 becomes evenly dispersed as it invades. Putting all this together, we see from Equation 9.10 that species 1 will invade if interspecific segregation grows sufficiently faster than aggregation and will fail to invade if aggregation grows sufficiently faster than segregation.

To expose only the spatial effects, it is convenient to consider the further restriction that $\lambda_{p1} = \lambda_{p2} = 0$. Then, the two species will coexist if $1 < S/A < \mu_1/\mu_2$; species 1 will invariably exclude species 2 if $1 < S/A$ and $\mu_1/\mu_2 < S/A$; species 2 will exclude species 1 if $1 > S/A$ and $\mu_1/\mu_2 < S/A$; and founder control will result if $\mu_1/\mu_2 < S/A < 1$. We have checked all of these outcomes both in the exact system 9.5 and in the approximation 9.6; spatial effects alone are capable of producing any of the four different outcomes of competition.

Two primary conclusions emerge from the stability analysis. First, long dispersal is not always beneficial. It facilitates persistence if the tendency to segregate is larger than the tendency to cluster but impedes persistence otherwise. Second, the interplay of local competition and dynamics is dynamically rich; it has the potential to transform any of the four quantitatively different outcomes of competition in the Poisson or mean-field limits (coexistence, founder control, exclusion of species 1, or exclusion of species 2) into any other outcome.

The Point Process

We now return to the point process that produced Figures 9.1–9.3. The point process is more difficult than the pseudospatial model above because there is no simple expression analogous to Equation 9.5. However, we can proceed directly to the moment approximations analogous to Equations 9.6.

Bolker and Pacala (1997) show that the equation for the mean abundance of species i in the point process converges to the first equation in Equations 9.6 with

$$\sigma_i^2 = U_{ii}(0)N_i + \int_{\substack{-\infty \\ y \neq 0}}^{\infty} U_{ii}(y)c_{ii}(y)\,dy, \qquad C = \int_{-\infty}^{\infty} U_{ij}c_{ij}(y)\,dy \tag{9.12}$$

where the U's are given by Equation 9.3. The function $c_{ij}(y)$ is the interspecific spatial covariance function. Consider many pairs of small quadrates of area A, with the members of each pair a distance y apart. The expected covariance between the *densities* of species 1 in one member of a pair and species 2 in the other is $c_{ij}(y)$ in the limit of small A. Similarly, $c_{ii}(y)$ gives the within-species spatial covariance.

To complete the model, Bolker and Pacala (1997) derived integro-partial differential equations for the dynamics of $c_{11}(y)$, $c_{22}(y)$, and $c_{12}(y)$. During the derivation, they closed the system by assuming that all third central moments were zero, exactly as in the derivation of Equations 9.6 (Bolker and Pacala 1997).

Pacala (1997) reported an equation for α^{est}, obtained by solving for the equilibrium values of N_1, N_2, $c_{11}(y)$, $c_{22}(y)$, and $c_{12}(y)$:

$$\alpha^{\text{est}} = \alpha\left(\frac{M\sqrt{1-\alpha} - W}{M\sqrt{1-\alpha} + W}\right), \qquad W = \frac{\beta}{f-u}\sqrt{\frac{2\mu}{f-u}}. \tag{9.13}$$

For simplicity, this expression is presented for the special case in which M is relatively large and α is close to one. The fact that α^{est} in Equation 9.13 may be negative for very small M is an artifact of this approximation. However, the plot in Figure 9.1 shows that Equation 9.13 generally provides a reasonably accurate approximation (note that M is relatively small in Figure 9.1).

Unlike the pseudospatial system (Eqs. 9.5–9.6), but consistent with the simulations of the point process, expression 9.13

predicts that the α–α^{est} relationship is humped even for large values of M. Note in expression 9.13 that the hump will always occur at a value of α between zero and one, with the hump closer to one the larger the ecological scales M.

The difference between the pseudospatial model and the point process is explained by the between-species covariance (recall expression 9.7). In the same case leading to equation 9.13:

$$\hat{c}_{ij}(y) \approx -\frac{\hat{N}}{2M\sqrt{(1-\alpha)\hat{N}/\mu}}\, e^{-(|y|/M)\sqrt{(1-\alpha)\hat{N}/\mu}}. \quad (9.14)$$

Because of the $1-\alpha$ term in the denominator of Equation 9.14, the interspecific spatial covariance becomes strongly negative as α approaches one. This implies that the two species spontaneously segregate into large monospecific patches in the point process if α is close to one (even if M is large), thereby causing the observed drop in α^{est} toward zero (Figure 9.1). The phenomenon is similar to the spatial clustering exhibited by some discrete interacting particle systems, such as the voter model (Durett and Levin 1994a).

The difference between the pseudospatial model and the point process is that extreme levels of spatial segregation occur only at very low levels of coupling (small m) in the pseudospatial model, but at large or small ecological scales (M) if α is near one in the point process (compare expressions 9.8 and 9.13). Because the spatial distribution in the point process breaks into ever larger and more monospecific patches as α approaches one, the effective level of between-patch coupling progressively decreases, ultimately causing dynamics analogous to those in the pseudospatial model with very low m's. Explicit space brings with it the potential for self-organization of a loosely coupled metapopulation.

CONCLUSIONS

Three primary results emerge from out analysis of the moment equations: (1) With finite dispersal and spatially local interactions, α^{est} is a humped function of α. The hump moves to the right as the scales of competition and dispersal increase. As a result, strong competition at the individual level leads to weak competition at the community level. This phenomenon is caused by an ecological analogue of genetic drift. As α approaches one, the deterministic forces opposing stochastic ecological drift decrease, and the tendency of either species to drift to local extinction (or fixation) increases, which gives distinct patches of each species. It is important to understand that the depression of community-level competition by this mechanism is potentially large relative to more widely recognized mechanisms. (2) Because α^{est} decreases as M decreases, short scales reduce community-level competition. Small competitive neighborhoods reduce competition because they facilitate the stochastic drift that leads to spatial segregation. Short dispersal further isolates local regions from one another and reinforces drift. (3) Interspecific spatial segregation can cause coexistence. In contrast to the advantages of long dispersal in fugitive-species models (e.g., Levins and Culver 1971), short dispersal facilitates persistence if a species' tendency to segregate exceeds its tendency to aggregate (condition 9.10).

Three independent lines of evidence provide empirical support for some of the above phenomena. First, spatial segregation is nearly ubiquitous and easy to observe. In any place where multispecies vegetation can be viewed from above, count individuals of a species close to randomly selected points (e.g., within a circle with radius equal to canopy height) and then close to randomly selected individuals of another species. A first estimate that is $(1 - b)\%$ of the second implies approximately a $b\%$ reduction in the strength of competition (if the neighborhood radius used is appropriate). If you try this, you will commonly observe biases in excess of 50%. Second, in three of four instances, observed levels of local spatial segregation were shown to have a large effect on community-level competition in field-calibrated models (yes in Pacala and Deutschman 1996; Cain et al. 1995; and Rees, Grubb, and Kelly 1996; no in Pacala and Silander 1990). Moreover, spatial segregation arises spontaneously in data-defined models (see Pacala et al. 1996; Pacala and Deutschman 1996). Third, as reported in Pacala (1996), Kelly and Tripler (unpublished) reviewed over three hundred published experimental field slides of competition. They found that plant-centered experiments in which neighbors were removed from quadrants centered on focal plants reported statistically significant competition at nearly three times the rate of experiments in which plots were established without reference to the locations of plants (72% versus 26% of experiments showed competition, respectively). Apparently, the fine-scale spatial segregation present in the plot-centered experiments but not in the plant-centered experiments caused the threefold reduction in the interspecific competition detected.

Obviously, the above evidence for the causes and consequences of spatial segregation is incomplete. Experiments are needed to test our four theoretical results against alternative hypotheses. The most likely alternative hypothesis is that habitats are spatially heterogeneous at fine scales and that species coexist and segregate, not because of finite dispersal and local interactions alone, but because each species is the dominant competitor in a different habitat type. One particularly worthwhile experiment would be to transplant plants from a natural community: (1) to a new location with scrambled spatial structure, (2) to a new location with preserved spatial structure, (3) to the same location with scrambled spatial structure, and (4) to the same location with preserved spatial structure (obviously with replicates). Comparisons of the resulting dynamics in 1 and 3 against 2 and 4 would show the importance of fine-scale spatial structure, whereas comparison of 1, 2, and 3 against 4 would test the alternative hypothesis of physically heterogeneous habitat. By following the treatments through time, one could also observe the dynamics of spatial pattern to gain further insight into the genesis of spatial segregation.

The methods presented here allow one to derive the macroscopic population dynamic equations implied by measurable microscopic rules governing births, deaths, and movement by individuals. These methods may be applied to models of virtually any ecological interaction. They promise to close the gap in the current explanatory theory containing the causes and consequences of biologically generated heterogeneity.

ACKNOWLEDGMENTS

We gratefully acknowledge the support of the Mellon Foundation, NASA (NAGW-3741, NAGW-468), the NSF (DEB-9221097), and the DOE (DE-FG04-94ER61815).

APPENDIX

Derivation of the Moment Equations 9.6

Let the random variable q_{it} be the number of species i individuals within a patch. By definition N_i is the mean of q_{it}, σ_i^2 is the variance, and C is the covariance between q_{1t} and q_{2t}.

During the small time interval Δt, q_{it} will change by the amount -1 with probability $\mu_i q_{it}\,\Delta t$ and by the amount $+1$ with probability:

$$\left[q_{it}(1 - m_i)F(q_{it} - q_{jt}) + m_i\overline{n_i F_i}\right]\Delta t$$

where

$$F(q_{it}, q_{jt}) = f_i - \beta_i q_{it} - \alpha_{ij}\beta_i q_{jt},$$

$\overline{n_i F_i}$ is defined by Equation 9.4.

If we define the operator $E[\;]$ as the expectation over the "ensemble" of all possible realizations of the model, then

$$E[\Delta q_{it}] = E\Big[(-1)(\mu_i q_{it})\,\Delta t + (+1)\Big((1 - m_i)q_{it}F_i(q_{it}, q_{jt}) + m_i E\big[q_{it}F_i(q_{it}, q_{jt})\big]\Big)\,\Delta t\Big].$$

The expression inside the brackets on the right-hand side is the expected change during Δt given the values of q_{it} and q_{jt} (the sum over all possible kinds of changes of the amount of the change times the probability of the change). The expectation of this conditional expectation over all possible values of q_{it} and q_{jt} (with the operator $E[\;]$) gives the total mean change.

Using the facts that $E[q_{it}] = N_i$, $E[q_{it}^2] = \sigma_i^2 + N_i^2$, and $E[q_{it}q_{jt}] = C + N_iN_j$, we have

$$\frac{\Delta N_i}{\Delta t} = -\mu_i N_i + N_i f_i - \beta_i(\sigma_i^2 + N_i^2) - \alpha_{ij}\,\beta_i(C + N_iN_j). \tag{9.A1}$$

After taking the limit as $\Delta t \to 0$, we have the first equation in Equations 9.6.

Turning now to the variance σ_i^2, if an individual dies in a patch then q_{it}^2 changes by an amount $-2q_{it} + 1$. Similarly, if an individual recruits into the patch then q_{it}^2 changes by an amount $2q_{it}^2 + 1$. Thus we have

$$\begin{aligned} E\left[\Delta q_i^2\right] = E\Big[&(-2q_{it} + 1)(q_{it}\,\mu_i)\,\Delta t \\ &+(2q_{it} + 1)\big(q_{it}(1 - m_i)F_i(q_{it}, q_{jt}) \\ &+m_i E\left[q_{it}F_i(q_{it}, q_{jt})\right]\big)\,\Delta t\Big]. \end{aligned} \tag{9.A2}$$

Note that $E[\Delta q_i^2] = [q_{it+\Delta t}^2 - q_{it}^2] = \Delta\sigma_i^2 + \Delta N_i^2$. Also, to evaluate the right-hand side of Equation 9.A2, we must evaluate $E[q_{it}^3]$ and $E[q_{it}^2 q_{jt}]$. We label the third moment $E[(q_{it} - N_i)^3]$ as T_{iii} and the third moment $E[(q_{iit} - N_i)^2(q_{ijt} - N_j)]$ as T_{iij}. Using these definitions

$$E\left[q_{it}^3\right] = T_{iii} + 3\sigma_i^2 N_i + N_i^3$$

$$E\left[q_{it}^2 q_{jt}\right] = T_{iij} + N_i^2 N_j + N_j\sigma_i^2 + 2N_i C$$

and

$$\begin{aligned} \frac{\Delta\sigma_i^2}{\Delta t} + \frac{\Delta N_i^2}{\Delta t} = &-2\,\mu_i(\sigma_i^2 + N_i^2 - N_i) \\ &+(1 - m_i)\big[f_i(2\sigma_i^2 + 2N_i^2 + N_i) \\ &-\beta_i(2T_{iii} + 6\sigma_i^2 N_i + 2N_i^3 + \sigma_i^2 + N_i^2) \\ &-\alpha_{ij}\,\beta_i\big(2T_{iij} + 2N_i^2 N_j + 2N_j\sigma_i^2 \end{aligned}$$

$$+4N_iC + N_iN_j + C)\big]$$

$$+ m_i\big[f_i(2N_i^2 + N_i)$$

$$-\beta_i(2N_i^3 + 2N_i\sigma_i^{\,2} + N_i^2 + \sigma_i^{\,2})$$

$$-\alpha_{ij}\beta_i\big(2N_i^{\,2}N_j + 2N_iC + N_iN_j + C\big)\big]. \qquad (9.A3)$$

Note that $\lim_{\Delta t \to 0} \Delta N_i^{\,2}/\Delta t = 2N_i\, dN_i/dt$. Passing to the limit of infinitesimal Δt and collecting like terms yields the second equation in Equations 9.6 plus two additional third moment terms on the right-hand side: $-2\beta_i T_{iii}(1 - m_i)$ and $-2\alpha_{ij}\beta_i T_{iij}(1 - m_i)$.

Proceeding in a precisely analogous way, we evaluate $E[\Delta q_1 q_2]$ to produce the third equation in Equations 9.6 plus two additional third-moment terms on the right-hand side: $-[\beta_1(1 - m_1) + \alpha_{21}\beta_2(1 - m_2)]T_{112}$ and $-[\beta_2(1 - m_2) + \alpha_{12}\beta_1(1 - m_1)]T_{221}$.

Now, why can we omit the third moment terms? If m_1 and m_2 both equal one, then the spatial distribution is Poisson (any non-Poisson initial distribution will die away as the plants initially present die; see Pacala and Silander 1985). We Taylor expand the ensemble distribution $P(n_1, n_2)$ about $m_1 = 1$ and $m_2 = 1$, yielding

$$P(n_1, n_2) \approx R(n_1, n_2) + (1 - m_1)W(n_1, n_2)$$
$$+(1 - m_2)Z(n_1, n_2)$$

where $R(n_1, n_2)$ is the bivariate Poisson distribution and $W(\ )$ and $Z(\ )$ do not depend on m_1 or m_2. Let T^R_{iii} and T^R_{iij} signify third moments of the Poisson distribution $R(n_1, n_2)$; $T^R_{iii} = N_i$ and $T^R_{iij} = 0$. Thus, the first third moment term above is

$$-2\beta_i(1 - m_i)T_{iii} \approx -2\beta_i(1 - m_i)N_i + O[(1 - m)^2]$$

where $O[(1 - m_i)^2]$ means of order $(1 - m_i)^2$ or $(1 - m_1)(1 - m_2)$, and the remaining three third moment terms above are simply $O[(1 - m)^2]$. We take m_1 and m_2 to be sufficiently close to one that $O[(1 - m)^2]$ terms are negligible. This leaves us with $-2\beta_i(1 - m_i)N_i$ in the equation for σ_i^2 as the sole remaining third moment term. One could analyze the model with this term included. However, if the average number of individuals per patch is large, then this term is negligible relative to other terms in the second equation in Equations 9.6 (e.g., the terms of order N_i^2, N_iN_j, σ_i^2, $(1 - m_i)N_i\sigma_i^2$, and $(1 - m_i)N_j\sigma_i^2$).

CHAPTER TEN

Habitat Destruction and Species Extinctions

David Tilman and Clarence L. Lehman

INTRODUCTION

The expansion of humans across the earth has caused a long string of species extinctions (see Ehrlich and Ehrlich 1981; Wilson 1992; May, Lawton, and Stork 1995; Pimm et al. 1995). Unfortunately, all of these past extinctions pale in comparison to the projected effects of recent and anticipated human habitat destruction (e.g., May et al. 1995; Pimm et al. 1995). Human conversion of natural ecosystems into managed ecosystems or into roads, rights-of-way, housing, and industry continues at an astounding rate. Already, about 43% of earth's terrestrial ecosystem area is directly harnessed for human benefit (Daily 1995) with some ecosystem types, especially those with fertile soils or favorable climates, much more heavily exploited. For instance, less than 1% of the original tallgrass prairie and less than 0.2% of the original oak savanna of Minnesota still remains. In this chapter we explore the potential effects of such habitat destruction by comparing and contrasting the predictions of spatially explicit models, spatially implicit models, and extensions of island biogeographic theory.

SPECIES-AREA AND SPECIES-ABUNDANCE RELATIONSHIPS

Early island biogeographers reported an empirical relationship between habitat size and species richness (Darlington 1957). The number of species in a given taxonomic group on islands of a given geographic region is a power function of island area:

$$S = cA^z, \quad (10.1)$$

where S is the number of species (species richness), A is the surface area of an island, z is an exponent observed to range from about 0.18 to 0.3, and c is a constant unique to each taxonomic group and set of islands. In this, as in all cases that follow, c is unique to each equation and situation. Similar relationships occur when areas of different sizes are sampled in mainland ecosystems, for which z values have been found to range from about 0.12 to 0.2. The lower z values of mainlands have been attributed to mainlands having a larger proportion of "transient" species (MacArthur and Wilson 1967).

The species-area relationship has been used to estimate how many species might be threatened with extinction following habitat destruction (Simberloff 1984; Wilson 1988, 1992). Consider a virgin habitat of area A_v that contains S_v species. If a proportion, D, were destroyed, the number of species, S_D, remaining would be

$$S_D = c[(1 - D)A_v]^z. \quad (10.2)$$

This means that the proportion of the original species that survived after destruction of a proportion D of the original virgin habitat would be

$$\frac{S_D}{S_v} = (1 - D)^z. \quad (10.3)$$

The proportion of the original species driven extinct would thus be

$$P_D = 1 - (1 - D)^z. \quad (10.4)$$

To evaluate the potential validity of this expression for P_D, consider what is encompassed in the empirical species-area relationship. In essence, it is a sampling function that shows how the number of observed species depends on the size of the area sampled. Why, though, might diversity increase with area sampled? Area encompasses at least three distinct factors that influence diversity. First, on average, larger areas contain more individuals. It is clearly impossible to observe more species than individuals, but the issue is deeper than this. The species in a given community have some characteristic distribution of abundances, such as a log-normal distribution (Preston 1948), a broken-stick distribution (MacArthur 1960), or a geometric distribution (May 1975). Species also have a minimal population size below which demographic stochastic fluctuations lead to extinction (May 1973b, 1975). May showed that these two relationships led to

$$S = c\left(\frac{I}{m_I}\right)^z, \tag{10.5}$$

where m_I is the minimum population size required for species survival, I is the total number of individuals of all species in the community of interest, and the value of z for this relationship depends on the underlying distribution (May 1975; Preston 1948).

A second reason for diversity to increase with area comes from the statistical distribution of areas required for the survival of a minimum population of each species. A species of small herbivorous mammal, for instance, can survive in a much smaller area than a large carnivore. Such species-specific minimum areas for survival are likely to be distributed, within a community, with log-normal, geometric, or broken-stick relationships, leading to a power function analogous to that above,

$$S = c\left(\frac{A}{m_A}\right)^z, \tag{10.6}$$

where m_A and z are derivable from the underlying distribution of minimal areas. This equation states that diversity increases with area because species differ in how much area they require to survive, and larger areas allow the survival of species that cannot survive in small areas.

A third reason why diversity should increase with area is heterogeneity. Habitats differ in their physical and biotic characteristics. Each species has some combinations of habitat characteristics at which it does best and some combinations in which it cannot maintain a viable population. Larger areas are likely to contain a larger range of habitat conditions and thus to encompass conditions conducive to the survival of more species. Habitat heterogeneity can be incorporated into Equation 10.6 by partitioning the effects of area into that caused by the minimum area requirement of species and the heterogeneity effect:

$$S = c\left(\frac{A}{m_A}\right)^{z_1}(Ah_A)^{z_2}. \tag{10.7}$$

Equations 10.5–10.7 predict the types of species expected to go extinct following habitat destruction, based on island biogeographic theory and its derivatives. First, the abundance distribution relationship behind Equation 10.5 predicts that as I is decreased by habitat destruction, rarer species have a much higher chance of extinction. In its most literal interpretation, Equation 10.5 states that species should go extinct in order from the least to the most abundant as D increases. Equation 10.6 predicts that species should go extinct in order from those with the largest to those with the smallest home ranges as habitat destruction increases. Equation 10.7 adds the expectation that some species will be driven extinct, independent of the first two factors, as their unique habitat requirements are lost following habitat destruction. This corresponds to the extinction of local endemics following destruction of their habitat. All three of these are predictions of biased extinction of rare species, with the latter two predictions including a more explicit reason for their rarity (i.e., large home range or unique but rare habitat requirements). Thus, the island biogeographic model leads to the general expectation that habitat destruction should lead to the biased extinction of rare species. The individual-based metapopulation-like models and explicitly spatial models that are analyzed next make a markedly different prediction (e.g., Nee and May 1992; Tilman et al. 1994; Tilman, Lehman, and Yin 1997).

INDIVIDUAL-BASED MODELS OF COEXISTENCE AND DIVERSITY

The metapopulation models of Levins (1969), Horn and MacArthur (1972), Levin and Paine (1974), Hastings (1980), Hanski (1983), Shmida and Ellner (1984), Cohen and Levin (1991), and others have been used to predict the effects of alternative management practices on survival of rare or endangered species (e.g., Harrison, Murphy, and Ehrlich 1988; Wootton and Bell 1992) and to predict the effects of habitat destruction on communities of numerous coexisting species (Nee and May 1992; Tilman et al. 1994; Moilanen and Hanski 1995; Loehle 1996; Tilman et al. 1997). We use such theory to explore the traits of species and numbers of species predicted to be threatened with extinction following habitat destruction.

The Analytical Model

As in Chapter 1, let an infinitely large habitat be divided into sites, each the size occupied by a single adult. Let the proportion of sites occupied by individuals of a species be p, the mortality rate m, the colonization rate c, and the proportion of permanently destroyed sites D. Assume that colonists entering a destroyed site do not survive, and that destroyed sites cannot produce colonists. As proposed by Nee and May (1992), this gives

$$\frac{dp}{dt} = cp(1 - D - p) - mp. \tag{10.8}$$

Note that the rate of propagule production, cp, is multiplied by the proportion of open and viable sites, $(1 - D - p)$, to give the rate of increase in site occupancy. The rate of mortality within sites, mp, is subtracted from this to give the rate of change in site occupancy. This is called a metapopulation model, but it is more correctly considered an implicitly spatial model of site occupancy by individuals and of competitive interactions among individuals. Here we use a multispecies extension of this model in which all species compete, as if for a single limiting resource (Tilman et al. 1994). This model was derived in Tilman et al. (1994) from the models of Hastings (1980), Nee and May (1992), and Tilman (1994). In our model, there is a simple competitive hierarchy, with species ranked from the best competitor (species 1) to the worst (species n). Species 1, the best competitor, displaces all other species from any site that it invades, and species 1 can invade any undestroyed site. The next best competitor, species 2, displaces all species except species 1, and so on to species n, which is displaced by all other species, can displace none, and only can invade empty undestroyed sites. The dynamics of site occupancy by the ith species, where subscripts i and j identify species, are (Eq. 10.9)

$$\frac{dp_i}{dt} = c_i p_i \left(1 - D - \sum_{j=1}^{i} p_j\right) - m_i p_i - \sum_{j=1}^{i-1} c_j p_i p_j. \quad (10.9)$$

Here D is a constant, which means that habitat destruction is permanent. The abundances of all species must be greater than or equal to zero. The competitive hierarchy means that a species is not affected by its inferior competitors. Interspecific trade-offs among competitive ability, colonization ability, and mortality are required to allow an unlimited number of species to stably coexist in a virgin, undestroyed habitat, even though they compete as if for a single resource (Tilman 1994). What are the effects of permanent habitat destruction on the diversity of such multispecies communities?

Consider the impact on the best competitor of destruction of a proportion, D, of the habitat. From Equation 10.9, at equilibrium (i.e., $dp_i/dt = 0$), the proportion of sites occupied by the ith species will be (Eq. 10.10)

$$\hat{p}_i = 1 - D - \frac{m_i}{c_i} - \sum_{j=1}^{i-1} \hat{p}_j \left(1 - \frac{c_j}{c_i}\right), \quad (10.10)$$

where equations are solved in order starting with species 1 and must have $p_i \geq 0$ for all i. When applied to the best competitor, species 1, this predicts that its abundance will be zero when $D_1 = 1 - m_1/c_1$. The equilibrium abundance of species 1 in a virgin, undestroyed habitat is also $p_1 = 1 - m_1/c_1$. This means that, independent of the abundances of inferior competitors, the best competitor is predicted to be driven extinct once a proportion, D_1, of habitat equal to its site occupancy in a virgin habitat is destroyed. Note that this is deterministic, not stochastic, extinction. Thus, if at equilibrium in a virgin habitat the best competitor occurred in 25% of the sites, this model predicts that permanent destruction of a randomly selected set of 25% of habitat sites would lead to its eventual extinction. The more abundant the best competitor, the greater is the random permanent destruction of sites required to drive it extinct. But, within the framework of this analytical model, habitat destruction of a randomly chosen proportion of D_1 sites has the same eventual effect—extinction—as if there had been the selective destruction of exactly those D_1 sites in which the species occurred.

What are the effects of habitat destruction on the extinction of inferior species? Tilman et al. (1994) derived results for a case in which species abundances form a geometric series, with the most abundant species being the best competitor, and increasingly less abundant species being increasing poorer competitors. This can be expressed as $p_i = q(1 - q)^{i-1}$, where q is the abundance of the best competitor and i refers to the species of competitive rank, i. It is possible to analytically calculate the D at which each species will go extinct, with D_i being the proportion of destroyed habitat beyond which species i can no longer survive (Tilman et al. 1994). In this case, $D_i = 1 - (1 - q)^{2i-1}$, which means that $D_1 < D_2 < D_3 \cdots < D_n$. Thus species are predicted to go extinct in order from the most abundant (best competitor and poorest disperser) to the least abundant (poorest competitor and best disperser) as habitat destruction increases. This prediction stands in direct contrast to the prediction of biased extinction of rare species made by derivatives of the species-area curve.

This surprising prediction seems counterintuitive at first. How could the most abundant species, which are also the best competitors, be the species most threatened with extinction following habitat destruction? How could deterministic extinction of species 1 occur when only a proportion, D_1, of an infinitely large habitat has been destroyed? An intuitive answer to these questions is provided by a consideration of the traits required of other species if they are to stably coexist in a virgin habitat with superior competitors. Within this model, a major way for an inferior competitor to coexist with a superior competitor in a virgin habitat is for the inferior competitor to be a better disperser. In contrast, the best competitors are the poorest dispersers of all, and their abundance is the most dispersal limited of all species. This makes them be the most susceptible to extinction from habitat destruction.

Tilman et al. (1997) explored three other analytical and several numerical cases to determine how general and robust the prediction of competitively ordered extinction was. In addition to the case in which poorer competitors were less abundant and all species had equal mortality, they considered cases in which (1) all species were equally abundant with equal mortality, (2) poorer competitors were more abundant with equal mortality, and (3) all species were equally abundant but poorer competitors had progressively greater mortality. Analytical solutions showed that, in all cases, species were predicted to go extinct in order of their competitive abilities, with the best competitors first, as habitat destruction increased. However, there are special cases in which some species go extinct out of this order (Tilman et al. 1997).

This biased extinction of the best competitors does not depend on the precise formulation of the model. For example, consider an analytical model that allows for the loss of seed at the edge of a habitat island. The length of habitat edge is proportional to $\sqrt{A}$, where A is the area of the habitat. This means that the fraction, L, of all seeds lost over the edge is proportional to $\sqrt{A}/A$, which is to say $L = K/\sqrt{A}$, where K is a constant. Thus each c_i in Equation 10.9 is replaced by $(1 - L)\ c_i = (1 - [K/\sqrt{A}])\ c_i$. When $A = \infty$, seed loss L is zero, just as assumed in Equation 10.9. When $S = K$, seed loss L is one, and all seeds are lost over the edge. Thus L in this model corresponds roughly to D in the original model, though it enters the equations differently:

$$\frac{dp_i}{dt} = (1 - L)c_i p_i \left(1 - \sum_{j=1}^{i} p_j\right) - m_i p_i - p_i \sum_{j=1}^{i-1} (1 - L)c_j p_j. \tag{10.11}$$

The equilibrium values for Equation 10.11 are

$$\hat{p}_i = 1 - \frac{m_i}{(1 - L)c_i} - \sum_{j=1}^{i-1} \hat{p}_j\left(1 + \frac{c_j}{c_i}\right). \tag{10.12}$$

The important questions are: How does the abundance $\hat{p}_i$ of species i change with L, and for what value of L does species i become permanently extinct? For the best competitor, setting Equation 10.12 to zero and solving for L yields $L_1 = 1 - m_1/c_1$. Notice that this L_1 for extinction is identical to D_1 for extinction in the original model. Indeed, for cases in which species have a geometric abundance series of a large habitat ($A = \infty$), simulations have shown that decreasing habitat size (A) leads to the same biased extinction as previously discussed. The first species to go extinct is the best competitor (poorest disperser), and it is followed, in order, by progressively better dispersers (poorer competitors) as the habitat has more and more edge, that is, as it is effectively smaller. Other models have also led to similar effects (Moilanen and Hanski 1995; Loehle and Li 1996).

In total our metapopulation-like models make a markedly different prediction than the models derived from the species-area relationship. The latter predict that extinction should be biased toward rare species, whereas the former predict that extinction should be biased toward superior competitors and poor dispersers, which could be among the most abundant species. These divergent predictions are caused by the different mechanisms of coexistence assumed in each model. The species-area models do not include any interspecific interaction and apply to species for which persistence depends only on the amount of available area. This causes species that require large areas (because they have large home ranges or small population sizes) to go extinct first. In contrast, the mechanism of coexistence in the metapopulation-like models comes from an interspecific trade-off among competitive ability, dispersal ability, and mortality. This model only applies to those species for which superior competitors are inferior dispersers (or have higher mortality) and thus predicts the biased extinction of superior competitors. It may be that aspects of both models operate in many natural habitats, causing the total number of species driven extinct by habitat destruction to be a weighted sum of metapopulation and species-area model predictions.

The number of species driven extinct by a given amount of habitat destruction depends on the traits of the species (Tilman et al. 1994; Stone 1995). If, in the virgin habitat, the best competitors are the most abundant species, and progressively poorer competitors are progressively less abundant, then the qualitative relationship of Figure 10.1 (curve B) would apply (Tilman et al. 1994), which is qualitatively similar to that predicted by the species-area relationship (Eq. 10.4). However, if, in the virgin habitat, the best competitors are rare, and poorer competitors are progressively more abundant, then the pattern will be more like that of curve A of Figure 10.1 (Stone 1995). Clearly the number of extinctions resulting from habitat destruction depends on the processes controlling species abundances in a community and may be markedly greater than that predicted by the classical species-area relationship (Stone 1995). This represents a second major difference between the species-area models and metapopulation-like models.

Extinction Dynamics

Although the ultimate outcome of habitat destruction is predicted to be the extinction of all species for which $D \geq D_i$, this gives no insight into the dynamics of these extinctions. All species that are driven extinct by destruction of D have an

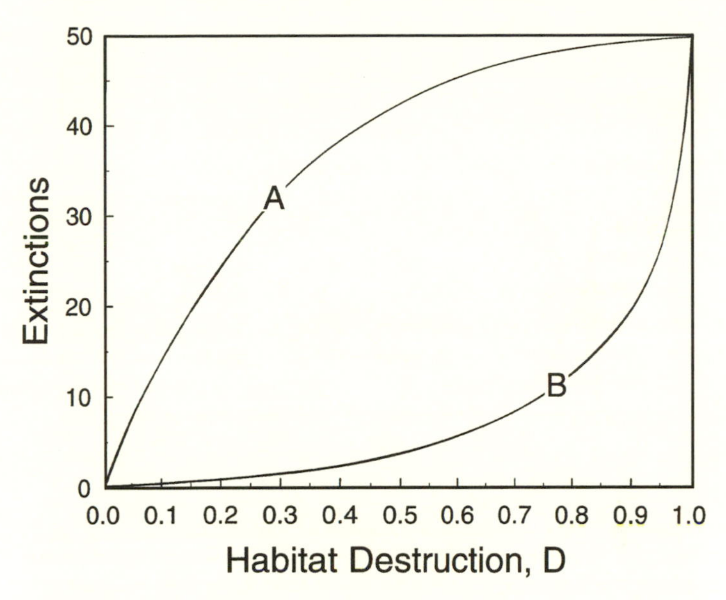

FIGURE 10.1. The number of extinctions predicted by the analytical multispecies version of Levins's model (Eq. 10.9). (A) Case in which the best competitor is the rarest species and progressively poorer competitors are more abundant (from Stone 1995). (B) Case in which the best competitor is the most abundant and progressively poorer competitors are less abundant (geometric abundance series of Tilman et al. 1994).

approximately negative exponential approach to an abundance of zero (Figure 10.2). The rate of this decay toward extinction depends on the c and m values of species relative to the amount of habitat destroyed. Values of c and m that seem realistic for grassland and tropical forest species yield a time delay between habitat destruction and species abundances falling to 1% of their original values that ranges from fifty to more than two thousand years.

Diamond (1972) reports time delays of this magnitude, and greater, between the formation of new islands by rising sea levels and the extinction of isolated bird populations. Terborgh (1974) and Case, Bolger, and Richman (1992) report similar cases of slow extinction following habitat isolation and fragmentation. Such time delays suggest that current habitat

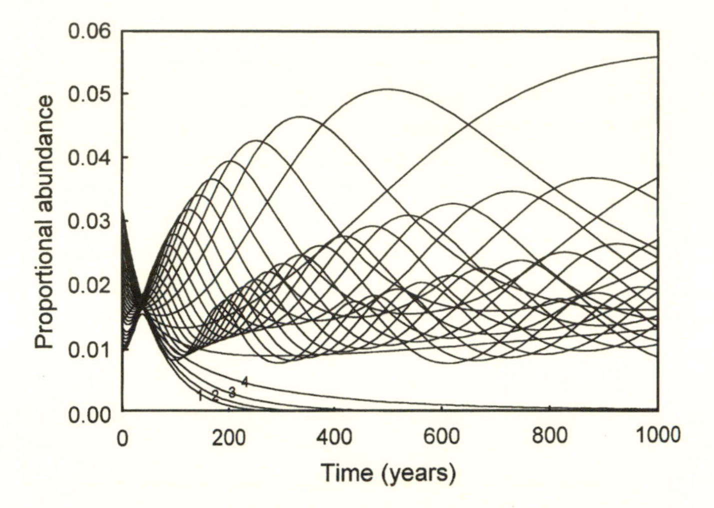

FIGURE 10.2. The dynamics of extinction following habitat destruction. Note that all twenty species coexisted in the virgin habitat. After permanent destruction of 25% of the habitat at time 0, the four best competitors (species 1–4) went extinct, and the other species persisted, eventually attaining new equilibrial abundances (from Tilman et al. 1994).

destruction is creating an extinction "debt." Many species that survive the initial act of destruction, and that may have been among the most abundant species in the remnant patches, are predicted eventually to go extinct. Thus, current acts of habitat destruction create a debt that is repaid by the delayed, selective extinction of species that are poor dispersers but superior competitors. Data regarding this idea are scarce. Bucher (1992) suggested that extinction of the passenger pigeons, which may have been the most abundant bird species in North America, was caused more by fragmentation of the oak forests upon which they depended, than by overhunting. However, their great dispersal abilities seem to argue against the metapopulation hypothesis proposed here. The African primates most harmed by habitat fragmentation may be those that are superior competitors (e.g., Skorupa 1986). Other studies of the effects of fragmentation on the loss of bird and mammal species also suggest biased extinctions (e.g., Lovejoy et al. 1984;

Laurance 1990, 1991; Stouffer and Bierregaard 1995). Diamond (1972) suggested that the species most threatened with extinction following island formulation from a mainland were poor dispersers. This possibility of biased extinction of formerly abundant species merits further study.

EXTINCTIONS IN EXPLICITLY SPATIAL HABITATS

Might any of the predictions we have discussed change if we were to make the metapopulation-like model explicitly spatial? To find out, we let each individual organism occupy a site within a hexagonal array of adjacent cells, with each cell being the size of the region occupied by an individual adult, as in Chapter 1. If any site is invaded by a propagule of a species that is a superior competitor to the current resident, the current resident is displaced by the invader. Residents produce propagules with a probability proportional to $c \cdot dt$ and experience mortality with a probability of $m \cdot dt$. Propagules disperse in the immediate neighborhood of a source cell, producing a "seed shadow" that can extend out one, two, three, or more hexagonal "rings" from the source cell. Our explicitly spatial simulations allowed us to determine the effects of relaxing the simplifying assumptions of the analytical model.

The analytical model assumed that dispersal was continuous, whereas many organisms reproduce seasonally. It assumed global dispersal of propagules across the entire habitat, whereas actual dispersal is local. It assumed immediate competitive displacement following invasion by a superior competitor, whereas actual competitive displacement occurs with some time delay. We relaxed all eight possible combinations of these three simplifying assumptions for a case in which four species coexisted as competitors in a virgin habitat. In all eight cases, as D increased the four competitors went extinct in order of their competitive abilities, just as in the analytical models (Tilman et al. 1997). However, the amount of habitat destruction required to drive a species extinct differed, sometimes markedly, from the analytical model and depended on which simplifying assumptions were relaxed. Thus, a more realistic treatment of space changed the quantitative nature of the effects of habitat destruction on extinction, but not the major qualitative feature, which is the selective loss of the superior competitors. This suggests that analytically predicted D_i values are of little, if any, quantitative value.

Spatial Patterning of Destruction

So far we only have considered destruction of randomly chosen individual sites in a spatial habitat, which is akin to destroying a habitat plant by plant. This was done for mathematical tractability but does not mimic the actual ways that humans destroy habitats. Most anthropogenic habitat destruction occurs in blocks. What might be the effects of having a habitat be destroyed in spatial blocks of various sizes? To explore this question, we again created a case in which four species coexisted in a virgin spatial habitat. We then destroyed this habitat in numerous alternative checkerboard patterns. For any given proportion, D, of habitat destroyed, we created alternative checkerboards ranging from 625 uniformly spaced small blocks of destruction to one large block of destruction. We then determined, for each type of checkerboard, the amount of destruction required to drive species 1, 2, 3, and 4 extinct. As might be expected, more destruction was required to drive a species extinct when large blocks were left intact than when the same undestroyed area was dispersed among many small blocks. However, except at the extremes, species went extinct in order of their competitive ability, with the best competitor being driven extinct first (Figure 10.3; Tilman et al. 1997).

The amount of habitat destruction required to drive a given species extinct was highly dependent on the size of the blocks into which a habitat was divided. The requisite amount of habitat destruction differed greatly from that predicted by the analytical model (i.e., for random destruction of single sites). For instance, the analytical model predicted that the best competitors would be driven extinct upon random destruction of 50% of all sites. However, in the spatially explicit cases the

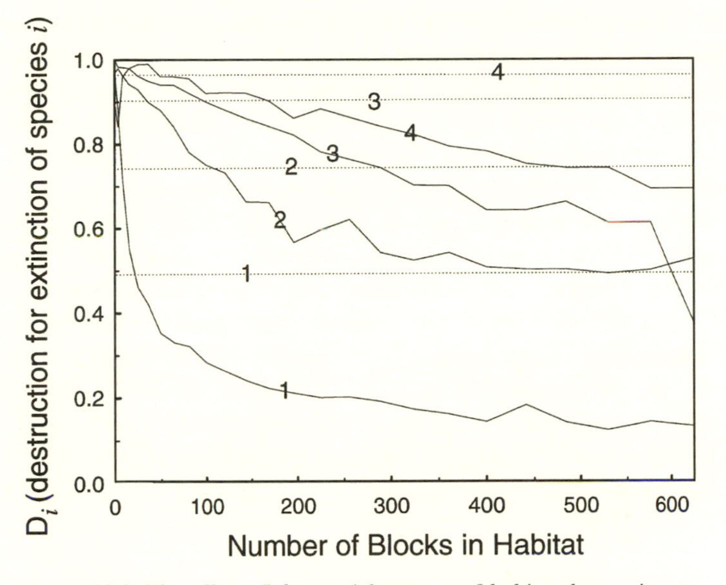

FIGURE 10.3. The effect of the spatial pattern of habitat destruction on the number of species driven extinct. In these simulations, a habitat was destroyed in a checkerboard pattern, with the total number of blocks into which a habitat was divided ranging from 1 to 625 (i.e., 25 × 25). The numbers refer to species, with 1 being the best competitor, 2 the next best, etc. The solid lines show the effect of the various patterns of habitat destruction on extinction by giving the proportion of each block that had to be destroyed to cause extinction of each species for any given total number of blocks. The dotted lines show the amount of habitat destruction required to cause extinction of these same species for the analytical model (Eq. 10.9). Figure from Tilman et al. (1997).

best competitors went extinct following anywhere from 95% to 15% habitat destruction, depending on the sizes of the blocks destroyed, and thus on the sizes of the areas left intact. For all four species, much more habitat destruction than analytically predicted was required to cause the extinction of a species if destruction left large intact blocks of habitat, and much less than analytically predicted was required if the intact habitat was left as many small but isolated patches. Thus, the analytical

D_i values predict the ordering of species extinctions, but not the level of habitat destruction at which this occurs. The latter only can be predicted by a more explicitly spatial consideration of fragment size, fragment locations, the patterning of corridors that link fragments, and movement patterns among fragments. The incidence function approach of Hanski (Chapter 2) provides one tool for such analyses. Detailed simulations that include spatial structure and incorporate knowledge of organismal movement patterns and behaviors are another approach that could be used to achieve this goal. However, as Wennergren, Ruckelshaus, and Kareiva (1995) have pointed out, the predictions of such models, especially when used to make management decisions about particular endangered species that live in highly fragmented habitats, may be highly sensitive to the model parameters, especially the estimated dispersal rate.

CONCLUSIONS

All organisms live in spatial habitats and interact mainly with neighboring organisms. Habitat destruction is also a spatial process, with some regions destroyed, others left in their pristine condition, and others modified intermediate amounts. Some of the effects of such habitat destruction can be understood in a qualitative manner by using nonspatial models, such as the species-area relationship and its derivatives. However, each incremental increase in the spatial realism in our models provided unique insights and predictions about the effects of habitat destruction on biodiversity. One such effect seems to be the tendency for biased extinction, following habitat destruction, of superior competitors that are poor dispersers. This effect, predicted by metapopulation-like models, was found to be remarkably robust when explored using explicitly spatial simulations. However, the level of habitat destruction required to drive a given species extinct was highly dependent on model assumptions, especially on the spatial patterning of destruction, and often differed dramatically from analytical D_i values derived from the metapopulation-like model. Such effects illustrate the limitation of the implicitly spatial metapopulation-like models and the need for greater spatial realism in considering the effects of habitat destruction on species extinctions.

ACKNOWLEDGMENTS

We thank Chengjun Yin and Chris Klausmeier for their discussions of these ideas and Adam Moshier, Angie Moshier, and Nancy Larson for their assistance.

CHAPTER ELEVEN

Local and Regional Processes as Controls of Species Richness

Howard V. Cornell and Ronald H. Karlson

INTRODUCTION

For the past forty years, a major goal in ecology has been to explain the evolution and maintenance of species diversity in ecological communities. This goal was prompted in part by the competitive exclusion principle, which predicted that no two species utilizing similar resources in similar ways could coexist indefinitely in the same habitat. The competitive exclusion principle raised an important question: If competition results in the exclusion of inferior competitors, how could so many species coexist in natural habitats? Efforts to answer this question focused almost exclusively on processes operating within the local community over relatively short periods of time. The assumption was that competition interacting with local abiotic conditions must determine the number of species coexisting within the habitat.

Mathematical models based on this assumption and the belief that communities were at competitive equilibrium predicted that many species could indeed coexist in a community by partitioning resources, but that there were limits to species richness set by the number of available limiting resources and physical factors (Tilman and Pacala 1993). Since the number of limiting factors within habitats must be finite, communities within habitats were most likely saturated with species, and only by subdivision of habitats would further coexistence be possible (MacArthur 1965).

The overwhelming emphasis on local processes prompted by these ideas engendered local explanations for diversity patterns over larger, regional spatial scales as well. Regional richness was viewed simply as the cumulative total of species occupying locally saturated habits within the region (Cornell and Lawton 1992; Ricklefs and Latham 1993). Differences in species richness among regions were attributed to differences in climate that modify the outcome of interactions within the habitat.

Four observations concerning the role of competition in structuring local communities challenge this view of local coexistence and the relationship between local and regional richness: (1) Many species residing in the same habitat and using resources in a similar way compete weakly if at all (Strong, Lawton, and Southwood 1984; Schoener 1983; Connell 1983; Gurevitch et al. 1992; Mahdi, Law, and Willis 1989; Compton, Lawton, and Rashbrook 1989). If interactions among potential competitors are weak or nonexistent, then there is no resource-based limit to the number of coexisting species in the community (e.g., Caswell 1976). (2) Nonequilibrial explanations for local richness might be more realistic in some communities than those based on competitive equilibrium (e.g., Connell 1978; Huston 1979). If communities are kept from reaching competitive equilibrium by disturbances or other perturbations, then competitive exclusion can be forestalled indefinitely (Caswell and Cohen 1991; Nee and May 1992; Dytham 1995; Moilanen and Hanski 1995). (3) Traditional mathematical models that assume competitive equilibrium do not predict saturated communities when certain simplifying assumptions are relaxed. Provided that these models incorporate environmental complexity and trade-offs in the ability of organisms to respond to environmental constraints, they predict the coexistence of a potentially unlimited number of species (Tilman and Pacala 1993). (4) Matched habitats with similar climate but in differentially rich regions support different numbers of species (Ricklefs and Latham 1993 and references cited therein). According to the principle of convergence (Schluter 1986), similar but independently developed communities controlled by competition should support similar numbers of species.

These four observations suggest on theoretical and empirical grounds that many communities might not be saturated and that other limits on local richness must be sought. Until recently, the strong emphasis on the local perspective has overshadowed the much older idea that regional patterns and processes might override local effects in setting local richness (Schluter and Ricklefs 1993). This older perspective holds that differences in regional richness do not derive from local interactions within habitats but are more likely to result from biogeographic and historical processes operating at larger spatial scales (Cornell and Lawton 1992; Cornell 1993; Schluter and Ricklefs 1993). Since local habitats are imbedded within larger geographic regions and thus exposed to dispersal from regional species pools, local richness can be inflated by processes operating at the regional scale, thus reversing the cause and effect relationship between local and regional richness patterns.

Single-species versions of this idea concerned mainly with the effects of dispersal and recruitment on local abundance and age structure of populations are embodied in the supply-side ecology literature (Lewin 1986; Roughgarden, Gaines, and Possingham 1988; Gaines and Bertness 1992; Doherty and Fowler 1994a,b; Downes 1995; Eggleston and Armstrong 1995). Recruitment limitation on population abundance is a well-studied phenomenon, and it is a small conceptual step to scale up this spatial mechanism to explain species enrichment at the level of the ecological community. The relationship between local and regional richness might be described by

$$\text{local richness} = p(\text{regional richness})$$

where p is the slope of the relationship (Hugueny et al. in preparation), that is, the average proportion of species in the regional pool found in a local community. Alternatively, it is the average proportion of local sites in a region occupied by a species. In a metapopulation context, this value is set by a balance between recruitment and extinction rates at sites within the region. If species enrichment is unabated by local processes, that is, if local richness shows no tendency to reach a clear upper limit in the most species-rich regions but, rather, increases proportionately with regional richness, then extinction rates would not increase relative to recruitment rates, p would be constant, and local communities would not be saturated in the equilibrium sense. Instead, species interactions would play a diminished role in setting limits on local richness, while processes that are responsible for differences in regional richness would play a more dominant role (Cornell and Lawton 1992).

This chapter has two goals. The first is to review case studies of the relationship between the number of species occupying a geographic region and the corresponding richness of local communities found within it. These studies provide insights into the processes controlling local richness and the scale at which they operate. The second goal is to demonstrate how local and regional perspectives can be synthesized with a study of regional and local controls on the richness of reef-building corals distributed throughout the world.

TESTING FOR SATURATION

The Method

The foregoing references to local versus regional richness relationships foreshadow a simple test for saturation in real communities (Cornell and Lawton 1992). The test requires estimates of species richness at local and regional spatial scales and a statistical evaluation of their interrelationship. If local richness is regressed on regional richness and the relationship is linear, the communities are unsaturated and are said to exhibit "proportional sampling" of the regional species pools (Figure 11.1, Type I). If the relationship is absent or if the slope is convexly curvilinear, the possibility of saturation arises (Figure 11.1, Type II). Since there can be no species occupying any habitat in a region with 0 richness, the curve must be constrained to transect the origin (0, 0).

Local versus regional richness regressions can be checked for curvilinearity by adding a quadratic term to the regression model (Cornell 1985a,b; Hawkins and Compton 1992; Caley and Schluter 1996; Cresswell, Vidal-Martinez, and Crichton 1995) or alternatively, by fitting any of the class of potential

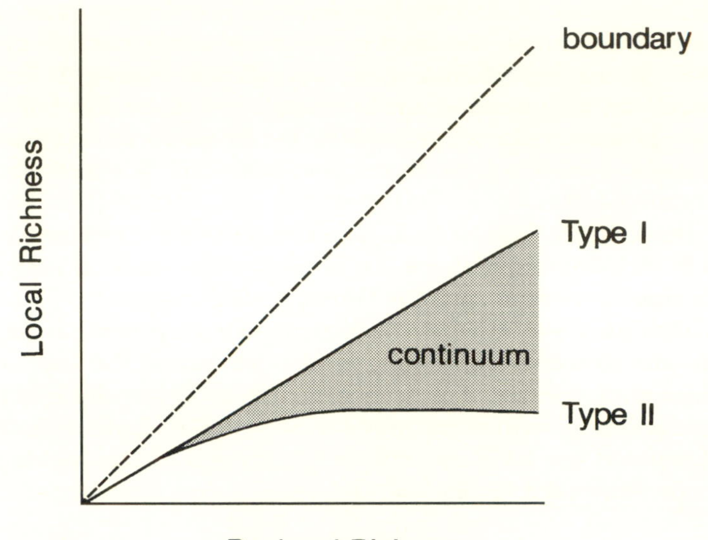

FIGURE 11.1. Possible relationships between local and regional richness in real communities. The relationship ranges from Type I where local richness increases proportionately with regional richness to Type II where local richness becomes independent of regional richness in the most species-rich regions (from Cornell and Lawton 1992).

saturating curves such as a power function (Griffiths 1997; Hawkins and Compton 1992; Cresswell et al. 1995; Kennedy and Guégan 1994; Richardson et al. 1995). We have used the former because the quadratic term in a second-order polynomial model can be tested for significance.

The Studies

The test in its formal statistical manifestation was first used to investigate saturation in assemblages of cynipid gall wasps on oak trees (Cornell 1985a,b). It has since been applied, with minor variations, to numerous taxa in terrestrial and aquatic environments (Caley and Schluter 1996; Hugueny and Paugy 1995; Griffiths 1997; Hawkins and Compton 1992; Kennedy and Guégan 1994; Dawah, Hawkins, and Claridge 1995;

Richardson et al. 1995). In several additional investigations, local richness was plotted but not actually regressed against regional richness (Ricklefs 1987; Lawton 1990; Aho 1990; Aho and Bush 1993; Gaston and Gauld 1993; Stuart and Rex 1994). Data from all of these studies have been replotted and checked for the Type I and Type II trajectories depicted in Figure 11.1 by fitting linear and second-order polynomial regression models. Lines were fitted to the data only if the first-order regression coefficient was significant, and curves were fitted only if the first-order and quadratic coefficients were significant.

The quality and completeness of the data are variable, but the results are instructive. Eight, or nearly one-half of the regressions, conform to the Type I relationship, indicating proportional sampling (Fig. 11.2A–H). An additional three regressions are significantly curvilinear but, with the possible exception of the fig wasp parasitoids, give no indication that they are close to saturation (Fig. 11.3A–C). Six regressions either show no relationship or roughly follow a Type II curve, suggesting saturation (Fig. 11.4A–F). However, this last result somewhat overstates the ubiquity of the Type II relationship since four of the regressions represent the same type of community (parasitic helminth worms in vertebrates). Thus, even with a conservative test for linearity, unsaturated patterns are common and widespread in natural communities.

Weak interspecific competition within an assemblage could permit unlimited local enrichment from the regional species pool. However, recent reviews of community theory have shown that strongly competitive assemblages can also be locally enriched as long as the environment is spatio-temporally heterogeneous (Cornell and Lawton 1992; see also the next section). If assemblages can be found that are both strongly competitive and routinely open to recruitment from the regional species pool, spatio-temporal heterogeneity provides a possible explanation and deserves to be investigated in real communities. Competitive interactions were not directly measured in any of the assemblages for which local and regional richness were compared. However, interspecific competition has been shown to be pervasive in ecological communities based on surveys of field-competition experiments (Schoener 1983; Gurevitch et al.

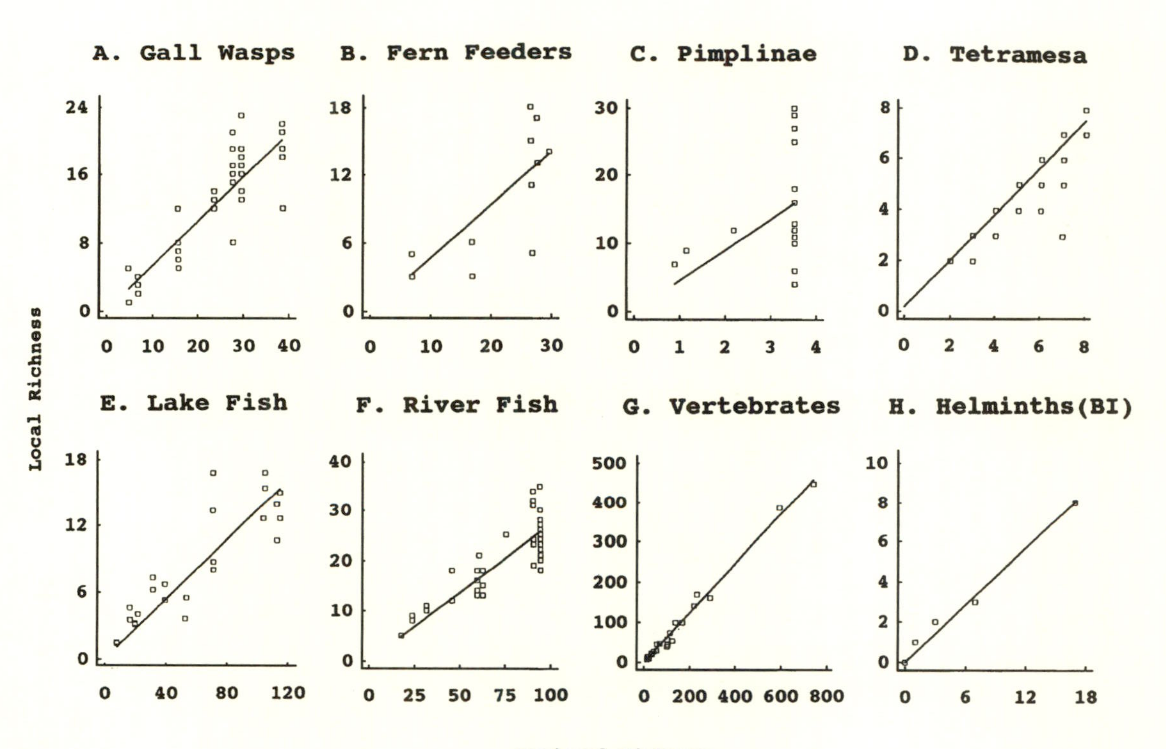
A. Gall Wasps
B. Fern Feeders
C. Pimplinae
D. Tetramesa
E. Lake Fish
F. River Fish
G. Vertebrates
H. Helminths(BI)
Local Richness
Regional Richness
24
16
8
0
0
10
20
30
40
18
12
6
0
0
10
20
30
30
20
10
0
0
1
2
3
4
8
6
4
2
0
0
2
4
6
8
18
12
6
0
0
40
80
120
40
30
20
10
0
0
25
50
75
100
500
400
300
200
100
0
0
200
400
600
800
10
8
6
4
2
0
0
6
12
18

1992; but see Connell 1983). Moreover, several assemblages in Figures 11.2–11.3 comprise the same taxa that consistently exhibit competitive effects in the surveys. Some of these assemblages express Type I patterns that are indistinguishable from those of noncompetitive taxa, suggesting that competitive impuissance is not the sole factor responsible for the unsaturated patterns.

Gall wasp and fern feeding herbivore assemblages are not saturated (Figure 11.2A,B) consistent with insect herbivores generally experiencing weak interspecific competition (Strong et al. 1984; Schoener 1983; Gurevitch et al. 1992; but see Damman 1993; Denno, McClure, and Ott 1995). However, bird assemblages have been model systems for the development of traditional competition theory, and, according to the surveys, bird species compete strongly among themselves (Connell 1983; Schoener 1983). Yet Caribbean and continental bird assemblages are both unsaturated despite some curvilinearity in the former (Figure 11.2G, 11.3B). Freshwater fish species have also been shown to compete without exception (Schoener 1983; Werner 1986), and their lack of saturation is even more dramatic (Figure 11.2E–G).

In a similar vein, it has been hypothesized that taxa at the carnivore, frugivore, detritivore, and producer trophic levels should compete more intensely than herbivores (Hairston, Smith, and Slobodkin 1960; Slobodkin, Smith, and Hairston 1967). If the hypothesis is correct, and if unsaturated patterns derive mainly from weak competition, herbivores should be less saturated than the other groups. As predicted, the two herbivore assemblages (gall wasps, fern feeders) are clearly

FIGURE 11.2. Assemblages exhibiting Type I relationships between local and regional richness. Data were taken from the following sources: (A) cynipid gall wasps (Cornell 1985a), (B) bracken fern herbivores (Lawton 1990), (C) pimpline parasitoid wasps (Gaston and Gauld 1993), (D) Tetramesa parasitoid wasps (Dawah et al. 1995), (E) North American lake fish (Griffiths 1997), (F) African river fish (Hugueny and Paugy 1995), (G) mixed terrestrial vertebrates (Caley and Schluter 1996), (H) helminth parasites of introduced British (BI) freshwater fish (Kennedy and Guégan 1994).

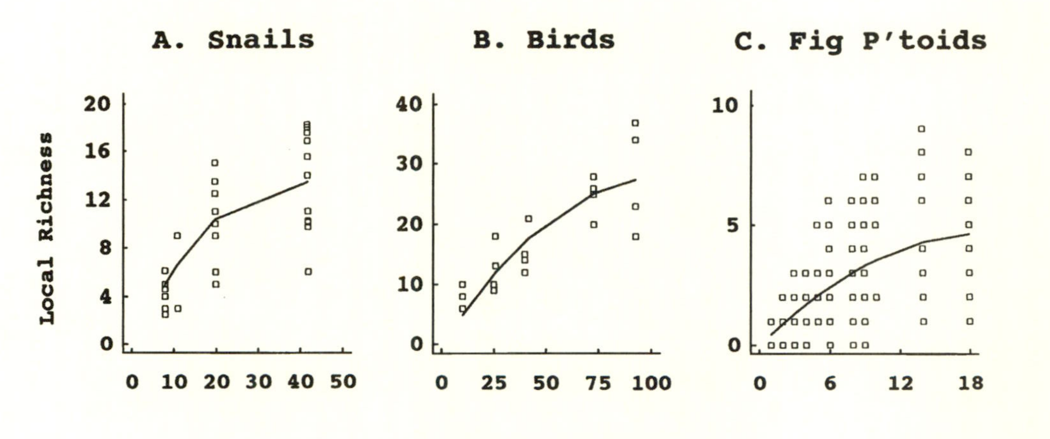
A. Snails
B. Birds
C. Fig P'toids
Local Richness
Regional Richness
20
16
12
8
4
0
0 10 20 30 40 50
40
30
20
10
0
0 25 50 75 100
10
5
0
0 6 12 18

Type I, and the producers (*Banksia*) and frugivores (fig wasps) are Type II. However, the pattern for carnivores is less consistent. Some (helminth worms, fig wasp parasitoids) are Type II as predicted, whereas others (carnivorous freshwater fish, *Tetramesa* parasitoids, Pimplinae parasitoids) are not. Deep sea prosobranch snails (presumably detritivores or carnivores) are also not saturated despite some curvilinearity in the graph. There is consistent survey evidence for competition in both fish and snails (Connell 1983; Schoener 1983; Werner 1986). There are no experiments from the surveys that explicitly examined parasitoid competition; however, competition between parasitoid species has been recently tested and confirmed in several instances (Patil et al. 1994; Wen, Smith, and Brower 1994; Wen and Brower 1995; Monge et al. 1995). In short, carnivore assemblages that in theory should be more competitive can elicit unsaturated patterns at least as frequently as saturated ones. For these and others where interactions may be intense, spatio-temporal heterogeneity should be given due consideration because of its theoretical plausibility (Cornell and Lawton 1992; see also the next section) and because experimental tests (e.g., Kneidel 1985; Ives 1991; Kouki and Hanski 1995) have confirmed that at least some forms of heterogeneity can facilitate competitive coexistence. Moreover, spatio-temporal heterogeneity is the rule rather than the exception in natural environments (Huston, 1994c; Hanski 1994).

In addition to the widespread occurrence of unsaturated assemblages, the clearest result to emerge from the analysis is that helminth parasites generally have a distinct local richness ceiling and thus might be good candidates for further study of limiting factors on communities (Figure 11.4C–F). The one helminth community that exhibits proportional sampling occurs on introduced fish species (Figure 11.2H). Nonnative fishes have probably not had time to accumulate sufficient

FIGURE 11.3. Assemblages exhibiting curvilinear relationships that do not level off. Data were taken from the following sources: (A) deep sea prosobranch snails (Stuart and Rex 1994), (B) Caribbean Island birds (Ricklefs 1987), (C) fig wasp parasitoids (Hawkins and Compton 1992).

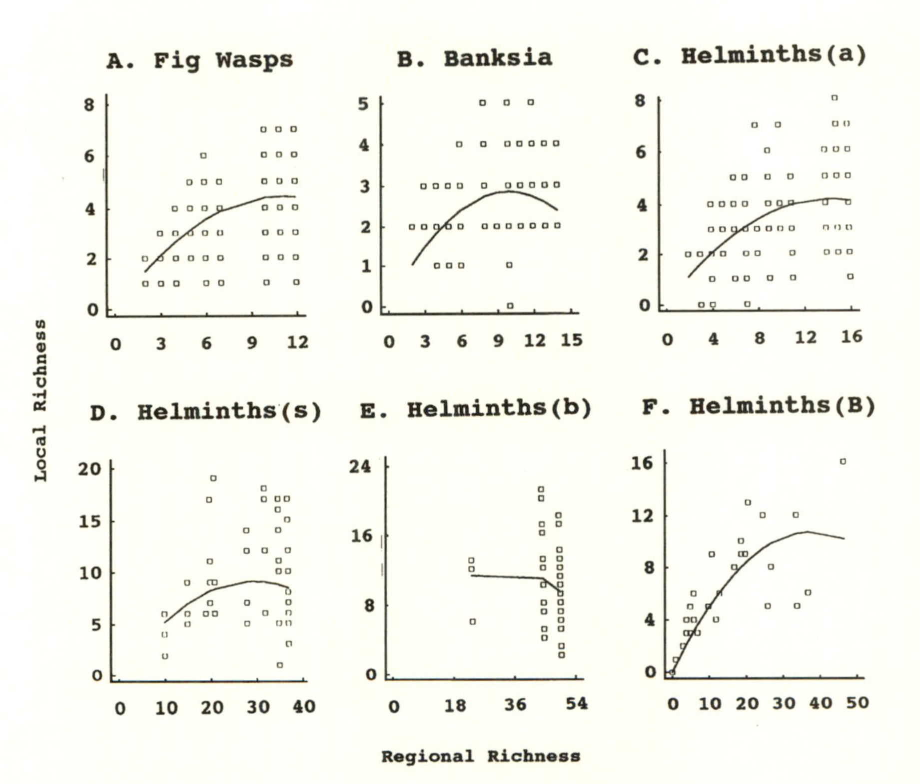
A. Fig Wasps
B. Banksia
C. Helminths(a)
D. Helminths(s)
E. Helminths(b)
F. Helminths(B)
Local Richness
Regional Richness

parasite species from the native pool to challenge any ecological limit on local richness (Kennedy and Guégan 1994). The depauperate fauna on nonnative fishes supports this notion and places these assemblages at the ascending part of the curve for native fish.

An important consideration for future analyses of this type is the spatial scale at which local richness should be measured. Communities are hierarchically structured assemblages of organisms that may be saturated at one scale but not others because the intensity of dynamic coupling among species varies with the unit of study. The studies in Figures 11.2–11.4 offer few clues about the relationship between scale and saturation mainly because local richness is evaluated at one scale in most analyses. There are two exceptions to this limitation, one that demonstrates scale-dependent saturation and one that does not. In the first case, parasite richness reached a hard upper limit of four species in individual eels (infracommunities) and remained constant even as component community richness (of the eel population) continued to increase (Kennedy and Guégan 1996). In the other case, Caley and Schluter (1996) demonstrated that vertebrate assemblages sampled at 1% and 10% of regional area both follow a Type I trajectory, with the only difference being that the slope is roughly 10% lower at the smaller scale. The small number of examples and the ambiguous results suggest that a more methodical exploration of scale is required. Nested sampling of matched habitats in several different regions combined with information on niche and habitat relationships will be necessary to make firm conclusions about the relationship between scale and saturation in different assemblages.

FIGURE 11.4. Assemblages exhibiting Type II or humped relationships. Data were taken from the following sources: (A) fig wasps (Hawkins and Compton 1992), (B) *Banksia* (Richardson et al. 1995), (C) helminths of amphibians (a) (Aho 1990), (D) helminths of sunfish (s) (Aho and Bush 1993), (E) helminths of bass (b) (Aho and Bush 1993), (F) helminths of native British (B) freshwater fish (Kennedy and Guégan 1994).

SYNTHESIS OF LOCAL AND REGIONAL PERSPECTIVES

The strong regional effects on local richness documented in these assemblages suggest that research emphasis should shift to regional processes and their effect on local diversity. However, this shift does not imply that consideration of local processes should be abandoned. Local and regional processes jointly influence local richness within habitats and thus should be examined simultaneously. The advantages of this broader perspective can be illustrated with a thought experiment. If physically identical habitats were transplanted into localities with different regional richness and allowed to settle into an equilibrial state transcending initial colonization history, by how much would per-locality richness vary among regions? The transplants would permit one to partition the contribution to per-locality richness among processes operating at local and regional spatial scales. By examining these processes simultaneously, such an experiment would synthesize regional and local perspectives and contribute fundamentally to a new multiscale understanding of community structure.

Although this kind of experiment is possible in principle, it is likely to be unwieldy and can be difficult to carry out. An alternative to habitat transplantation is to artificially increase the species pool to experimental treatments with untreated controls in the same habitat and to monitor short- and long-term changes in community richness. This approach is particularly attractive for plant communities where seed influx of potential colonists can easily be boosted. Two recent examples of this approach have demonstrated that, in the short term, plant communities are open to invasion and are unsaturated (Tilman 1994; Robinson, Quinn, and Stanton 1995).

Where both kinds of experiments are intractable, the problem can be addressed using a statistical comparative approach. For example, Schluter and Ricklefs (1993) used multivariate procedures to partition the influence of regional and local phenomena on local richness. Likewise, we have conducted an extensive review of the literature to quantify and partition the relative effects of local and regional variables on the local richness of reef-building scleractinian corals (Cornell and Karlson 1996).

We compiled quantitative data on local coral richness from sixty-three primary literature sources published from 1918 to 1993. The data were collected using quadrat, line transect, and point intercept samples of varying size at over a hundred sites worldwide. We corrected for these differences in sampling effort and method prior to analysis by conducting separate regressions of $\log_{10}$-transformed local richness against Box-Cox–transformed sample size (Sokal and Rohlf 1981). The residuals of local richness from each regression were then pooled over sampling method and standardized for further analysis.

We compiled quantitative data on regional richness using thirty-five primary sources from the systematics literature in addition to the sources noted above; in most cases, sources for regional richness were independent of those used for local richness estimates. The criterion of independence has not always been adhered to in past studies, but it is important to a critical assumption of regression analysis, namely that the errors of estimate will be independently distributed. We estimated regional richness as the number of species within 10^2–10^4 kilometers of each local sample site; this suboceanic scale is small enough to permit several regions within a single ocean basin to be distinguished. At this scale, regional richness varied between nineteen species in the eastern Pacific (Porter 1972) and 411 species in the most speciose region of the central Indo-Pacific (Veron 1993).

Since nearly all ecological variables known to influence coral richness at individual sites vary across unequivocally local scales (10^0–10^3 meters), we used depth (relative to mean low water) and habitat (ranked by relative distance from shore) as proxy variables for relevant local environmental variation. Five habitat categories encompassed inner flat, mid- and outer flat, crest and upper slope, midslope, and lower slope environments. Although depth and habitat heterogeneity may not subsume all local factors operating in the community, and some local processes may operate independently of the depth or spatial scale we chose, we know that light levels, sedimentation,

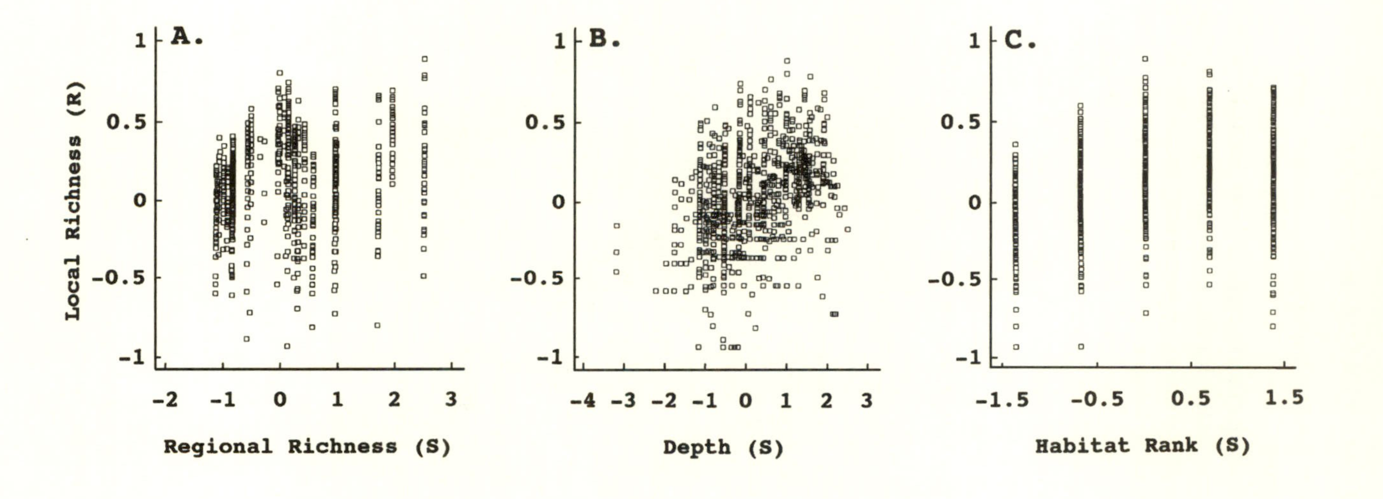
A.
B.
C.
Local Richness (R)
1
0.5
0
-0.5
-1
-2
-1
0
1
2
3
Regional Richness (S)
-4
-3
-2
-1
0
1
2
3
Depth (S)
-1.5
-0.5
0.5
1.5
Habitat Rank (S)

temperature, wave energy, food availability, tidal exposure, storm damage, predation, and other factors all vary along depth and habitat heterogeneity gradients (Huston 1994). The relationships between the pooled local richness residuals and standardized variables representing regional richness, depth, and habitat are depicted in Figure 11.5.

We used stepwise multiple regression analysis to identify the most important local and regional variables influencing variation in local richness, and we tested ten terms representing the linear and quadratic effects of each variable as well as the two-way and three-way interactions among the three independent variables. The linear effect of depth was the most significant term followed closely by the linear effect of regional richness and the quadratic term for habitat (Table 11.1). Although the combined effect of the local variables is more than twice that of regional richness, the individual effects of depth, habitat, and regional richness are nearly equal (Table 11.1). The quadratic term for regional richness was weak but significant, indicating curvilinearity in the effect of regional richness on local richness. This curvilinearity is subtle but provides at least a suggestion that local coral richness approaches a ceiling in the most species-rich regions.

Nevertheless, the primary result of our work is the detection of a significant degree of regional enrichment of local scleractinian biotas on a global scale (Figure 11.5A). The magnitude of this enrichment relative to the effects of the local variables might at first seem surprising. Corals are well known to engage in strong biological interactions, and the structure of these tropical communities has been widely believed to result primarily from local factors (Huston 1985; Done et al. 1991).

FIGURE 11.5. Local scleractinian diversity residuals (R) plotted against standardized variables representing (A) regional richness, (B) depth, and (C) habitat ranked by distance from shore (modified after Cornell and Karlson 1996). Independent variables are given in standardized units as indicated by (S).

TABLE 11.1. Multivariate stepwise regression of the pooled residuals of $\log_{10}$ local richness versus depth (1^0), regional richness (1^0 and 2^0), and habitat (2^0). Habitat values were ranked by distance from shore, whereas depth and regional richness were $\log_{10}$-transformed. All variables were standardized prior to analysis. The four terms appear in the order in which they entered the stepwise model (from Cornell and Karlson 1996)

Source of Variation	df	SS	MS	F
Model (R^2 = 37%)	4	336.55	84.14	133.78**
1. Depth (1^0)	1	125.38	125.38	199.38**
2. Regional Richness (1^0)	1	102.22	102.22	162.54**
3. Habitat (2^0)	1	100.92	100.92	160.48**
4. Regional Richness (2^0)	1	8.04	8.04	12.79*
Error	901	566.60	0.63	
TOTAL	905	903.15		

*$P < 0.0005$, **$P < 0.0001$

However, reef environments are spatio-temporally heterogeneous and subject to periodic disturbance (Connell 1978; Huston 1985); thus, according to theory, they should be open to regional influences. Even if niche space is uniform, spatio-temporal heterogeneity provides opportunities for competitively inferior species entering the community from the colonization pool to persist in probabilistic patchworks where they can avoid otherwise inevitable exclusion by competitive dominants (Skellam 1951; Levin and Paine 1974; and Shorrocks 1990; Hanski 1990; Chesson 1991; Tilman 1994 for reviews). As previously noted, such patchworks can be created by spatio-temporally unpredictable disturbance (see the Introduction), but they also derive from random mortality and recruitment (Tilman 1994 and references cited therein), by aggregation of individuals on fragmented resources (Ives 1988, 1995; Shorrocks 1990; Shorrocks and Sevenster 1995), by local neighborhood dispersal (Hassell et al. 1994; Holmes et al. 1994), or by shifting spatial variation in the risk of attack by natural enemies on population recruits (Armstrong 1989). Even though strong interspecific competition is possible in such heterogeneous environments, local richness can increase with increased regional richness, often without theoretical limit (e.g., Tilman 1994) because of the existence of probabilistic refuges which reduce competitive contacts.

In a recently developed patch occupancy model, Caswell and Cohen (1993) make specific predictions about the relationship between local and regional species richness for interactive communities subject to varying levels of disturbance. According to their model, regional enrichment should not occur in highly competitive undisturbed communities; such communities should reach upper limits to local richness that are independent of the size of the regional species pool. However, they found that small increases in disturbance frequency can result in significant regional enrichment at local scales. Although species interactions may be necessary, they are not sufficient for saturation in local communities. Under the right conditions, regional processes can penetrate even strongly interacting communities (such as those found on coral reefs) to increase local richness.

SUMMARY

Several conclusions can be drawn in reference to the goals set forth at the beginning of this chapter. First, many natural communities appear unsaturated despite the conservative nature of the test for saturation and are thus subject to strong regional controls. Even highly competitive assemblages such as freshwater fish and songbirds are unsaturated, suggesting that interspecific competition is not sufficient to produce a ceiling on local richness. Various forms of spatio-temporal heterogeneity can forestall competitive exclusion indefinitely and may provide the best explanation for Type I patterns in many communities. Second, we demonstrate a direct comparison of the relative contribution of local and regional processes on the local richness of reef-building corals. Even though corals experience strong biological interactions, regional processes are just as important as local ones in setting levels of richness on reefs. The strong influence of local processes on local richness suggests that there should be continued emphasis on experimental work at local spatial scales. However, the significant effects of regional richness confirm that investigations on reefs should also expand the scale at which local variation in richness is examined (Karlson and Hurd 1993). Integration of systematics, biogeography, and oceanography with local dynamics is essential for a complete understanding of the local and regional processes affecting natural communities.

PART IV

The Final Analysis: Does Space Matter or Not? And How Will We Test Our Ideas?

Throughout this book models with and without a spatial dimension have been compared and contrasted. For virtually any ecological dynamic or species interaction it is certainly possible to produce a model that reveals some striking effect of space. But this does not address some fundamental questions surrounding spatial processes: (1) When does the presence of a spatial dimension profoundly alter population dynamics or opportunities for coexistence in the real world? and (2) To what extent must our models be spatially explicit? Questions as basic as these often elicit starkly contrasting answers, and in this final part we present three contributions that take very different views on the subject of "space in ecology." These highlight a debate ecology needs to resolve before space is adequately included in its framework.

In a fascinating analysis of so-called production functions, Roughgarden argues that explicit representation of space can usually be abstracted away, so that implicit treatments of space suffice. Roughgarden's argument is noteworthy because production functions are one of the fundamental concepts in ecology—these are functions that describe how population growth changes with population density. They are central to fisheries biology, to agriculture, and to basic population biology. If these production functions can indeed always be abstracted to be nonspatial, then spatially explicit models may provide more detail, and entail more work, than necessary for issues related to productivity of managed ecosystems.

In striking contrast, Levin and Pacala argue that analyses of spatially explicit models are key to our attempts to comprehend the patterns of nature. The need for spatially explicit models is justified on the basis of their importance in deriving scaling laws that allow us to see how dynamics at one scale are manifest at other scales. This does not mean Levin and Pacala are wedded to complex spatially explicit models, but rather that a thorough understanding of such models is needed to know the limitations of simpler nonspatial theory.

Finally, in a totally orthogonal direction, Steinbek and Kareiva explore the challenges an empiricist faces when testing spatially explicit models in ecology. The constraints are mundane—spatial extent makes replication more costly—but the implications can be profound. If testing "spatial theory" is going to be hard, then we need to think carefully about what questions are addressed. Steinbek and Kareiva assert that issues of coexistence and extinction should take precedence over those of stability, pattern formation, and invasion.

When views as discordant as the chapters in this part arise, it does not mean that someone is "correct" and someone is "wrong." The differences come from the divergent interests and problems the authors want to address. Two points of consensus are these. First, every ecologist needs to think about the consequences of space before proceeding along a research path that is wholesalely nonspatial. Second, simple spatial models may give as much insight, with much less effort, as complex ones. Space may not always matter, but when it does matter, to neglect it could produce major misunderstandings.

CHAPTER TWELVE

Theories of Simplification and Scaling of Spatially Distributed Processes

Simon A. Levin and Stephen W. Pacala

INTRODUCTION: THE CHALLENGE OF SCALING

The problem of scaling is at the heart of ecological theory, the essence of understanding and of the development of a predictive capability (Levin 1992). The description of any system depends on the spatial, temporal, and organizational perspective chosen; hence it is essential to understand not only how patterns and dynamics vary with scale, but also how patterns at one scale are manifestations of processes operating at other scales (Haury, McGowan, and Wiebe 1978; Levin 1992). Evolution has shaped the characteristics of species in ways that result in scale displacement: Each species experiences the environment at its own unique set of spatial and temporal scales and interfaces the biota through unique assemblages of phenotypes. In this way, coexistence becomes possible, and biodiversity is enhanced. By averaging over space, time, and biological interactions, a genotype filters variation at fine scales and selects the arena in which it will face the vicissitudes of nature. Variation at finer scales is then noise, of minor importance to the survival and dynamics of the species, and consequently of minor importance in any attempt at description.

In attempting to model ecological interactions in space, contributors throughout this book have struggled with a trade-off between simplification and "realistic" complexity and detail. Although the challenge of simplification is widely recognized in ecology, less appreciated is the intertwining of scaling questions and scaling laws with the process of simplification. In the context of this chapter, simplification will in general mean the use of spatial or ensemble means and low-order moments to capture more detailed interactions by integrating over given areas. In this way, one can derive descriptions of the behavior of the system at different spatial scales, which provides the essentials for the extraction of scaling laws by examination of how system properties vary with scale.

PATTERN FORMATION

The search for scaling laws is ineluctably intertwined with pattern and process: the detection and description of pattern, and its relation to process. Statistical methodologies are invaluable tools for discovering relationships and suggesting mechanisms, but they will not be discussed further in this limited review. The problem of pattern formation and maintenance, however, is central to our theme and will reassert itself repeatedly throughout this chapter. There is no pattern without variation, and the theory of pattern is a theory of how diversity arises in potentially uniform systems, of what limits diversity, and of the apportionment of the causes of diversity among exogenous and endogenous causes. Hence theories of biological diversity echo themes from the literature on pattern (Levin 1978a, 1981, 1988). Again, as in so many aspects of religion, philosophy, and science, order arises from an opposition of forces, stimulatory and repressive: *Short-range activation* enhances deviations from uniformity, breaking symmetry and instigating pattern; *long-range inhibition* constraints the spread of novelty. The forces eventually may achieve a sort of balance, such that patterning is sustained. This theme is evident in a number of chapters in this book (e.g., Lehman and Tilman, Chapter 8).

The simplest ecological example of pattern forming in this way comes from considering predator-prey systems in which prey move on more limited spatial scales than do predators. Prey are activators: A small increase in prey density will stimulate growth of the predator population, and also the prey if an

Allee effect exists. But those stimulatory influences will be short-range, limited to the scale on which prey move. On the broader scale of predator movement, the inhibitory influence of predators tends to stabilize pattern, leading to alternating regions of high and low density (Segel and Jackson 1972; Levin 1974).

Similar effects can be seen in the sinusoidal front patterns sometimes seen in wildebeest herds (Gueron and Levin 1993), although the mechanisms are different. Animals have "personal zones" about them and tend to move away from others that intrude into that space (Hediger 1950). In the model of Gueron and Levin (1993), the consequence is that animals in the lead speed up, and trailing animals slow down, if they are in each other's personal space. This local effect can serve to destabilize an otherwise uniform front, causing inhomogeneities to begin to develop. However, as distances between individuals increase, repulsion is replaced by attraction, as individuals seek to remain with the herd. The result is the maintenance of stable wavelike patterns, again from the balance between short-range activation and long-range inhibition.

The story is, as in these two examples, most easily told in space (Gierer and Meinhardt 1972), which therefore will be the focus of the rest of this chapter, but the issue is the same in other dimensions (Levin and Segel 1985; Holling 1992). In particular, through such mechanisms, the dynamics of independent components become entrained and systems self-organize into units that interact strongly within, and weakly (and on longer timescales), with other units (Simon and Ando 1961; Paine 1980). Recognition of the structures that develop becomes key to understanding the dynamical organization of these self-organized networks (Iwasa, Levin, and Andreasen 1987, 1989; Holling 1992; Kauffman 1993).

In standard usage (see Barenblatt 1994), scaling laws are power laws relating variables and reflect self-similarity: recapitulation of patterns across scales of space and time. Such self-similar patterns have been observed often in ecological systems and the output of models (Levin and Buttel 1987; Levin 1992; Moloney et al. 1992) and motivate a search for explanation. Scaling laws may be derived phenomenologically based on observations, or through dimensional or other analyses of constitutive laws (e.g., Taylor 1950). To develop understanding, it is best to be able to proceed from mechanistic laws that capture basic interactions, through analysis of how properties change with scale. In physics, for example fluid mechanics, those laws may be conservation laws or other relations derived from first principles, and involving properties such as densities. In ecology, first principles are typically expressed at small scales or the level of individuals, and the first step in deriving scaling laws is to simplify the morass of information to derive macroscopic relationships. The next several sections describe ways to do this, in the hope of guiding the search for scaling laws through simplification.

FROM MEAN-FIELDS TO INDIVIDUALS

The classical approach to modeling ecological systems (Volterra 1926) simplifies by ignoring space completely and in essence assumes that every individual is equally accessible to every other individual; the result is a system of differential or difference equations for the mean abundances of the various types. Because every individual is assumed to see the average or mean-field, such equations comprise the mean-field description of the system.

The mean-field approach is a sensible place to begin, but it ignores much of what is important about the dynamics of ecological communities. In reality, interactions typically are restricted to a subset of the individuals in the population; more generally, there is a likelihood or preference structure for the probability of interactions between two individuals as a function of the distance between them, either in real space or in some abstract character space.

Recognition of the inadequacy of mean-field models is not a recent phenomenon. In epidemiology and population genetics, it was acknowledged in the first part of this century (Brownlee 1911; Fisher 1937; Haldane 1948), and the literature is extensive on the use of diffusion models for the spread of advantageous alleles (Fisher 1937; Kolmogorov, Petrovsky, and Piscunov 1937) and other dynamic aspects of population genetics (Dobzhansky and Wright 1947). In ecology, it was Skellam's seminal paper that stimulated the application of diffusion approximations to the spread of species, and a variety of important ecological issues were addressed through this framework (Skellam 1951; Kierstead and Slobodkin 1953; Okubo 1980; see review in Levin 1976). But such diffusion models, and the reaction-diffusion systems that were introduced by simply adding diffusion terms (plus possibly advection) to the equations for interacting species, are limited in their range of application. In particular, because diffusion models imply infinite speeds of propagation, they prevent the transient isolation that often is essential to persistence of competitively inferior types. Furthermore, the deterministic approach of the reaction-diffusion systems precludes explicit consideration of stochastic terms, such as localized disturbances. As will be discussed later, diffusion limits often can be derived to reflect the mean behavior of stochastic systems; but it is in the limiting process that leads to these approximations, which will differ from the simplistic ones obtained by simply appending diffusion terms to reaction equations, that the rules of scaling become apparent.

Recognition of these limitations has led to a variety of other formulations, including those where space is discrete (patchy) (Levin 1974); among these are the metapopulation models that have attracted considerable interest in recent years in a wide variety of applications (Levin and Paine 1974; Paine and Levin 1981; Pickett and White 1985; Chesson 1986; Gilpin and Hanski 1991). In the last few years, encouraged by rapidly increasing computational capabilities, attention has shifted to individual-based models (DeAngelis and Gross 1992) and spatial stochastic models such as interacting particle models (Durrett and Levin 1994b), in which detailed information concerning spatial distributions is retained. But such models by themselves generate cartoons, and much more detail than we have a right to know. In Book VI of Plato's *Republic*, Socrates says, "The many, as we say, are seen but not known, and the ideas are known but not seen." It is those unseen ideas that must be extracted in order to determine the essence of what is responsible for fundamental patterns; all else is commentary.

The extraction process involves the derivation of the scaling laws themselves.

The differences among the various modeling approaches expose the consequences of different modeling assumptions and of the different scales of biological interactions they represent. Durrett and Levin (1994b) compare mean-field, reaction-diffusion, metapopulation, and interacting particle models of a variety of types of interactions; no two are identical in their predictions. In particular, for systems in which a successional hierarchy is assumed, coexistence is found under certain conditions in the interacting particle and metapopulation models when it would not be possible in the mean-field or reaction-diffusion descriptions. The reason is apparent: The first two models allow the early colonist to remain isolated sufficiently long that it can build up its local population to a critical size; the infinite speed of propagation in the diffusion models makes this impossible in those representations. It is obvious that the nature of movement and the scale of interactions play fundamental roles in the persistence of species and the maintenance of biodiversity.

RENORMALIZATION AND SCALING

The scale of a process is the range over which it varies according to some criterion, or simply the range over which measurements are averaged. Scaling laws, in their simplest form, are power-law relationships among measurements made at different scales, as seen in subjects from autecology (Peters 1983; Harvey and Pagel 1991) to physics (as in Kepler's laws) to clouds and coastlines (Mandelbrot 1977). Kadanoff (1966) suggests that they occur when a system, over some range, "looks the same on all length scales." Goldenfeld (1992) argues that this is not quite correct, but a good starting point for discussion. The formal approach is through renormalization groups (Wilson 1983; Goldenfeld 1992), which makes rigorous the scaling process through the derivation of equations for blocks of cells in terms of the units that make them up, then iterates the process. Thus, the methods discussed in the next section may be seen as steps toward a theory of renormalization for spatial stochastic systems in ecology.

The theory of renormalization groups arose to deal with the phenomenon of critical phenomena and phase transitions in physics (Ma 1976); similar phase transitions occur in many spatial problems in ecology and suggest parallel approaches, see, for example, Solé et al. (1996). Consider, for example, the introduction of a species into an environment where a competitor is already established. Our starting point (Durrett and Levin 1994a) is the interaction between two species whose dynamics are specified by the interaction matrix

$$M = \begin{bmatrix} a & b \\ c & d \end{bmatrix}. \tag{12.1}$$

The terms of the matrix M are assumed to specify the net payoffs to each type when interacting with individuals of a particular type. Thus, a is the payoff for type 1 when dealing with type 1, b is the payoff for type 1 versus type 2, etc. In the mean-field version, this leads to the equations

$$\begin{aligned} \frac{dH}{dt} &= H\left[a\,\frac{H}{H+D} + b\,\frac{D}{H+D} - k(H+D)\right] \\ \frac{dD}{dt} &= D\left[c\,\frac{H}{H+D} + d\,\frac{D}{H+D} - k(H+D)\right] \end{aligned} \tag{12.2}$$

in which a, for example, now should be interpreted as the mean per-capita growth rate for type 1 in a region in which type 2 is rare, etc. Here, k indicates a density-dependent death rate (a density-independent component of mortality is included in the "game" matrix M). For definiteness, we assume first

$$M = \begin{bmatrix} 0.7 & 0.4 \\ 0.4 & 0.7 \end{bmatrix}, \tag{12.3}$$

which reflects the fact that individuals do better in the company of their own type. For this system, then, initial conditions will have an important influence on which type prevails.

Gandhi, Levin, and Orszag (1998), considering a variant on this system in which space is a continuum and every randomly walking individual carries an interaction neighborhood with it, show that, at low initial densities of the invader, the system specified by matrix 12.3 is well described by the mean-field approximation; at higher initial densities, however, the situation is more complicated, and a phase transition (shifting dominance to the invader) occurs when the initial density exceeds the density of the resident. Close to the critical point, power-law behavior prevails; that is, the time to extinction of the losing type increases as a power of the initial density.

Simulations show that, away from the critical point, the density of the introduced type is low everywhere. Every invader individual is surrounded by individuals of the opposite type, and the invasion is quelled before it can get started. No beachheads are established, and the system remains essentially spatially homogeneous; hence, mean-field theory is applicable. Beyond a threshold introduction density, however, the rare type forms monospecific clusters within a sea of the common type on a rapid timescale (Figure 12.1). On a much longer time scale, those clusters shrink in proportion to their surface curvature and diffusion (Gandhi et al. 1998), leading to a relationship of the form

$$\frac{dR}{dt} \sim -\frac{\mu}{R} \tag{12.4}$$

in which R is the radius of a cluster and μ is the diffusion constant associated with individual movement.

Integration shows that the extinction time of a cluster is thus proportional to R_0^2/μ, where R_0 is the initial size of a cluster. From this, it follows that the extinction time satisfies a power law scaling in relation to initial density. The agreement of theory and simulation can be seen in Figure 12.2 where, for large grid size, the time to extinction is shown as a function of initial density. Note the agreement between mean-field theory and the simulations for low initial density and the scaling near the threshold x_c, where a phase transition takes place.

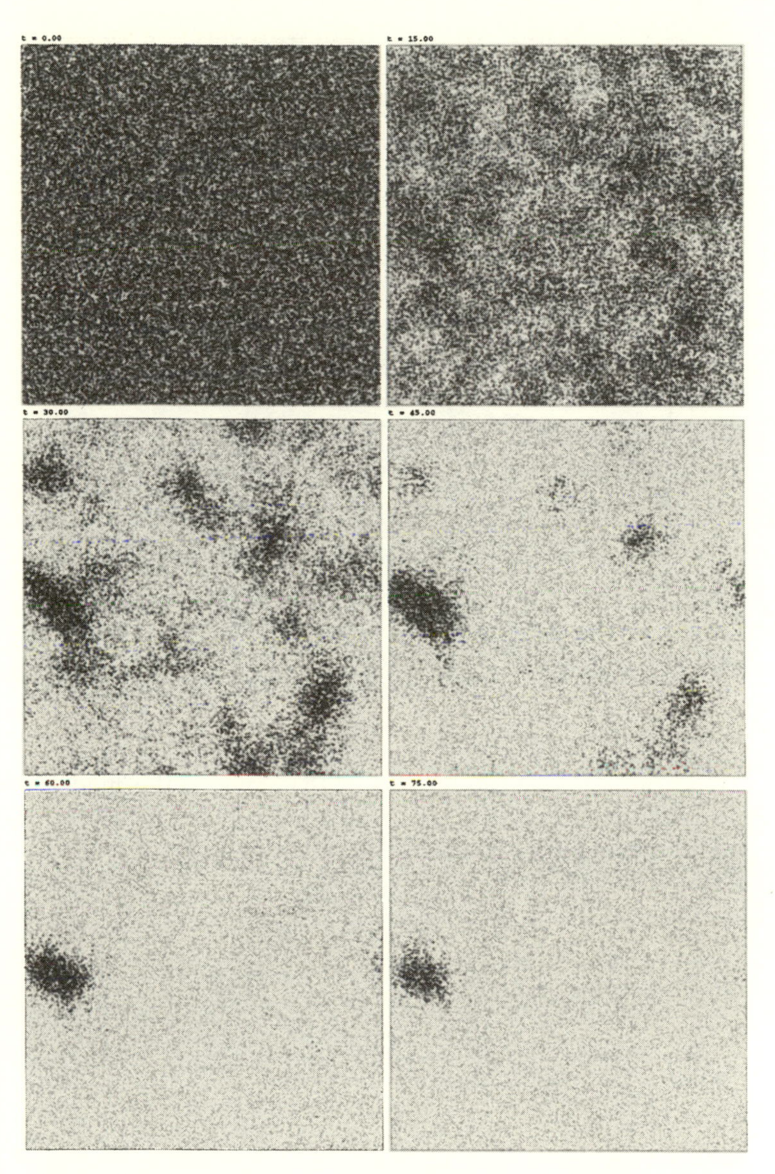

FIGURE 12.1. Snapshots from simulation at $t = 0, 15, 30, 45, 60, 75$ (see text). S_1 particles are light gray, and S_2 particles are black. The run required a million particles on an 128×128 grid (from Gandhi et al. 1998).

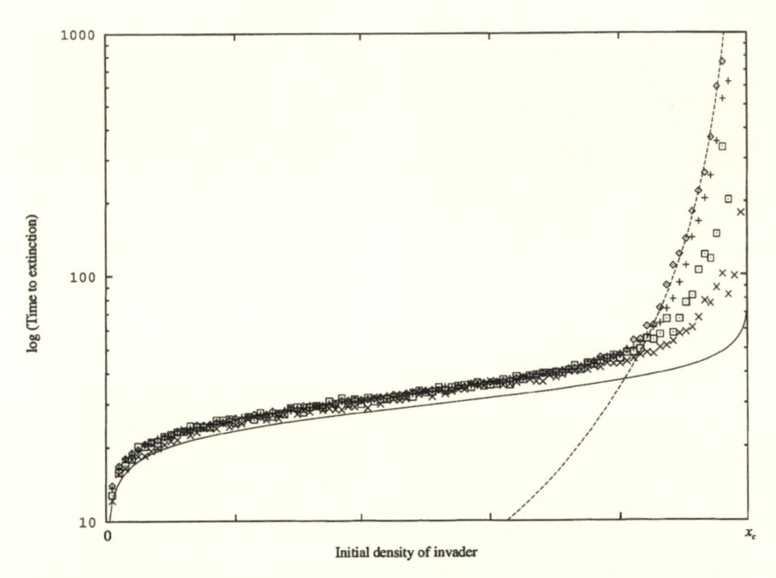

FIGURE 12.2. Time to extinction as a function of initial density; x_c is density of resident type. Solid line is mean-field approximation; dotted line is approximation from surface-effect theory. Linear dimension of grid is 256 (◇), 128 (+), 64 (□), 32 (×) (modified from Gandhi et al. 1998).

One can approximate the spatial dynamics of this model in the traditional manner, by appending diffusion terms to the mean-field Equation 12.2 to yield the classical reaction-diffusion system

$$\frac{\partial H}{\partial t} = \Delta u + \left[H \left(a \frac{H}{H+D} + b \frac{D}{H+D} - k(H+D) \right) \right]$$

$$\frac{\partial D}{\partial t} = \Delta v + \left[D \left(c \frac{H}{H+D} + d \frac{D}{H+D} - k(H+D) \right) \right]. \tag{12.5}$$

Here space has been normalized so that the diffusion coefficients (assumed to be identical) equal unity. Simulations of these equations produce patterns (Figure 12.3) that capture the

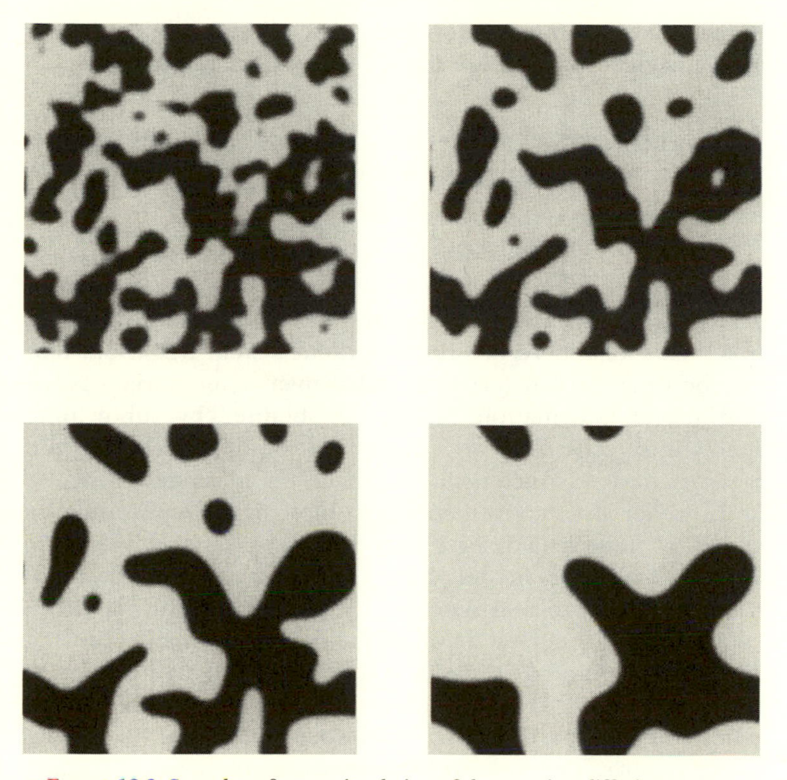

FIGURE 12.3. Snapshots from a simulation of the reaction-diffusion system at $t = 50, 100, 200, 650$. Initial conditions correspond to setting initial density of invader at $0.99x_c$, grid size at 256×256. Black corresponds to areas occupied by S_2, and white corresponds to areas occupied by S_1. Notice that the PDE captures the formation of clusters and their evolution through surface effects (from Gandhi et al. 1998).

mean features of the spatial dynamics (see Gandhi et al. in preparation; Durrett and Levin 1994a) and assure eventual extinction of one type (depending on initial conditions). However, this phenomenological approach to deriving a diffusion approximation for the spatial stochastic process does not work in general (Durrett and Levin 1994a), and we do not have clear rules for when it will. The more appropriate way to this end is to derive a hydrodynamic limit, starting from an individual-based formulation and proceeding formally. An example is given in a later section.

APPROACHES TO SIMPLIFICATION AS ILLUSTRATED BY THEORIES OF RANDOM WALKS AND DIFFUSION

The variety of ways discussed earlier to capture the role of space provide alternative descriptions for the same systems. That systems are composed of individuals, for example, does not mean that there are not emergent system behaviors that can be described in terms of global means and variances, or that useful information cannot be obtained by subdividing space into cells that are treated as homogeneous within. Systems have characteristic dynamics on a range of scales, and, either for convenience or biological importance, a variety of these may deserve attention.

Individual-based models have the advantage that they are closest in detail to real systems; that advantage is also a disadvantage in that they retain all of the detail that may hide what is really important at broader scales. The challenge then is to understand how those collective dynamics arise from the detailed interactions, that is, to build a statistical theory of collective dynamics. This is a problem that is at the core of inquiry throughout science. How is information transferred across scales? In this section we illustrate a variety of approaches to this question for a single example, the movement of individuals under density independence. Pedagogically this is valuable because random walks are easily visualized; more importantly, the structures of the underlying models embody the essential features of a wide range of stochastic processes and provide the foundation for their approximation through "diffusion limits." The approach in this section, to simplify by deriving equations for aggregate dynamics that integrate over a given region, obviously provides a mechanism for scaling among regions of different sizes.

The simplest model of movement is the random walk. In particular, we introduce here three familiar approaches to deriving a diffusion limit of a random walk, since these will motivate the three main approaches we exploit for more complicated problems. The simple random walk assumes that individuals do not interact, so the resulting equations are linear in form and fairly simply derived; the fundamental issues are retained, however, when nonlinearities enter.

1. *Hydrodynamic limits of interacting particle models.* Imagine that individuals are confined to move from point to point on a line, with points being equally spaced at distances δx apart. Assume further that the net flux from site x to site $x + \delta x$ in a small time δt, beginning at time t, is proportional to the difference $N(x,t) - N(x + \delta x, t)$, where $N(x,t)$ is the (expected) number of individuals at x at time t. If the constant of proportionality $D\,\delta t$ is the same for all pairs of sites, then it follows that N changes according to the rule

$$N(x, t + dt) = N(x,t) + D\,\delta t[N(x + \delta x, t) + N(x - \delta x, t) - 2N(x,t)]. \quad (12.6)$$

This defines the random walk for a given grid size and given time step. If one allows the time step and lattice size to shrink to zero in a compatible way (that is, so that the series of models with finer and finer steps represent approximations to the same basic process), then Equation 12.6 gives way in the limit to the familiar diffusion equation

$$\frac{\partial N}{\partial t} = D\frac{\partial^2 N}{\partial x^2} \quad (12.7)$$

(see, e.g., Murray 1990 or Okubo 1980). This process works equally well in any number of dimensions and can be extended to cover situations in which movement is biased or spatially variable, either of which may lead to inclusion of a first derivative in x, an "advection" term. Equation 12.7 is a special case of what is known as a Fokker-Planck, or forward Kolmogorov, diffusion equation for the density function of a stochastic process.

2. *Lagrange to Euler.* The approach above focuses attention on a particular location and simply counts the number of

individuals at that site and quantifies the fluxes among sites. An alternative, and equivalent, method is to follow each individual as it makes its way through space. In fluid dynamics, this approach bears the name of Lagrange. Again, one may assume that movement is among points on the line separated by distances δx, and occurring in time steps δt. $N(x, t)$ now reflects the probability that an individual will be found at x at time t; for example, for a point release of individuals, the probability distribution will take a binomial form. In the limit, as δx and δt tend to zero in harmony with one another, the binomial distribution becomes the normal, with variance $2Dt$, and the distribution is again seen to be governed by the Equation 12.7 for any initial distribution. Here D enters through the limit of the ratio of $(\delta x)^2/\delta t$ as both δx and δt go to zero. The end result is the same as in the earlier section, but the process is fundamentally different. Again, extensions to higher dimensions and biased diffusion are straightforward. The limiting Equation 12.7, in the jargon of fluid dynamics, is known as the Eulerian formulation of the problem, after Euler.

3. *Moment expansions*. The previous two approaches are familiar to anyone who has dealt with the diffusion equation. A third approach, less familiar, uses very different methods and at the same time makes explicit exactly what is being ignored in the diffusion approximation. In this approach, one assumes that the probability that an individual will be at x at time $t + \delta t$ is given by the integral representation

$$N(x, t + \delta t) = N(x, t)(1 - k\,\delta t) + k\,\delta t \int N(x - y, t) P(y)\, dy, \tag{12.8}$$

where the integral is taken over all space.

This reflects the assumption that an individual moves in time δt with a certain probability, and that the distance of movement is not limited to nearest neighbors; rather, it is governed by a probability distribution $P(x)$ for moving a distance x. In the simplest case, $P(x)$ will be a symmetric function decreasing from its (maximum) value at zero; but biased diffusion may be reflected in a nonsymmetric P, and movement by "great leaps" may mean that $P(x)$ does not have its peak at zero (Okubo and Levin 1989), or that long tails or even secondary peaks are possible (Mollison 1977). This approach may be thought of either as providing an Eulerian description of the expected number of particles at x, or as a Lagrangian description of the probability that an individual particle will be at x, at time t.

In this case, the limiting process bears some unique features that distinguish it from the previous two examples. $N(x - y, t)$ must be expanded in a Taylor series about x before δt is allowed to shrink to zero. In the absence of bias, a diffusion term emerges as the lowest-order term in the expansion (with coefficient proportional to the variance of P), but the next nonzero term (that of fourth order) modifies that based on the kurtosis of $P(x)$; higher-order terms also appear. The resulting equation, if left undisturbed, has infinite order, so the usual assumption is that all terms beyond the leading term are negligible, resulting in the standard diffusion equation. More generally, however, one might retain the fourth-order term as a correction (Cohen and Murray 1981; Levin and Segel 1985), or seek to represent it in terms of the variance so that a modified diffusion equation results. If the fourth-order term is not discarded, one can obtain phenomena (such as Mollison's "great leaps forward" or stabilization of the front of advance) that are not possible in the second-order (diffusion) approximation.

Equation 12.7 provides detailed spatial information regarding the dynamics of a homogeneous spatial stochastic process. If one integrates with respect to space, one finds reassuringly that

$$\left\langle \frac{dN}{dt} \right\rangle = 0, \tag{12.9}$$

where $\langle \ \rangle$ denotes the expectation at any point in space; this confirms the fact that movement is a conservative process. Adding the possibility of birth (or death) anywhere at net per-capita rate r yields the equation

$$\frac{d}{dt}\langle N\rangle = \left\langle\frac{dN}{dt}\right\rangle = r\langle N\rangle, \tag{12.10}$$

implying exponential growth of the mean.

Nonlinearities introduce another level of complexity, however. If, for example, deaths depend on the density in some neighborhood, the averaging process involves variance and spatial correlations; the dynamics of $d\langle N\rangle/dt$ are not closed. In this case, one must either "close" the system by representing the higher moments in terms of the mean, or complement the basic equation with equations for the moments, ultimately closing the system at a later point. We return to this "moment closure" problem in a later section.

HYDRODYNAMIC LIMITS OF PARTICLE SYSTEMS

The process of taking hydrodynamic limits is a standard one, but often difficult when interactions are nonlinear. To illustrate this point, we modify the matrix 12.1 from the form 12.3, replacing it by

$$M = \begin{bmatrix} -0.6 & 0.9 \\ -0.9 & 0.7 \end{bmatrix}. \tag{12.11}$$

This is usually termed "Prisoner's Dilemma," but we more generally use the nomenclature "hawks" and "doves" for the two species.

As will be described below, for simplicity we will also modify the scheme described in the previous section to assume a fixed spatial grid divided into cells. For this example, type 1 (termed hawks) outcompetes type 2 (termed doves) under any conditions, but hawks drive themselves to extinction when isolated. Doves, in contrast, always lose out in competition but are self-sustaining in the absence of hawks (see Durrett and Levin 1994a).

In the mean-field version of the dynamics, the outcome is straightforward and unequivocal: Hawks eliminate doves, then drive themselves to extinction. Spatial localization can be achieved naively by again adding diffusion terms to Equation 12.7, but the result is the same. Under random initial spatial conditions, some areas initially are taken over by doves, others by hawks; on a longer timescale, however, the clusters of hawks form advancing fronts or blobs that take over the system, followed by waves of death as the hawks drive themselves to extinction.

A more biologically realistic way to introduce space is to treat individuals as individuals, which is both politically and mathematically correct. Space (two dimensions) is divided up into cells centered at the nodes of the integer lattice $S = Z^2$, where Z is the set of all integers. Let $\eta(t, x)$, $\zeta(t, x)$ represent the numbers of hawks and doves, respectively, in the unit square centered at the lattice point x; η and ζ are both integers. The dynamics are reflected by three processes, each with its own characteristic range:

Migration: Each individual, at rate μ, changes its location to a randomly chosen site within a neighborhood N_1. The notion of rate means that time is treated as continuous, and that $\mu h + o(h)$ is the probability that the event (movement) will occur in a small time step h. The notation $o(h)$ indicates a term of an order negligible relative to h for h small.

Mortality: Death may occur either due to the density-independent term buried in M, or due to density dependence at rate $k\{\eta(t, x) + \zeta(t, x)\}$, where here the braces indicate that η and ζ are averaged over a neighborhood N_2.

Game step: The interactions specified by M are also localized, with their own characteristic neighborhood N_3. Dynamics are as in the mean-field version, except that the relevant densities used in determining growth rates are the averages of η and ζ taken over N_3; the average can be chosen to be a weighted one, for example, with nearer neighbors bearing greater influence. We ignore that complication here.

The interacting particle system specified above has very different dynamics than the mean-field or reaction-diffusion systems. Locally, the dynamics are exactly as for the mean-field system: Hawks eliminate doves, then themselves. But if N_3 is sufficiently small, that dynamic simply creates gaps—targets of opportunities for doves to reestablish themselves. The system has the potential to persist indefinitely (the expected persistence time, of course, depends on grid size in any simulations). Although phrased in terms of hawks and doves (or more strictly, players in "Prisoner's Dilemma,") the model works equally well for epidemiological systems (susceptibles and infectives), or for obligate successional series.

As already mentioned, adding diffusion terms to the mean-field equations does not lead to fundamentally different dynamics and hence does not provide a good approximation to the interacting particle system (ips). There is a way, however, to derive a "hydrodynamic" diffusion limit that does capture the essential feature of the particle system: persistence. Recall that, in the ips, individuals are units. Imagine, as in the previous section, that each such individual is performing an independent random walk on the grid; it may be established then that the joint distribution of the number of individuals in any finite subset of the grid will converge to one specified by independent Poisson variables. This is the critical step in the limiting process, and the same assumption will reappear in the next sections. If migration occurs on a much faster timescale than births and deaths, sites near x at time t may be treated as independent, so that the numbers of hawks and doves are Poisson with means $u(x,t)$ and $v(x,t)$, respectively. If the spatial step is now shrunk to zero, as before (specifically, individuals on a lattice with spacing ϵ perform random walks at a rate that scales with $1/\epsilon^2$), u and v may be shown in the limit (see Durrett and Levin 1994a) to satisfy a set of reaction-diffusion equations

$$\frac{\partial u}{\partial t} = \Delta u + u\left[r + g\left(a\frac{u}{u+v} + b\frac{v}{u+v}\right) - k(u+v)\right]$$

$$\frac{\partial v}{\partial t} = \Delta v + v\left[s + g\left(c\frac{u}{u+v} + d\frac{v}{u+v}\right) - k(u+v)\right], \tag{12.12}$$

where Δ is the two-dimensional Laplacian (sum of second partial derivatives), r and s are the net birth rates of isolated hawks and doves, and

$$g = g(u,v) = 1 - \exp[-[N](u + v)] \qquad (12.13)$$

where $[N]$ is the number of points in the neighborhood N_3. Note that the effect of the "proper" limiting process is a set of equations that introduces the standard diffusion term but modifies the dynamic equations to minimize the importance of the hawk-dove interaction at low densities (g vanishes as $u, v \to 0$). The significance of this is that small isolated dove populations are protected from invasion by "nanohawks" (Mollison 1977) moving at faster than the speed of light and are able to build up local densities. The modified dynamical system, with diffusion suppressed, now has a globally attracting equilibrium and so will persist.

FROM LAGRANGE TO EULER

The second method introduced for deriving macroscopic laws for nonlinear systems begins by following an individual as it moves through space, leading to a Lagrangian description of the dynamics. In this approach, typically, one deals with effects either on the acceleration of an individual (e.g., Sakai 1973; Okubo 1986; Grünbaum 1992) or the velocity (Aoki 1982; Huth and Wissel 1992; Gueron and Levin 1993). For the purposes of this discussion, we will use the Newtonian momentum focus on accelerations, leading to an equation for the acceleration of an individual in terms of the forces acting upon it. These forces will include density-dependent ones, for example, inherited from the fluid dynamics, or representing taxis with regard, say, to thermal or chemical signals that are not generated by other individuals; they will also include the nonlinear responses of an individual to the location, densities, or velocities of other individuals. Models of this sort have been used widely to simulate animal grouping (see, e.g., Heppner and Grenander 1990; Grünbaum 1992, 1994).

To proceed to the Eulerian description of this system (in one dimension), the nonlinear equivalent of Equation 12.2, one must determine the flux $J(x,t)$ and average velocity $U(x,t)$ of individuals, with the objective of deriving an equation for the density distribution $p(x,t)$. Typically, one seeks a conservation equation

$$\frac{\partial p}{\partial t} = -\frac{\partial J}{\partial x}, \qquad (12.14)$$

where $J = pU$ (Grünbaum 1992); the challenge is to relate U to individual behaviors. As in the preceding section, one way to do that is to assume that the distribution of neighbors is Poisson, with the mean given by the density distribution. Making this assumption allows one to proceed (nontrivially) to an Eulerian description, typically including integral terms, reflecting the fact that individuals are responding to the dynamics of other individuals in a nonvanishing neighborhood of their location (Grünbaum 1992, 1994). A similar partial-differential integral formulation will appear in the next section, in a quite different context, but for similar reasons.

MOMENT EXPANSIONS AND MOMENT CLOSURE

The application of moment methods to spatial stochastic processes is well illustrated with a simple model, the contact process, which then can be generalized to a widely used forest growth simulator. In the basic contact process (Durrett and Levin 1994b), cells are either occupied or empty. Occupied cells may become extinct at rate δ, or colonize empty neighbors at rate λ, so that the dynamics of the mean probability of occupancy is (Levin and Durrett 1996)

$$\frac{du(1)}{dt} = -\delta u(1) + n\lambda u(01), \qquad (12.15)$$

where n is the number of neighboring cells considered, $u(1)$ is the probability a cell is occupied, and $u(01)$ is the probability that an ordered pair of cells will be in the configuration unoccupied-occupied. Equation 12.12 specifies the dynamics of the mean but does not provide a closed description. To close it, one either must make some assumption about how $u(01)$ relates to $u(1)$—in the mean-field approximation, for example, $u(01) = u(1)[1 - u(1)]$—or write an equation for the dynamics of $u(01)$. The trouble with this is that the latter equations will include terms like $u(001)$ and other triples. The system of equations for the moments, essentially for the moment-generating function, is infinite, unless truncated by a closure rule at some point. The hope is that higher-order approximations (truncation at a later stage) will lead to greater accuracy, but that depends upon the particular dynamics. Equation 12.15, the second-order approximation, say, with $n = 4$, appends a second equation

$$\dot{v} = -(\lambda + \delta)v + \delta(u - v) - 3\lambda v \frac{(2v + u - 1)}{(1 - u)} \tag{12.16}$$

to Equation 12.15, where $u = u(1)$, $v = u(01)$.

This decoupling approximation, called pair approximation (Matsuda et al. 1992), has been carried out for many models of population dynamics of plants (Harada and Iwasa 1994; Harada et al. 1995) and sometimes predicts qualitative results accurately when mean-field dynamics fail (e.g., Sato, Matsuda, and Sasaki 1994). Although the pair approximation is not accurate in the basic contact process near critical values, it is surprisingly accurate if the system includes a small random long-range dispersal that makes the spatial pattern closer to random (Harada and Iwasa 1994). The same tendency has been seen for point processes, such as the forest model SORTIE (Bolker and Pacala 1997).

The second-order approximation not only gives improved estimation of the actual dynamics (Figure 12.4) but also provides information on biologically important clustering (Figure 12.5). The contact process is a prototypical model for forest growth, species invasion, epidemic spread, and forest fire (see, e.g., Durrett 1988). Its wide-ranging applicability is reflected in its appearance (in modified form) throughout this book (see,

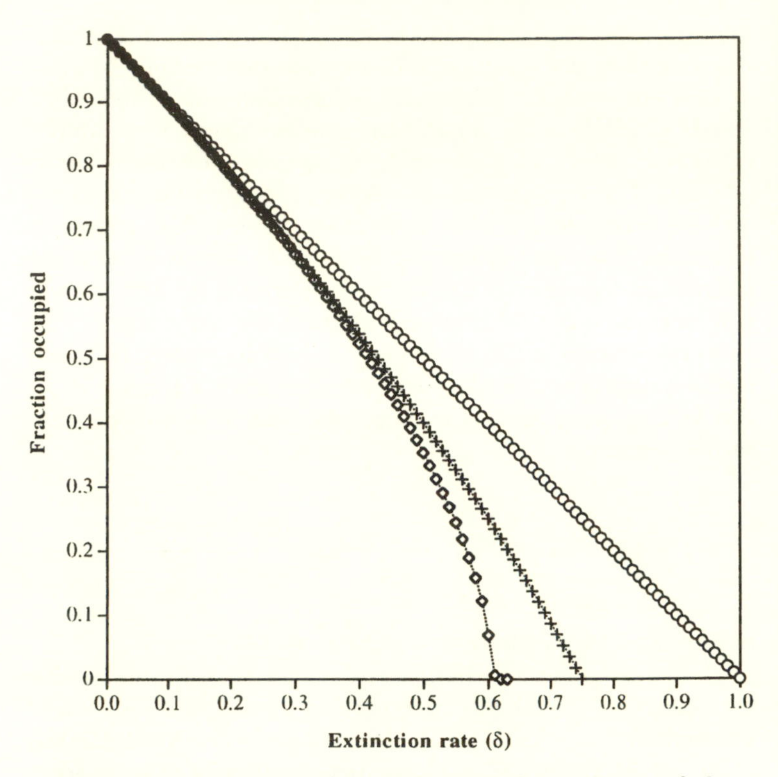

FIGURE 12.4. Occupancy fraction as a function of extinction rate δ, for contact process. ◇ is simulation, ○ is mean-field approximation, + is pair approximation (from Levin and Durrett 1996).

e.g., Chapters 5, 8, and 9). Furthermore, extending ideas of Hubbell (1995, 1997), Durrett and Levin (1996) use generalizations of this model to derive species-area relationships, prototypical scaling laws.

Moment closure methods can be applied whenever there is variation in some characteristic, be it with regards to space, functional properties, or the uncertainty associated with a particular realization of a stochastic process. For example, consider the problem of competition between two species, distributed across two patches. It is easy to show (Levin 1978a), that under local Lotka-Volterra dynamics, and conservative coupling, the equations for the spatial means of the two

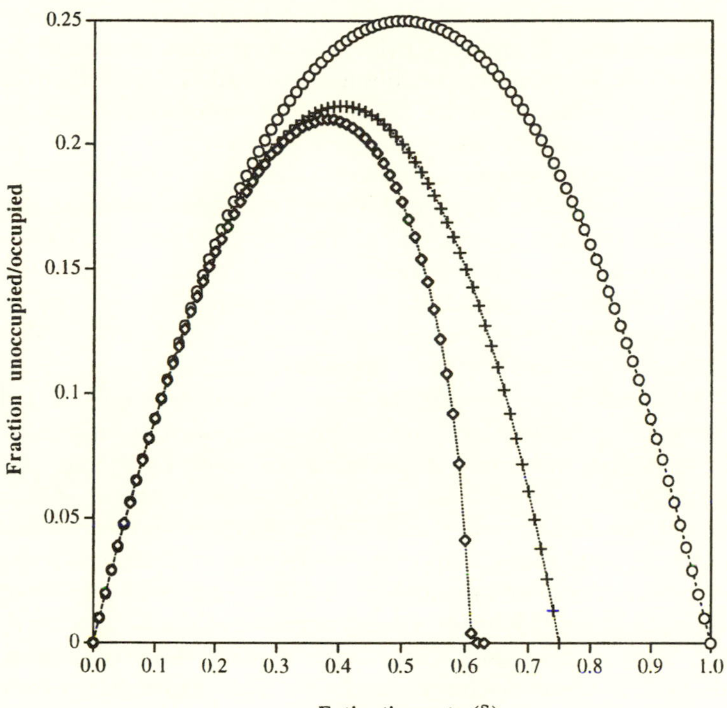

FIGURE 12.5. Correlation structure, expressed as proportion of pairs in unoccupied/occupied configuration, versus extinction rate for contact process. ◇ is simulation, ○ is mean-field approximation, + is pair approximation (from Levin and Durrett 1996).

species are

$$\frac{d\bar{u}}{dt} = r\bar{u}(k - a\bar{u} - b\bar{v}) - ra\sigma_u^2 - rb\,\mathrm{cov}(u,v)$$

$$\frac{d\bar{v}}{dt} = s\bar{v}(L - c\bar{u} - d\bar{v}) - sc\,\mathrm{cov}(u,v) - sd\sigma_v^2 \quad (12.17)$$

in which σ_u^2 and σ_v^2 are the spatial variances of u and v. If the densities in the two patches are identical, mean-field theory works perfectly, and the variance, covariance terms vanish. More generally, however, the system 12.17 is not closed and must be extended to include closure rules or equations for the variances and covariances (which in turn must be closed or extended). When more patches are involved and movement is not purely local, the entire spatial correlation function may be implicated, as we discuss in the next example.

For a spatial stochastic process, one may consider either spatial or ensemble moments, where "ensemble" refers to the set of all possible realizations of the stochastic process. Under appropriate assumptions, including those discussed here, the process is ergodic, and there is an equivalence between spatial and ensemble moments; other examples discussed in this chapter would not, however, simplify in this way.

The same approach can be applied to more sophisticated "contact" processes, in continuous space, such as neighborhood models with finite dispersal or the forest growth simulator SORTIE. For example, in a single-species version of this model (see Bolker and Pacala 1997), discussed in more detail in Pacala and Levin (Chapter 9), births to an individual at a site in continuous space are determined by a weighted function of densities of individuals in a neighborhood of the focal individual, with a weighting that decreases with distance. Again, all processes are stochastic, and one takes expectations over the infinite ensemble of possible realizations, obtaining a description of the form

$$\frac{dn}{dt} = H(n) - \alpha \int_{-\infty}^{\infty} U(r)c(r)\, dr, \tag{12.18}$$

where n is mean occupancy.

Here $H(n)$ is a term that depends on n alone (the "mean-field" term), $U(r)$ is the *competition kernel*, and $c(r)$ is the covariance at lag r. Once again, Equation 12.18 gives the dynamics of n in terms of variables other than n, so the system is not closed. The approach then is to complement Equation 12.18 with a second equation, a partial differential integral equation, for the dynamics of $c(r)$, and then to close the series by making appropriate assumptions about the higher moments. This powerful approach is developed further in Pacala and Levin (Chapter 9).

SUMMARY

In this chapter we have introduced a number of approaches to describing simplification and scaling laws for spatial stochastic ecological models. Scaling laws are power-law relationships that relate measurements on one scale to another; but the techniques of simplification provide more general ways to develop quantitative relationships among phenomena on difference scales. Through a series of examples, we introduce the classical notions of hydrodynamic limits of particle systems and Eulerian descriptions of individual-based (Lagrangian) models. Although different in detail, there are remarkable similarities in how the limiting processes are carried out.

In the last quarter century, especially since the work of Kadanoff and Wilson, renormalization methods have provided a powerful approach to developing scaling laws. The use of moment closure techniques bears relation to renormalization and is discussed in considerable detail. In general, approaches to simplification and scaling hold the potential to revolutionize the modeling of spatial dynamics in ecology. Further examples are given in Pacala and Levin (Chapter 9); the ultimate goal is to show, through scaling, what detail, either in real systems or individual-based simulations, is essential to understanding pattern and process on higher scale. Without such rules, the modeling process lacks robustness; with it, we may achieve genuine understanding of spatial dynamics.

ACKNOWLEDGMENTS

We are pleased to acknowledge the support of the National Aeronautics and Space Administration under grant NAGW-4688 to Princeton University, of the Andrew W. Mellon Foundation, and of the Office of Naval Research through its support of the University Research Initiative Program at Woods Hole Oceanographic Institution under grant ONR-URIP N00014-92-J-1527. Thomas Powell, Yoh Iwasa, David Tilman, and Peter Kareiva provided very useful criticisms and discussions.

CHAPTER THIRTEEN

Production Functions from Ecological Populations: A Survey with Emphasis on Spatially Implicit Models

Jonathan Roughgarden

INTRODUCTION

The search for generalization in ecology was delayed in the 1970s when Smale (1976), among others, demonstrated that models for ecological systems were inherently not robust if many strongly interacting populations were involved. Models of four or more competing species produced a variety of qualitatively different predictions depending on seemingly small details of the functional form of the equations. This nonrobustness is a consequence of nonlinearity, strong interactions, and high dimension. The response of ecologists during the 1980s was to devise population models specific to certain classes of systems, so we now have models for competition among plants, marine invertebrates of the rocky intertidal zone, and so forth. These new models begin at a scale, usually that of the individual organism, at which the model's assumptions and parameters can be directly measured and justified, and we can predict the consequences for population dynamics. It is perhaps now time to ask if predictions from the new, relatively specific models share features in common with one another and with the earlier, rather nonspecific theory. Of course, such generalizations may not exist, and I for one would be perfectly happy if theoretical ecology eventually yielded a geographic atlas of models, each model special to the processes and species at some spot on the globe, and thereby achieved generality through exhaustive coverage of a finite world. But it will make life easier for everyone if there are some simple and general take-home messages that apply to all, or nearly all, systems.

A common, though not universal, feature of the new models is an attention to spatial position—because in many kinds of species, interaction between individuals takes place only when organisms are physically close to one another. Thus, space is fundamentally important. Nonetheless, it may be helpful to remove the space, making models spatially implicit rather than spatially explicit. This surgical outlook is motivated in part by the difficulty of deriving scientifically useful results from spatially explicit models. This chapter therefore emphasizes models that begin as spatially explicit formulations and that can be converted to spatially implicit equations instead.

In this chapter, models for a single population are compared in terms of the "ecological production function" implied by the population model. The production function is simply the graph of dN/dt versus N which, for a logistic equation, is a downward-opening parabola whose peak is at $K/2$ and whose roots are at 0 and K.[1] The peak of the production function is the point of maximum sustainable yield, though usually not of maximum sustainable revenue. For a logistic equation, the slope of the production function at $N = 0$ is r and at $N = K$ is $-r$ because a parabola is symmetric about its peak.

The production function is the most useful aspect of a population model in economic contexts. The production function can be thought of as indicating the interest or return, in organisms per time, earned on a stock of organisms. Because of ecological density dependence, the interest earned depends on the stock size. In ecology little attention is focused on the production function—it is integrated to predict the trajectory of population size through time and then forgotten. But the production function itself is needed for analyzing the eco-

[1] The logistic equation is $dN/dt = rN(K - N)/K$, where the state variable, N, is the number of organisms in the population, r is the intrinsic rate of increase, and K is the carrying capacity.

nomic value of ecological services because with it the return on natural stocks can be compared with the return on financial stocks (cf. Roughgarden and Smith 1996).

The production function is a highly aggegrated representation of a population's dynamics that depends on a plethora of biological and geographic details. The policy mechanisms that influence environmental use are often coarse and broad in scope—do we cut down a thousand hectares of forest, drain a thousand hectares of wetland, harvest 500,000 tons of fish? The policy mechanism is rarely spatially explicit or focused on individual species because of the infeasibility and cost of implementing detailed policies. Therefore, we need to take models well founded in ecological mechanism and reduce them to a point where they are useful for policy analysis. Of course, most models in ecology are intended for other ecologists, cooked up by scientists for consumption by scientists. To transfer ecological knowledge outside of ecology, and particularly into the economic design of environmental policy, we should deliver models whose degree of aggregation matches the available policy mechanisms.

NEIGHBORHOOD MODELS OF PLANTS

The importance of spatial proximity in bringing about interactions between individuals is particularly obvious in terrestrial plants, where interaction depends on overlapping canopies, overlapping root systems, or diffusable chemicals released into the soil. Models called "neighborhood models" have been studied by Pacala and Silander (1985) to represent plant population dynamics based on the spatial proximity of the members of the population.

The population consists of annuals and is censused each year at the seedling stage. S_t is the number of seeds per area at year t. The basic setup is an equation whereby the seed density at time $t + 1$ equals the seed density at time t multiplied by the germination probability, g, to yield the seedling density; this is then multiplied by the probability of surviving from seedling to maturity, $Z(S_t)$, and the fecundity, $F(S_t)$:

$$S_{t+1} = (gS_t)Z(S_t)F(S_t). \tag{13.1}$$

This setup has been analyzed by taking $g = 1$ and $Z(S_t) \equiv P$, leaving all the density dependence in the fecundity, $F(S_t)$. The fecundity of each plant is then developed in terms of the average number of "neighbors" that a plant has—a plant's neighbors are those whose centers are contained in a unit circle around its stem. If the dispersal is sufficiently far relative to the unit circle, the plant centers are approximately Poisson distributed on a plane, and the equation for S_{t+1} becomes

$$S_{t+1} = (gS_t)P \sum_{n=0}^{\infty} \frac{e^{-PgS_t}(PgS_t)^n}{n!} f_a(n) \tag{13.2}$$

where $f_a(n)$ is the seeds produced by an adult plant with n individuals overlapping its radius, that is, n neighbors. The expression $f_a(n)$ is called a "fecundity predictor," and several functional forms have been considered.

Exponential Fecundity Predictor

If fecundity declines exponentially with the number of neighbors

$$f_a(n) = Qe^{-\nu n} \tag{13.3}$$

then

$$S_{t+1} = gPQS_t e^{-gP(1-e^{-\nu})S_t}. \tag{13.4}$$

To express this as a production function, subtract S_t from both sides (and suppress the subscript t), yielding

$$\Delta S = (gPQe^{-gP(1-e^{-\nu})S} - 1)S. \tag{13.5}$$

This production function is illustrated in Figure 13.1,[2] and the formulae for the equilibrium, denoted as K, the slope of the

[2] Parameters are $P = 0.2$, $g = 1$, $\nu = 0.2$, and $Q = 10, 20, 100$.

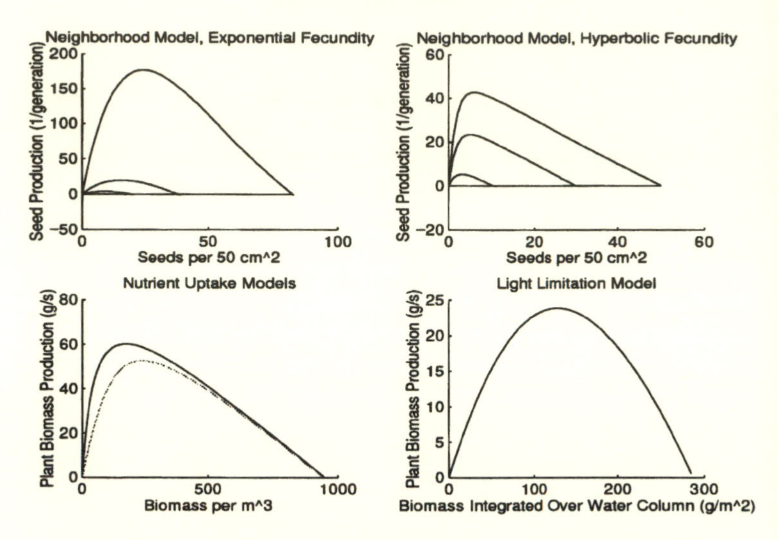

FIGURE 13.1. Production functions for plant populations. Top left and right: From "neighborhood models" in which an individual's seed production, probability of survival, and distance it was dispersed before germination are assumed to depend on the number of other individuals within some circle around it. Bottom left: From a model in which plant growth is limited by nutrients. Bottom right: From a model in which growth is limited by light.

production function at equilibrium, denoted as $-r$, and the slope of the production function at the origin, denoted as r_0, are all recorded in Table 13.1.[3]

Hyperbolic Fecundity Predictor

Fecundity may decline hyperbolically with number of neighbors, as suggested by data on *Arabidopsis thaliana*. If so, the potential seed of a plant, M, may be thought of as divided among the plant and its neighbors, leading to the following expression for the fecundity predictor:

$$f_a(n) = \frac{M}{1 + cn}. \tag{13.6}$$

[3] Parameters are $P = 0.2$, $g = 1$, $\nu = 0.2$, and $Q = 10, 20, 100$.

TABLE 13.1. Formulae for equilibrium population size and initial rate of increase.

Name of Model	K	r	r_0
Exponential Fecundity Neighborhood	$\frac{\ln(gPQ)}{gP(1-e^{-\nu})}$	$\ln(gPQ)$	$gPQ - 1$
Hyperbolic Fecundity Neighborhood	$\approx M$	≈ 1	$gPM - 1$
Linear Fecundity Neighborhood	$\frac{\gamma gP - 1}{\mu(gP)^2}$	$\gamma gP - 1$	$\gamma gP - 1$
Linear Nutrient Uptake	$\frac{uS - da}{dwu}$	$\frac{(uS - da)d}{us}$	$\frac{uS - da}{a}$
Michale-Menton Nutrient Uptake	$\frac{(v-d)S - dah}{(v-d)\,dw}$	$\frac{[(v-d)S - dah]\,d(v-d)}{v^2S - 2vSd + d^2ah + d^2S}$	$\frac{(v-d)S - dah}{S + ah}$
Light Limitation			$\frac{p_{\max} I_{\text{in}}}{H + I_{\text{in}}} - 1$
Space Limited Metapopulation	$\frac{cA(m-\mu) - \mu\nu}{ca(m-\mu)}$	$\frac{(m-\mu)[cA(m-\mu) - \mu\nu]}{m\nu}$	$\frac{cA(m-\mu) - \mu\nu}{cA + \nu}$
Benthic/Oceanic Metapopulation	$\frac{p}{2ac(m-\mu)q}$	$\frac{(m-\mu)p}{m}$	$\frac{p}{1 + 2Acq}$
Terrestrial Insects	$(1/\alpha)(\lambda^{(1/\beta)} - 1)$	$\beta(1 - \lambda^{-1/\beta})$	$\lambda - 1$
Optimal Feeding Territories	$\frac{q_3}{2q_2}\frac{L}{v}$		$-\nu + b\frac{\left(e_i - \frac{e_w}{ar_s} - \frac{e_p r_s}{v}\right)}{\left(\frac{1}{ar_s} + \frac{r_s}{v}\right)}$

With $c = 1$, the production function is

$$\Delta S = M(1 - e^{-gPS}) - S. \tag{13.7}$$

See Figure 13.1[4] for an illustration and Table 13.1 for the formulae. Numerical analysis of this model suggested that a positive equilibrium existed if $gPM - 1 > 0$ and that this equilibrium was stable for $c \leq 1$ and possibly unstable if $c > 1$.

Linear Fecundity Predictor

The neighborhood model has also featured a fecundity predictor that declines linearly with the number of neighbors

$$f_a(n) = \gamma - \mu n. \tag{13.8}$$

The corresponding production function is

$$\Delta S = (\gamma gP - 1)S - \mu(gPS)^2 \tag{13.9}$$

which is identically logistic. Formulae appear in Table 13.1.

NUTRIENT UPTAKE MODELS OF PLANTS

The alternative approach to modeling the population dynamics of plants has been to focus on how growth explicitly depends on the amount of available resources, specifically, the concentration of limiting nutrients in the soil and the light intensity. In this approach, spatial proximity is largely ignored. These models were originally developed for the growth of algae suspended in a well-mixed water column and have now been applied quite successfully to terrestrial plant populations as well. In this section we review the basic model pioneered by Tilman (1977, 1982) for nutrient-limited growth of biomass. The basic model has two state variables: B, the biomass of plant material in some area, and R, the amount of some nutrient (or resource), usually nitrate, in the area.

[4] Parameters are $P = 0.6$, $g = 1$, $M = 10, 30, 50$.

If the kinetics of resource uptake are modeled as a rectangular hyperbola, in accordance with the Michale-Menton formulae from chemistry, the biomass growth is

$$\frac{dB}{dt} = B\left(\frac{vR}{h + R} - d\right) \tag{13.10}$$

where v is the maximum uptake velocity per unit biomass and h is the resource level that produces half the maximum uptake rate. The corresponding resource dynamics are

$$\frac{dR}{dt} = S - aR - wB\frac{vR}{h + R}. \tag{13.11}$$

As before, if the nutrient uptake dynamics are fast, the equilibrium resource level for a given biomass can be found. Let $q(B)$ be defined as

$$q(B) \equiv S - ah - wvB. \tag{13.12}$$

Then

$$\hat{R}(B) = \frac{q(B) + \sqrt{q(B)^2 + 4aSh}}{2a} \tag{13.13}$$

and the production function for biomass becomes

$$\frac{dB}{dt} = B\left[\frac{v\hat{R}(B)}{h + \hat{R}(B)} - d\right]. \tag{13.14}$$

The production function and formulae appear in Figure 13.1[5] and Table 13.1.

Figure 13.1 and Table 13.1 also record the production function and formulae for K, r, and r_0 where the uptake kinetics are linear.[6] The graph of the production function for Michale-Menton uptake kinetics is very close to that for linear uptake kinetics.

[5] Parameters are $v = 1$, $h = 10$, $d = 0.1$, $S = 10$, $a = 0.5$, and $w = 0.1$.

[6] Parameters are $u = v/h$ where $v = 1$, $h = 10$, and $d = 0.1$, $S = 10$, $a = 0.5$, and $w = 0.1$.

LIGHT-LIMITATION MODELS OF PLANTS

In this section we review a model for light-limited growth of plant biomass, following Huisman and Weissing (1994, 1995). Consider a water column with a vertical light gradient, and assume the water column is well stirred from top to bottom. Light with intensity $I(s)$ at depth s is given by the Lambert-Beer law

$$I(s) = I_{\text{in}} e^{-k(W/z)s} \tag{13.15}$$

where I_{in} is the incident light at the water surface, k is the extinction coefficient of light, W is the total biomass integrated over the water column, and z is the depth of the water column. Photosynthesis per unit biomass is assumed to follow Michale-Menton kinetics

$$p(I) = p_{\max} \frac{I}{H + I} \tag{13.16}$$

with $p_{\max}$ denoting the maximum photosynthesis rate and H the intensity producing half the maximum photosynthesis rate. The respiratory loss per unit biomass is a constant, l. The growth per unit biomass at depth s is photosynthesis minus respiration, $p(I) - l$, which, when integrated over the water column, leads to the following equation for the growth of the total biomass:

$$\frac{dW}{dt} = \frac{p_{\max}}{k} \ln\left(\frac{H + I_{\text{in}}}{H + I_{\text{in}} e^{-kW}}\right) - lW. \tag{13.17}$$

The production function is illustrated in Figure 13.1,[7] and the formula for r_0 appears in Table 13.1; simple formulae for K and r were not found.

[7] Parameters are $p_{\max} = 10$, $H = 100$, $l = 1/2$, $I_{\text{in}} = 10$, and $k = 0.005$.

MODELS OF SPACE-LIMITED BENTHIC INVERTEBRATES

Marine invertebrates found at the sea shore typically have a life cycle consisting of egg, pelagic larva, and benthic adult. Together with colleagues, I have developed population models for this situation (Roughgarden and Iwasa 1986; Iwasa and Roughgarden 1986; Roughgarden, Pennington, and Alexander 1994; Alexander and Roughgarden 1996). One formulation views the larvae as contained in an unstructured "larval pool" and the adults as residing on benthic substrate. The larvae "settle" or recruit to unoccupied substrate. This formulation, called the "metapopulation" version, applies to places where the larval pool quickly mixes across the benthic habitat. The rocks and hard surfaces that line an estuary such as San Francisco Bay are an example. The water of the bay mixes in days, whereas the larval life spans weeks, so all the surface where the adults live communicates with a common larval pool. The alternative formulation, called the "cross-shelf transport" version, applies to the open coast where the larval pool is structured by horizontal advection and diffusion, and where recruitment is caused by episodes of onshore currents that return larvae to the coastal benthic habitat where they were born.

Metapopulation Formulation

The number of adult organisms in the benthic site is B, the number of larvae in the larval pool is L, and the area of the benthic habitat is A. The free space, F, is defined as the area not occupied by organisms,

$$F \equiv A - aB \tag{13.18}$$

where a is the basal area covered by a single individual and A is the total area of the benthic habitat. Then the dynamics within the benthic site are

$$\frac{dB}{dt} = cLF - \mu B \tag{13.19}$$

where c is the settlement coefficient that describes the accessibility of the benthic site to the larval pool, and μ is the benthic mortality rate. The input to the benthic site is the term cLF, which represents recruitment or settlement onto empty space —the recruitment rate is assumed to be proportional to both the amount of empty space and the number of larvae, as though recruitment takes place by random contacts of larvae with empty space. The larval population dynamics are

$$\frac{dL}{dt} = mB - cLF - \nu L \tag{13.20}$$

where m is the fecundity rate of adults, and ν is the mortality rate in the larval pool. The input to the larval pool is the reproduction by adults, and the loss from the larval pool is from both mortality and from settlement into the benthic habitat.

To obtain a production function for the adults (which would be the life stage of commercial or recreational interest) the larval population may be assumed to come to equilibrium relative to the number of adults (called the stock). The equilibrium larval population for a given adult stock is

$$\hat{L} = \frac{mB}{cF + \nu} \tag{13.21}$$

leading to adult dynamics, providing the larval abundance tracks the adult population size, of

$$\frac{dB}{dt} = \left[\frac{m}{1 + \nu/c(A - aB)} - \mu\right]B. \tag{13.22}$$

This production function is illustrated in Figure 13.2[8] and the formula for K, r, and r_0 in Table 13.1.

[8] Parameters are $A = 50$, $a = 0.0001$, $\mu = 0.1$, $m = 1$, $\nu = 0.5$, and $c = 0.01$.

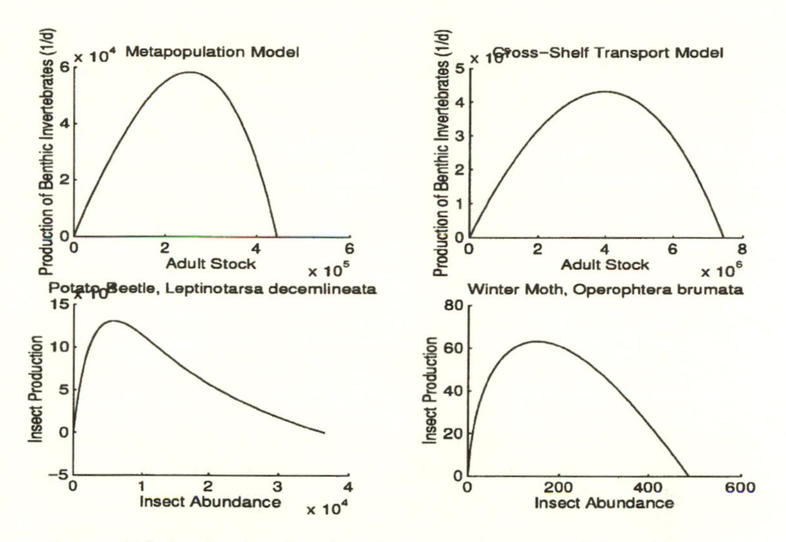

FIGURE 13.2. Production functions for invertebrate populations. Top left and right: These are for marine invertebrates with a bottom-dwelling (benthic) adult phase and an ocean-going (pelagic) larval phase. Top left assumes an unstructured "larval pool" and is called a metapopulation model; top right assumes the larval pool occupies a band along the coast within which larvae are moved about by currents and by diffusion. Bottom left and right: These pertain to insect populations.

This production function is unimodal, as have been those illustrated previously. However, with two or more benthic sites communicating with a common larval pool, if some of the benthic sites are net sources of larvae and others net sinks for larvae, multiple equilibria may exist, and the production function is multimodal. In this situation, a threshold propagule must be introduced for the population to establish at the highest equilibrium. Although multiple equilibria are theoretically possible if the benthic sites include both larval sources and sinks, there is no evidence that this situation actually occurs in nature.

Cross-shelf Transport Formulation

In this formulation the larval pool is viewed as being contained in the ocean between the coast and an offshore front.

The coast is located at $x = 0$ and the front at $x = x_f$. The larval pool and the population of benthic adults are coupled via a boundary condition at $x = 0$, and the front at $x = x_f$ is a reflecting boundary. Thus the model for the larval pool is a reaction-diffusion equation, and that for the benthic adults an ordinary differential equation, and that for the benthic adults an ordinary differential equation. The dynamics of adults, $B(t)$, at the coast are

$$\frac{dB(t)}{dt} = cF(t)L(0,t) - \mu B(t) \tag{13.23}$$

where F is the free space, $L(0,t)$ the larval concentration at the coast, c a coefficient of larval settlement, and μ the adult mortality rate. The total space, A, identically equals

$$A \equiv F(t) + aB(t) \tag{13.24}$$

where a is the average basal area of an adult. The dynamics of larvae, $L(x,t)$, in the ocean are

$$\frac{\partial L(x,t)}{\partial t} = -u\frac{\partial L(x,t)}{\partial x} + k\frac{\partial^2 L(x,t)}{\partial x^2} - \nu L(x,t) \tag{13.25}$$

where x_f is the location of an offshore reflecting boundary, u the cross-shelf advection rate, k the eddy-diffusion coefficient, and ν the larval mortality rate. The boundary condition at $x = 0$ is

$$uL(0,t) - k\left.\frac{\partial L(x,t)}{\partial x}\right|_{x=0} = mB(t) - cF(t)L(0,t) \tag{13.26}$$

where m is the fecundity rate, and at $x = x_f$ it is

$$uL(x_f,t) - k\left.\frac{\partial L(x_f,t)}{\partial x}\right|_{x=x_f} = 0. \tag{13.27}$$

To obtain a production function for B as before, we solve for an L that is at equilibrium for fixed B because benthic processes are slow compared to water-column processes (i.e., solve for $L(x, t) = L[x, B(t)]$),

$$L(x, B) = \frac{2Bm(e^{q_1 x/2k} q_3 - e^{q_2 x/2k} q_4)}{[2(A - aB)c + q_2]q_3 - [2(A - aB)c + q_1]q_4} \tag{13.28}$$

and specifically,

$$L(0, B) = \frac{2Bm(q_3 - q_4)}{[2(A - aB)c + q_2]q_3 - [2(A - aB)c + q_1]q_4} \tag{13.29}$$

where the water-column parameters group as

$$\begin{aligned} q_1 &= u + \sqrt{4k\nu + u^2} \\ q_2 &= u - \sqrt{4k\nu + u^2} \\ q_3 &= e^{q_2 x_f/2k} q_1 \\ q_4 &= e^{q_1 x_f/2k} q_2. \end{aligned} \tag{13.30}$$

The the dynamics of the adult stock with an implicit larval pool become

$$\frac{dB}{dt} = B\left\{ -\mu + \frac{2(A - aB)cm(q_3 - q_4)}{[2(A - aB)c + q_2]q_3 - [2(A - aB)c + q_1]q_4} \right\}. \tag{13.31}$$

This production function is illustrated in Figure 13.2[9] and the formulae in Table 13.1, in terms of two lumped parameters, p,

[9] Parameters are $x_f = 10{,}000$, $u = 10$, $k = 100{,}000$, $A = 1{,}000$, $a = 0.0001$, $\mu = 0.1$, $m = 1$, $\nu = 0.5$, and $c = 0.1$.

the positivity requirement,

$$p = -\mu + 2Ac(m - \mu)q \tag{13.32}$$

and q, a water-column descriptor

$$q = \frac{q_3 - q_4}{q_2 q_3 - q_1 q_4}. \tag{13.33}$$

This formulation has been used to model population fluctuations caused by changing ocean conditions. For example, take $u = u(t)$ and $x_f = x_f(t)$, and consider a scenario in which the front moves sinusoidally between one and twenty-five kilometers offshore with a period of sixty days, and the rate of advection varies from 0.2 m/s at maximum upwelling to 0 m/s at maximum relaxation. This idealizes how changes in the California Current control a coastal benthic ecosystem.

INSECT POPULATION MODELS

Hassell, Lawton, and May (1976) have used the following phenomenological single-species model for insects:

$$N_{t+1} = \frac{\lambda N_t}{(1 + \alpha N_t)^{\beta}} \tag{13.34}$$

where λ is the finite net rate of increase (the density-independent component of population growth), and α and β characterize the density-dependent feedback. Examples include the Colorado potato beetle, *Leptinotarsa decemlineata*, for which $\alpha = 0.00007$, $\beta = 3.4$, and $\lambda = 75$, and the winter moth, *Operophtera brumata*, with $\alpha = 0.6$, $\beta = 0.3$, and $\lambda = 5.5$. This model has the property that if $\lambda < 4$ the equilibrium is stable, and if $\lambda > 4$ and if also β is high enough (say, > 3) then limit cycles and chaos occur.

Although originally introduced as a phenomenological model, de Jong (1979) and Hassell (1980) subsequently showed that the model could be derived in a spatial conceptualization. The total population of adult insects is distributed into patches according to a negative binomial distribution. Within each patch survival is exponentially distributed, with a parameter that is density dependent. These assumptions lead precisely to the model above. Other models too may be derived assuming the distribution across patches is Poisson or a positive binomial.

To obtain the production function corresponding to these dynamics, subtract N from both sides,

$$\Delta N = \left(\frac{\lambda}{(1 + \alpha N)^{\beta}} - 1 \right) N. \qquad (13.35)$$

The formulae for K, r, and r_0 appear in Table 13.1. Figure 13.2 illustrates production functions for the Colorado potato beetle, *Leptinotarsa decemlineata*, and the winter moth, *Operophtera brumata*.

FISHERY POPULATION MODELS

The fishery literature has long had a collection of phenomenological models for population dynamics, as summarized by May et al. (1978). N represents the stock as a fraction of the unexploited equilibrium population size, and other scaling conventions are used, such as a unit death rate, and a fecundity whereby the recruitment to a stock of one-half the unexploited size is three-fourths of that in an unexploited stock. The models are listed in Table 13.2. Figure 13.3 (left) offers a plot of the production functions listed in Table 13.2.

MODEL FOR ORGANISMS DISTRIBUTED IN FEEDING TERRITORIES

The model of this section is a prototype for the population dynamics of *Anolis* lizards, which occupy feeding territories as adults. This model is an animal counterpart of the neighborhood model for plants in that the organisms interact with their adjacent neighbors. Population increase leads to a shrinking of the territory size, and a consequent reduction in the foraging yield from those territories, until a point at which direct

TABLE 13.2. Phenomenological models from fisheries biology.

Name of Model	dN/dt
Beverton-Holt	$N\left(\frac{1}{b+(1-b)N}-1\right)$ with $b = 1/3$
Chapman	$b(1-e^{-aN})-N$ with $b = 9/8$ and $a = \ln(9)$
Ricker	$aNe^{-bN}-N$ with $a = 9/4$ and $b = \ln(9/4)$
Pella-Tomlinson	$bN(1-N^2)$ with $b = 2/3$
Pella-Tomlinson	$bN(1-N^3)$ with $b = 4/7$
Logistic	$N(1-N)$
Fox	$-bN\ln(N)$ with $b = 1/\ln(4)$
Cushing	$N^b - N$ with $b = 0.415$

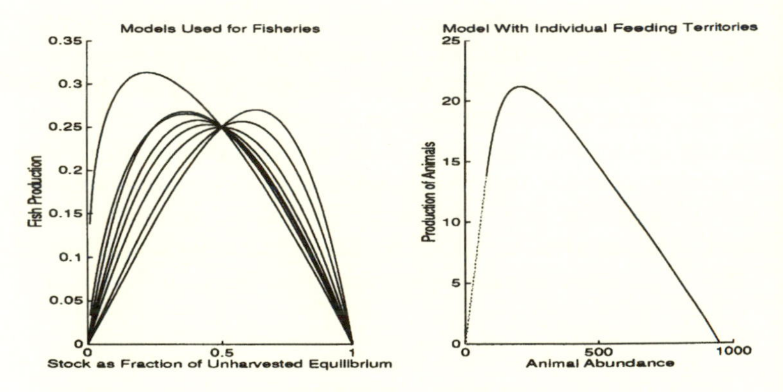

FIGURE 13.3. Production functions for vertebrate populations. Left: The many forms found in the literature on fisheries management. Right: A neighborhood model for a terrestrial vertebrate where individuals each have feeding territories that become increasingly compressed and less profitable as the population increases. The initial part of the curve is a straight line, where the population dynamics are density independent. The curve begins to bend when the desired territories intersect, and the density dependence begins.

replacement occurs. This model is an extension to interacting foragers of the foraging theory developed in chapter 1 of my recent book on *Anolis* lizards (Roughgarden 1995).

For simplicity, consider foraging on a line for one prey type without prey escape. We begin with a solitary forager on a line ($r \geq 0$), with a single prey type. The rate of appearance of prey in the interval from r to $r + dr$ is

$$\rho(r)\,dr = a\,dr \tag{13.36}$$

with a in units of insects per meter per second. If the cutoff distance is r_s, the average waiting time is

$$t_w(r_s) = \frac{1}{\int_0^{r_s} \rho(r)\,dr} = \frac{1}{\int_0^{r_s} a\,dr} = \frac{1}{ar_s} \tag{13.37}$$

and the average pursuit time, to the insect and return, is

$$t_p(r_s) = \frac{\int_0^{r_s} \left(\frac{2r}{v}\right) \rho(r)\,dr}{\int_0^{r_s} \rho(r)\,dr} = \frac{\int_0^{r_s} \left(\frac{2r}{v}\right) a\,dr}{\int_0^{r_s} a\,dr} = \frac{r_s}{v} \tag{13.38}$$

where v is the forager's sprint velocity. The foraging yield is then

$$E(r_s) = \frac{e_i - e_w t_w(r_s) - e_p t_p(r_s)}{t_w(r_s) + t_p(r_s)} \tag{13.39}$$

where e_i is the energetic content of an insect, e_w is the energy per second of waiting time, and e_p is the energy per second in pursuit. The optimal cutoff distance for a solitary forager is then the r_s that maximizes $E(r_s)$:

$$r_s = \frac{-(e_p - e_w) + \sqrt{(e_p - e_w)^2 + ave_i^2}}{ae_i}. \tag{13.40}$$

There is no density dependence until the crowding is sufficient to force the territory sizes to be less than r_s.

Now suppose the two foragers facing each other are close enough to want the food in each other's home range. We must now predict the foraging yield as a function of the interindividual distance. We assume the territory boundary comes to lie halfway between the positions of the two foragers. If we assume the interindividual distance, d, is less than $2r_s$ so that the desired home ranges overlap, then the average waiting time, pursuit time, and yield are

$$t_w(d) = \frac{2}{ad}$$

$$t_p(d) = \frac{d}{2v}$$

$$E(d) = \frac{e_i - 2e_w/(ad) - e_p\, d/(2v)}{2/(ad) + d/(2v)} \qquad (13.41)$$

assuming each forager uses half the distance between it and its neighbor.

The spatial pattern of the foragers equilibrates after a few days, and for the given number of animals in a finite site, foraging yield of these individuals can be determined. To determine the population dynamics, other parameters needed are the mortality rate, the time taken to reach maturity, the energetics of converting foraging yield into eggs, and the average fraction of the day that is spent foraging.

Suppose the individuals are arrayed as pairs of foragers facing each other on a circle of length L. (A circle avoids edge effects, but if desired, one could work with explicit left and right boundaries.) With N individuals, the interindividual distance, d, is $L/(N/2)$. Then the population dynamics are

$$\Delta N = \left(-\nu + bE\left(\frac{L}{N/2}\right)\right)N \qquad (13.42)$$

where $E[L/(N/2)]$ means the foraging yield per forager with an interindividual distance of $L/(N/2)$. Here ν and b are lumped parameters,

$$\nu = \mu + m(\tau - \tau_f)e_r$$

$$b = m\tau_f \tag{13.43}$$

where μ is the daily mortality rate, τ is the total number of seconds in twenty-four hours, τ_f is the average number of seconds spent foraging per day, e_r is average metabolic rate while not foraging, and m is the quantity of eggs produced per joule of foraging yield times the probability of surviving to adult. If we substitute the expression for $E[L/(N/2)]$ a more explicit form of the production function is obtained:

$$\Delta N = \left[-\nu + \frac{b\left(e_i - \dfrac{e_w}{ar_s} - \dfrac{e_p r_s}{v}\right)}{\dfrac{1}{ar_s} + \dfrac{r_s}{v}} \right] N \quad \text{for } N \le \frac{L}{r_s}$$

$$\Delta N = \left[-\nu + \frac{b\left(e_i - \dfrac{e_w N}{aL} - \dfrac{e_p L}{N_v}\right)}{\dfrac{N}{aL} + \dfrac{L}{N_v}} \right] N \quad \text{for } N > \frac{L}{r_s}. \tag{13.44}$$

For $N \le L/r_s$ the population dynamics is density independent, and when N is greater than this value the density dependence begins because at this point the territories begin to overlap. The production function is illustrated in Figure 13.3.[10]

[10] Parameters are $a = 0.005$ insects/m/s, $e_i = 10$ j, $v = 1$ m/s, $e_w = 0.01$ j/s, $e_p = 0.1$ j/s, $t_f = 5 \times 60 \times 60$ s, $m = (1/e) \times 0.0001$ eggs/j, $\mu = 1/365$/d, $L = 1{,}000$ m, $e_r = e_w$, $b = mt_f$, $\nu = \mu + m(24 \times 60 \times 60 - t_f)e_r$.

The equilibrium population size may be written in terms of the quantities

$$q_1 = \nu + be_p$$

$$q_2 = \nu + be_w$$

$$q_3 = be_i av + \sqrt{b^2 e_i^2 a^2 v^2 - 4av(\nu + be_p)(\nu + be_w)}\,. \quad (13.45)$$

The formulae for K and r_0 appear in Table 13.1.

DISCUSSION

Figures 13.1–13.3 illustrate natural production functions for population-dynamic models developed for organisms extending from plants through invertebrates to vertebrates. The figures all show unimodal downward-opening curves that usually are not as symmetric as the logistic, leaning either to the right or the left. The logistic seems middle-of-the-road in comparison.

The models based on spatially explicit accounts of how density dependence happens do not differ as a class from those lacking a spatial basis. Taking account of the spatial aspects of the density dependence has made models more biologically credible but has not led to a different mathematical form of model. Of course, an experienced modeler can easily incorporate space in a way that will lead to mathematical outcomes not possible in a nonspatial counterpart. This is particularly easy in metapopulation formulations with spatially nonuniform parameter distributions coupled with asymmetric migration tendencies. It is not clear empirically, though, if spatial considerations often affect the aggregate properties of a population in ways fundamentally different than predicted by nonspatial models. If spatial pattern, per se, is of primary interest, then obviously a spatially explicit model cannot be replaced with a nonspatial simplification. However, for aggregate properties, the importance of spatial explicitness may be less than the gain in utility obtained with a nonspatial approximation.

In the logistic, the slope at $N = 0$ is r, and the acceleration there is negative—that is, the first unit of stock earns the highest interest, and each successive unit of stock earns less and less because of the density dependence. (Economists would say that the marginal value of the stock is a decreasing function of stock size.) This is true of all but one of the models of Figures 13.1–13.3 too. The model to the right in Figure 13.3, for a population whose individuals have feeding territories, has a zero acceleration for a while. What happens is that density dependence begins only after enough animals have accumulated in the habitat for their territories to intersect. The situation has been termed a "density-vague" population regulation and is represented as a production function that begins with a straight line that bends once the abundance attains the threshold density at which the density dependence occurs. None of the models has a positive acceleration at a stock of zero, called "positive density dependence" or the "Allee effect."

For the resource-based models, r_0, r, and K vary together. The formulae for these parameters in terms of component processes often show that they are proportional to one another and therefore vary up and down together as environmental conditions change in space or time.

SUMMARY

Production functions are surveyed from single-species population-dynamic models in ecology. Many are relatively new models that incorporate biological detail appropriate to particular classes of organisms, including nutrient-, light-, and neighborhood-limited plants, marine and terrestrial invertebrates, and marine and terrestrial vertebrates. The production functions are all quite similar, and the logistic model's production function is a middle-of-the-road model compared to others in the literature. Production functions from models based on spatial assumptions do not differ as a class from those based on nonspatial assumptions.

CHAPTER FOURTEEN

Challenges and Opportunities for Empirical Evaluation of "Spatial Theory"

Eleanor K. Steinberg and Peter Kareiva

INTRODUCTION

This book is based on the premise that the character of population dynamics and species interactions can be influenced profoundly by spatial structure. This is supported by an overwhelming number and diversity of mathematical models demonstrating important consequences of space. Unfortunately, although empiricists may appreciate this new "spatial theory," they have been slow to explicitly test its predictions. Indeed, perhaps the greatest challenge facing "spatial ecology" is wedding the pertinent theory to empirical research. We know "in theory" that space can make a huge difference, but we do not know how often "in fact" it does make a difference. In particular, given that our resources are limited, it is not clear how much we should invest in field studies focusing on the spatial dimension of ecological processes. Collecting spatially structured data, much less conducting "spatial experiments," is costly.

We begin this chapter with a computer simulation that is intended to illustrate how difficult it will be to devise "clean" manipulative experiments that test specific aspects of spatial theory. With this cautionary tale as a jumping-off point, we discuss in general how we think spatial theory should be addressed via empirical programs and then list specific issues that are especially amenable to immediate empirical progress. Our goal is to point out likely pitfalls when pursuing the empirical assessment of spatial theory, as well as to highlight opportunities for breakthroughs.

A SIMULATION MODEL

A widely cited explanation for the stability of host-parasitoid interactions is the idea that a combination of population subdivision, environmental variability, and dispersal can be stabilizing when summed over an entire ensemble of subpopulations. One of the most illuminating theoretical examinations of this issue (Reeve 1988) argued that definitive evidence could be obtained by experimentally isolating subpopulations from the exchange of dispersers, and then comparing population fluctuation and persistence between isolated and nonisolated (control) ensembles of patches. This clear advice has led at least two population ecologists to do Reeve's recommended experiment—isolate subpopulations (by either distance, or actually enclosing them with huge nets) and then compare population fluctuations between ensembles of subpopulations that are isolated and ensembles interconnected by dispersal. In one experiment involving scale insects attacking grapefruit and the parasitoids of the scale insects, grapefruit trees were either enclosed with netting (thereby cutting off dispersal) or left open. The coefficient of variation for repeatedly sampled host or parasitoid populations did not differ between isolated and nonisolated collections of subpopulations (Murdoch et al. 1996). This experiment ran for three generations of the host and included five replicates.

In a second experiment that involved predatory mites and their mite prey, the researcher manipulated the degree of isolation by arranging apple trees either in clusters of four or sixteen trees or as single isolated trees (Walde 1994). This experiment uncovered an effect of isolation on persistence but not on the coefficient of variation. However, the significant persistence effect did not arise when patches were simply isolated versus unisolated—it required artificial augmentation of prey populations at the beginning of the experiment (Walde 1994). This experiment ran for three to five generations and included four replicates.

An important question is whether absence of "significant effects" in these elegant spatial experiments is due to a true absence of any spatial effect, or simply limited experimental power. The power of experiments is always an issue in ecology but attains even greater importance where the experiments are spatially complex, because opportunities for replication are limited and expenses are much greater with experiments that involve spatially extensive sets of populations.

To formally examine the issue of power we use Reeve's model as an illustration. This host-parasitoid model takes the form

$$H_i(t+h) = (1-\mu_H)H_i(t) + \frac{\mu_H[\Sigma H_1(t)]}{n} \quad (14.1a)$$

$$P_i(t+h) = (1-\mu_P)P_i(t) + \frac{\mu_P[\Sigma P_1(t)]}{n} \quad (14.1b)$$

$$H_i(t+1) = F_i(t-h)H_i(t+h)\left[1 + \frac{\alpha_i(t+h)P_i(t+h)}{k}\right]^{-k} \quad (14.2a)$$

$$P_i(t+1) = H_i(t+h)\left\{1 - \left[1 + \frac{\alpha_i(t+h)P_i(t+h)}{k}\right]^{-k}\right\} \quad (14.2b)$$

where host and parasitoid populations are divided into n separate subpopulations, with $H_i(t)$ and $P_i(t)$ representing the densities of hosts and parasitoids in the ith subpopulation at time t. In this model, all subpopulations are assumed to be equally accessible to the pool of dispersers, represented by a fraction of the hosts (μ_H) and parasitoids (μ_P).

The model breaks the population dynamics into two separate processes: (1) migration and (2) reproduction and mortality. First, migration occurs in the time interval between t and $t + h$, as represented in Equations 14.1a and 14.1b, describing how host and parasitoid populations change as a result of dispersal. The subscript i ranges from 1 to n. Then, after dispersal, each subpopulation changes according to May's (1978) classical negative binomial model, as presented in Equations 14.2a and 14.2b. In May's model, the parameter k describes the degree of parasitoid clumping independent of host density. Variability is entered into the model by allowing the parasitoid attack rate (α_i) and the host (or prey) reproductive rate (F_i) to be random variables, drawn independently for each patch and each generation. The reproductive rate was taken to be a log-normal random variate, and the attack rate was assumed to be a gamma random variate.

Reeve's simulation model ran for a hundred generations and involved a hundred replicates of each run, with a hundred patches in each ensemble of subpopulations. He contrasted coefficients of variation and persistence time (the number of generations an ensemble of patches maintained at least one population of hosts and one population of parasitoids) as a function of dispersal versus no dispersal. Of course, a field counterpart of Reeve's simulation experiment is unimaginable. More realistically, one would expect field experiments to involve no more than five to ten replicates, and to run for no longer than five to ten generations—indeed, these estimates of a reasonable duration and number of replicates exceed the effort either Murdoch et al. or Walde put into their tests of Reeve's model. It is not clear whether a logistically feasible field experiment could ever include enough replication or run for enough generations to detect the effect Reeve obtained by running simulations for a hundred generations and over a hundred replicates.

To address exactly this issue we reran Reeve's original model, using the same equations and the same reproductive values, search efficiencies, and negative binomial aggregation parameters that he used. Instead of the large system, run for a long time period and many replicates, we ran it for limited time periods with fewer replicates and smaller numbers of total patches in each ensemble (Figure 14.1). What we learned was that it is very unlikely to detect a strong signal for stabilization

COEFFICIENT OF VARIATION

		10 "Natural" Patches			100 "Natural" Patches			10 "Controlled" Patches		
		LH/LP	HH/LP	LH/HP	LH/LP	HH/LP	LH/HP	LH/LP	HH/LP	LH/HP
5 gen	**5 rep**	0.47	0.16	0.80	0.67	0.20	0.29	0.76	0.25	0.86
10 gen	**5 rep**	0.34	0.22	0.53	0.10	**0.04***	0.44	0.22	**0.00***	0.61
5 gen	**10 rep**	0.87	0.19	0.50	0.54	**0.02***	0.41	0.24	0.35	0.34
10 gen	**10 rep**	0.86	0.73	0.49	0.34	0.11	0.43	0.58	0.44	0.93

PERSISTENCE TIME

		10 "Natural" Patches			100 "Natural" Patches			10 "Controlled" Patches		
		LH/LP	HH/LP	LH/HP	LH/LP	HH/LP	LH/HP	LH/LP	HH/LP	LH/HP
5 gen	**5 rep**	0.55	0.09	0.28	1.00	1.00	1.00	0.37	1.00	0.37
10 gen	**5 rep**	0.49	**0.02***	0.09	**0.05***	**0.02***	0.10	0.06	**0.01***	**0.02***
5 gen	**10 rep**	0.32	**0.00***	0.84	1.00	1.00	1.00	1.00	1.00	1.00
10 gen	**10 rep**	0.14	**0.00***	0.08	**0.05***	0.08	**0.00***	**0.00***	**0.00***	**0.00***

FIGURE 14.1. Summary of simulation results involving Reeve's (1988) host-parasitoid model as a means of assessing the power of different experimental designs. One set of simulations began with "controlled" patches that were initiated under equilibrium conditions, with all patches starting out with the exactly same numbers of hosts and parasitoids (as though the researcher artificially enforced identical initial populations in all patches). Other simulations started with what we call "natural" patches in which we did not enforce identical initial populations. These "natural" populations were initiated by running the simulation for one generation before beginning sampling (to produce the sort of background variability one would expect in the field). Experimental conditions included ten or one hundred patches, five or ten generations, five or ten replicates, or all of these. In addition, varying dispersal rates (represented as fraction of total population entering dispersal pool) for both hosts and parasitoids were simulated at the following levels: Low Host (LH) = 0.01, High Host (HH) = 0.1, Low Parasitoid (LP) = 0.01, and High Parasitoid (HP) = 0.1.

assuming the limitations of a realistic field experiment (Figure 14.1)—specifically, we detected a significant effect of dispersal only if there were either ten generations or ten replicates (an effort rarely achieved by field ecologists). Thus, even though we know the elimination of dispersal destabilizes host-parasitoid dynamics that are governed by Reeve's model, for practical numbers of replicates and durations of experiments, it may be impossible to detect such an effect. It is worth noting that persistence time was generally more sensitive to the presence or absence of dispersal than was coefficient of variation.

A second obvious field experiment for patchy host-parasitoid interactions would involve the removal of patches. Although the effect of patch removal has been examined in models, it has not generally been a focus of field experiments (but see Wennergren, Ruckelshaus, and Kareiva 1995). We also used Reeve's model to simulate experiments aimed at detecting the effect of removing patches. To do this we started with one hundred patches in a system governed by Reeve's model and eliminated patches until there were fifty, twenty-five, ten, and five patches remaining in the system. Interestingly, in contrast to the simulated experiments pertaining to the elimination of dispersal, reduction in the number of patches did appear to yield a consistent effect on persistence time with as few as five replicates and generations (Figure 14.2)—although the reductions in patch number required to see this effect were quite severe (by at least 50%). Again, coefficient of variation did not provide a sensitive measure for the effect of a reduction in patch number on the dynamics of the host-parasitoid interaction (Figure 14.3). Indeed, there was no trend whatsoever relating the coefficient of variation for host populations to the number of patches remaining in the system following habitat destruction (Figure 14.3).

Obviously, our simulated field experiments based on Reeve's simulation model do not comprise a general or robust examination of all possible spatial experiments. But they do make the point that a theoretically striking result may be difficult to demonstrate experimentally, even though the theory is "correct." The general lesson is that empiricists interested

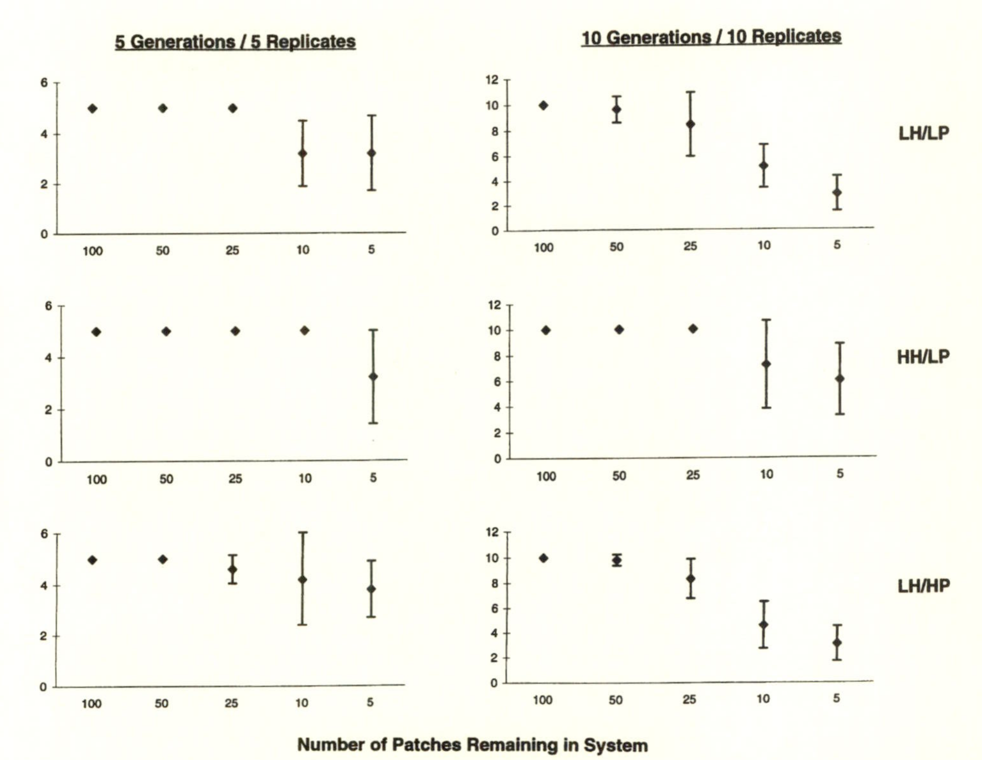
5 Generations / 5 Replicates
10 Generations / 10 Replicates
LH/LP
HH/LP
LH/HP
Persistence Time (generations)
Number of Patches Remaining in System

in testing spatial theory should pay careful attention to the power of their experimental designs, given the processes being examined. It is going to be difficult to replicate large spatially distributed systems as effectively as nonspatial experiments.

TWO APPROACHES TO TESTING SPATIAL THEORY: MANIPULATIVE EXPERIMENTS AND "MODEL" FITTING

There are two distinct routes by which we can test key ideas from spatial models in ecology. One approach is to invest our faith in a particular model, parameterize the model in detail, and then test quantitatively the prediction of the model. This approach generally has not been adopted, probably because few experimentalists are willing to put much faith in any single model. A more common approach is to test some general prediction (such as the hypothesis that eliminating dispersal will destabilize a predator-prey system), without worrying about a specific model. These tests of "general predictions" seem simple and straightforward. They are not. When one returns to the models and the actual dynamics and considers the uncontrollable variation typical of the natural world, it is clear that attaining adequate replication, duration, and spatial extent may become prohibitively expensive. Given this dilemma, we believe that before spatial questions are addressed experimentally, ecologists should first frame their questions in terms of explicit models and then invest extensive effort into parameterizing these models. We also point out that manipulative experiments need not be as low in power as our simulated experiments on host-parasitoid interactions might suggest. Instead, by astute choice of which aspects of spatial theory to pursue via a model/observation study, and which aspects

---

FIGURE 14.2. Summary graphs of patch-removal simulation results for persistence time. As above, simulations were run for five or ten generations and replicated five or ten times for low and high host and parasitoid dispersal parameters. The one-hundred-patch system was considered "intact," and the fifty-, twenty-five-, ten-, and five-patch systems were simulated to represent removal of habitat.

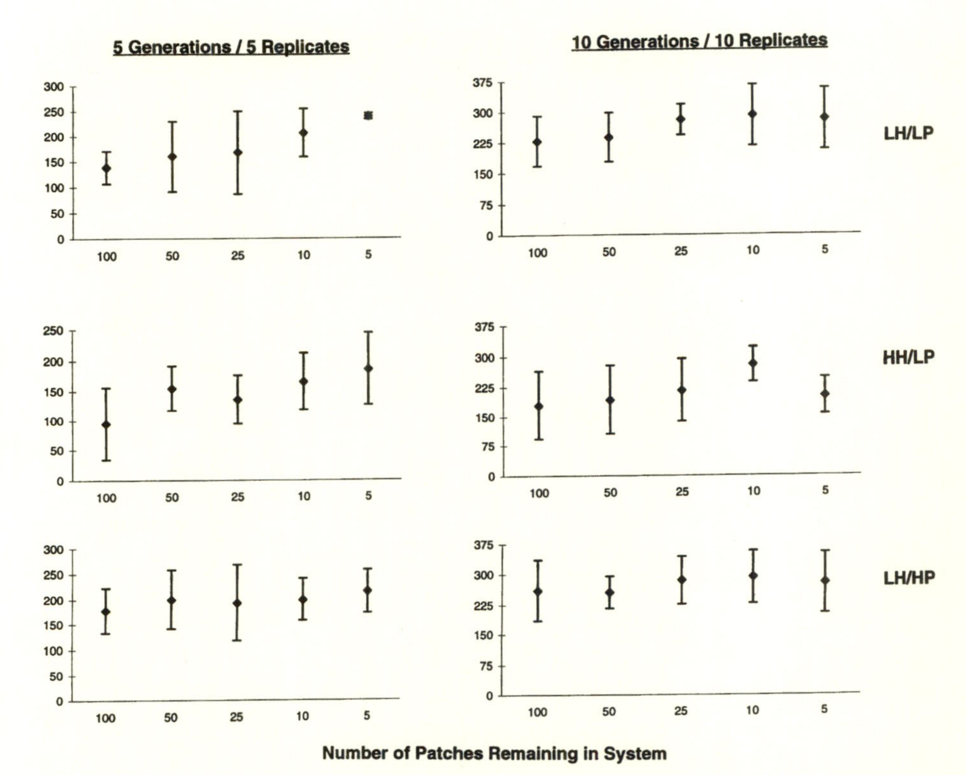
5 Generations / 5 Replicates
10 Generations / 10 Replicates
LH/LP
HH/LP
LH/HP
CV of Host Density
Number of Patches Remaining in System
100
50
25
10
5

to pursue via manipulative experiments, we can look forward to great progress in our understanding of spatial processes in the near future.

This book explores the implications of space for five major ecological phenomena: (1) stability, (2) coexistence, (3) variation in biodiversity, (4) invasions, and (5) pattern formation. Not all of these phenomena are equally accessible to experimental scrutiny. First, clear-cut manipulative experiments that test how space affects stability will be unavoidably plagued by low statistical power and the ambiguities surrounding how we define stability. Thus, documentation of compelling consequences of space for the stability of species interactions is likely to require tightly controlled microcosm experiments (*sensu* Lawton 1996) or the good fortune to have a long-term dataset interrupted by a natural disaster such as a drought (see Tilman and Downing 1994; Tilman 1997). Second, spatially explicit models of ecological invasion are similarly risky foci for field experiments, because as the theory points out, rare long-distance dispersal events can drive the speed and shape of invasions (Chapter 3)—and such events will not be a predictable feature of any experimental study of invasion. Finally, pattern formation, which arises so beautifully and clearly in many spatially explicit models, will be difficult to demonstrate in natural systems because the inevitable soil and microclimate heterogeneities of all field situations will drive their own patterns. Moreover, we are lacking theoretical studies that investigate the nature of pattern formation in heterogeneous environments (since most modelers focus on pattern formation in homogeneous environments). The spiral waves and checkerboard patterns that capture our imagination as outputs from spatial models (see Hassell and Wilson, Chapter 4) are likely to become very blurry if they are overlaid on a heterogeneous environment. This implies that pattern formation will best be studied in laboratory or greenhouse settings that homogenize the environment as much as possible. Of

---

FIGURE 14.3. Summary graphs of patch-removal simulation results for coefficient of variation. See legend of Figure 14.2 for details.

course, the fact that one can obtain pattern formation in a petri dish does not give the field biologist any insight into the more pertinent question: How much of the patchiness and patterning that one sees among natural plant and animal populations is due to *internal* (e.g., diffusive instability as discussed in Chapters 1 and 13) versus *external* causes (e.g., soil heterogeneity)?

In general, then, we suggest that issues of stability, invasions, and pattern formation will not be amenable to the classical "treatment/control" manipulative experiment but will best be examined by combining specific models with large-scale or long-term datasets. It would be a major mistake if ecologists decided spatial models were somehow less scientific just because it was impossible to design a simple clear-cut experimental "test"—many processes in ecology cannot be studied by a strictly experimental approach and require a combination of observation and model fitting.

EVALUATING PREDICTIONS ABOUT PATTERN FORMATION, INVASIONS, AND STABILITY USING SPECIFIC MODELS

The first example of how spatial theory could be "tested" using observational data takes advantage of the fact that most pattern-generating mechanisms in spatial models have a characteristic wavelength or scale that can be calculated if one knows the time rates of local dynamics and the length scales of dispersal (Chapters 4 and 13; Kareiva 1990). By applying a specific model to a patchy distribution of predators and prey, for example, and estimating parameters in the model, one could ask whether the typical "length" of patches matches the length predicted by the theory. Clearly, this enterprise will have the greatest chance of success if alternative causes of patchiness (such as physical heterogeneities) are minimized. Second, it would be ideal if one discovered a system whose scale of patchiness varied along some environmental gradient, and if that variation could be related to predicted changes in the characteristic length of self-generated pattern. Since rates of population growth for the "catalyst" (e.g., the prey in a predator-prey system) influence the scale of pattern formation, any environmental gradient that altered prey reproductive rate should alter the scale of patchiness. Insect predator-prey or host-parasitoid systems are thus likely candidates for this type of inquiry. In particular, the intrinsic rates of increase for most insect species are a predictable function of temperature, and as a result any pattern formation involving such insects might well change predictably in accord with this temperature-driven effect.

A second opportunity for combining models and observational data involves invasions. Indeed, as was already discussed in Chapter 3, several biologists have been able to predict rates of spread using simple invasion models and estimates of mean dispersal rates and intrinsic rates of population growth (i.e., Andow et al. 1990; Viet and Lewis 1996). Of course, there are just as many examples of invasion models failing to aptly predict the velocity with which a species range limit expands, presumably because of overly simplistic descriptions of dispersal and unpredictability of long-range dispersal events. Fortunately, we now have theory (see Chapter 3) and methods for estimating long-distance movement (see Assuncao and Jacobi 1996). These theoretical and methodological advances suggest that we might be able to predict rates of invasive spread for a wide variety of organisms, given a concerted effort to quantify dispersal and demography.

The question of stability in species interactions and spatial models is probably the most challenging aspect of spatial theory. This is because stability reveals itself only after long time series of population censuses. One way around this limitation is to force the issue—perturb a system artificially and thereby expose its stability properties. The complication to this approach is that one does not know immediately what aspect of a system's transient response to perturbation ought to be used as a surrogate for stability. For instance, does the fact that the combination of space and dispersal stabilizes host-parasitoid models mean that in the short run a perturbed host population will return to its mean density more rapidly in systems with dispersal than it would in systems lacking dispersal? We may not know the answer to that question now, but it could be addressed by exploring specific models with the aim of identifying transient behaviors that are good indicators of long-term stability properties.

MANIPULATIVE EXPERIMENTS AIMED AT THE EFFECTS OF SPACE ON COEXISTENCE AND THE MAINTENANCE OF BIODIVERSITY

Whereas stability, invasions, and pattern formation may not lend themselves to elegant experiments, the influence of space on coexistence and diversity is eminently tractable. We list here some obvious experimental approaches that should be given a high priority:

1. All metapopulation models predict a decline in species richness and a decline in the fraction of habitat patches occupied by particular species, as parcels of habitat are removed from the landscape (Chapters 3, 10). Spatially explicit models, such as those developed by Tilman and Lehman (Chapter 10), even indicate what *types* of habitat loss will produce the most striking effects. Following the lead of this theory, ecologists should begin experimentally removing habitat for plant-insect associations; such experiments could be designed to yield results within two to four years.

2. Many models indicate that the spatial dimension is key because it allows coexistence through spatial segregation. One could easily test whether such spatial segregation were indeed a critical factor underlying plant coexistence by simply broadcasting the seeds from successive generations of competing plants in either a random pattern, or a spatially segregated pattern, and asking whether the clustering of seeds halted the march to competitive exclusion.

3. Diseases are expected to run a different course through populations if the populations are well mixed and thus effectively nonspatial, as opposed to comprising individuals with limited movement (Chapters 5–7). By experimentally adjusting the degree of mixing in a host-pathogen system it would be straightforward to assess the importance of space and dispersal

in constraining the pattern of epidemics. For example, one feasible approach might involve moving potted plants around to varying degrees in a plant-pathogen system and then recording the fate of disease innocula.

4. Certain types of host-parasitoid coexistence require a sufficiently large spatial arena for the development of spatial patterning (Chapter 4). If this "size effect" plays a major role in a particular host-parasitoid system, then there should be a threshold "arena size" below which the parasitoid and host cannot coexist. It should be possible to experimentally identify such thresholds for a wide variety of pathogens that attack insects or plants. Huffaker's (1958) report of a prolonged species coexistence associated with increasing the number of oranges on which predatory and prey mites interacted is a crude example of this type of experiment. Unfortunately Huffaker did not analyze or present the data in such a way as to allow one to see whether spatial segregation was lost as the number of oranges was reduced.

5. Theory allows us to identify situations in which nonspatial "mean-field" descriptions of population dynamics are likely to do a poor job of describing population (integrated over space) change (Chapters 9, 12). In a greenhouse or common garden, one should then be able to create experimental populations or assemblages of species that have the properties associated with the inadequacy of mean-field descriptions and then show that nonspatial models invariably fail. Even if we lack a specific spatial model, it should be possible to experimentally determine whether systems possessing the same mean densities, but contrasting spatial distributions of organisms, proceed along markedly different population trajectories. If contrasting spatial dispersions do not produce dramatic effects in terms of aggregate properties such as rates of increase or total population change, then it is hard to argue that space is very important.

These are just a sampling of the sorts of obvious and feasible experiments that need to be launched in ecology. Only after such experiments are conducted will we have clear evidence for the importance of spatial processes in nature, as opposed to in theory.

SUMMARY

One of the purposes of this book is to spark interest among field ecologists and experimental ecologists in exploring spatial theory. In this chapter we have first alerted empiricists to the challenge of testing spatial models. We showed that one could easily produce models that yield clean effects (such as noting that dispersal in a patchy world is stabilizing to host-parasitoid interactions) yet have a hard time uncovering such an effect empirically because of the need for a level of replication that is hard to achieve in a spatial system. Given that cautionary tale, we then argued that one class of spatial issues should be best attacked by applying specific spatial models to specific systems. The issues we think are ideally addressed by linking specific models to observations involve pattern formation, ecological invasions, and stabilization (or destabilization) due to spatial effects. A second group of issues, primarily involving the maintenance of biodiversity and mechanisms of coexistence, is ripe for direct experimental assessment (perhaps even without any models as guides). The experiments we call for in these systems typically represent direct manipulations of habitat availability or interventions that alter dispersal syndromes, both of which are practical for a limited suite of systems, which we identify. Ultimately, we hope that the availability of this book will accelerate experimental programs that pursue such empirical assessments of spatial theory. Without an enormous intensification of empirical research concerning spatial ecology, our now rich theoretical framework will never fulfill its promise.

References

Aho, J. M. 1990. Helminth communities of amphibians and reptiles: comparative approaches to understanding patterns and processes. In G. W. Esch, A. O. Bush, and J. M. Aho, eds., *Helminth Communities of Amphibians and Reptiles: Comparative Approaches to Understanding Patterns and Processes*, 157–195. New York: Chapman & Hall.

Aho, J. M., and A. O. Bush. 1993. Community richness in parasites of some freshwater fishes from North America. In R. E. Ricklefs and D. Schluter, eds., *Species Diversity in Ecological Communities: Historical and Geographical Perspectives*, 185–193. Chicago: University of Chicago Press.

Akcakaya, H. R. 1994. Ramas/GIS. *Linking Landscape Data with Population Viability Analysis*. Setauket, N.Y.: Applied Biomathematics.

Alexander, H. M. 1989. An experimental field study of anther-smut disease of *Silene alba* caused by *Ustilago violacea*: genotypic variation and disease incidence. Evolution 43:835–847.

Alexander, H. M. 1990. Epidemiology of anther-smut infection of *Silene alba* caused by *Ustilago violacea*: patterns of spore deposition and disease incidence. Journal of Ecology 78:166–179.

Alexander, H. M., and J. Antonovics. 1995. Spread of anther-smut disease (*Ustilago violacea*) and character correlations in a genetically variable experimental population of *Silene alba*. Journal of Ecology 83:783–794.

Alexander, H. M., J. Antonovics, and A. Kelly. 1993. Genotypic variation in disease incidence: integration of greenhouse and field studies on *Silene alba* and *Ustilago violacea*. Journal of Ecology 81:325–334.

Alexander, S., and J. Roughgarden. 1996. Larval transport and population dynamics of intertidal barnacles: a coupled benthic/oceanic model. Ecological Monographs 66:259–275.

Allee, W. C. 1938. *The Social Life of Animals*. New York: W. W. Norton.

Anderson, R. M. 1989. Populations and infectious diseases: ecology or epidemiology? Journal of Animal Ecology 60:1–50.

Anderson, R. M., B. T. Grenfell, and R. M. May. 1984. Oscillatory fluctuations in the incidence of infectious disease and the impact of vaccination: time series analysis. Journal of Hygiene of Cambridge 93:587–608.

Anderson, R. M., and R. M. May. 1983. Vaccination against rubella and measles: quantitative investigations of different policies. Journal of Hygiene 90:259–325.

Anderson, R. M., and R. M. May. 1985a. Age-related changes in the rate of disease transmission: implications for the design of vaccination programmes. Journal of Hygiene of Cambridge 94:365–436.

Anderson, R. M., and R. M. May. 1985b. Vaccination and herd immunity to infectious disease. Nature 315:323–329.

REFERENCES

Anderson, R. M., and R. M. May. 1991. *Infectious Diseases of Humans: Dynamics and Control*. Oxford: Oxford University Press.

Andow, D. A., P. M. Kareiva, S. A. Levin, and A. Okubo. 1990. Spread of invading organisms. Landscape Ecology 4:177–188.

Antonovics, J. 1992. Toward community genetics. In R. S. Fritz and E. L. Simms, eds., *Plant Resistance to Herbivores and Pathogens*, 426–449. Chicago: University of Chicago Press.

Antonovics, J. 1994. The interplay of numerical and gene-frequency dynamics in host-pathogen systems. In L. Real, ed., *Ecological Genetics*, 129–145. Princeton: Princeton University Press.

Antonovics, J., Y. Iwasa, and M. P. Hassell. 1995a. A generalized model of parasitoid, venereal, and vector-based transmission processes. American Naturalist 145:661–675.

Antonovics, J., D. Stratton, P. H. Thrall, and A. M. Jarosz. 1995b. An anther smut disease (*Ustilago violacea*) of fire-pink (*Silene virginica*): its biology and relationship to the anther smut disease of white campion (*Silene alba*). American Midland Naturalist 135:130–143.

Antonovics, J., and P. H. Thrall. 1994. The cost of resistance and the maintenance of genetic polymorphism in host-pathogen systems. Proceedings of the Royal Society of London B 257:105–110.

Antonovics, J., P. H. Thrall, A. M. Jarosz, and D. Stratton. 1994. Ecological genetics of metapopulations: the *Silene-Ustilago* plant-pathogen system. In L. Real, ed., *Ecological Genetics*, 146–170. Princeton: Princeton University Press.

Aoki, I. 1982. A simulation on the schooling mechanism in fish. Bulletin of the Japanese Society of Scientific Fisheries 48:1081–1088.

Armstrong, R. A. 1976. Fugitive species: experiments with fungi and some theoretical considerations. Ecology 57:953–963.

Armstrong, R. A. 1989. Competition, seed predation, and species coexistence. Journal of Theoretical Biology 141:191–194.

Aron, J. L., and I. B. Schwartz. 1984. Seasonality and period-doubling bifurcations in an epidemic model. Journal of Theoretical Biology 110:665–679.

Askew, R. R. 1975. The organisation of chalcid-dominated parasitoid communities centred upon endophytic hosts. In P. W. Price, ed., *Evolutionary Strategies of Parasitoids*, 130–153. New York: Plenum Press.

Assuncao, R., and C. Jacobi. 1996. Optimal sampling design for studies of gene flow from a point source using marker genes or marked individuals. Evolution 50:918–923.

Atkinson, W. D., and B. Shorrocks. 1981. Competition on a divided and ephemeral resource: a simulation model. Journal of Animal Ecology 50:461–471.

Babad, H. R., D. J. Nokes, N. J. Gay, E. Miller, P. Morgan-Caper, and R. M. Anderson. 1995. Predicting the impact of measles vaccination in England and Wales: model validation and analysis of policy options. Epidemiology and Infection 114:319–344.

REFERENCES

Bailey, N. J. T. 1975. *The Mathematical Theory of Infectious Diseases and Their Application*. London: C. Griffin.

Bailey, V. A., A. J. Nicholson, and E. J. Williams. 1962. Interaction between hosts and parasites when some host individuals are more difficult to find than others. Journal of Theoretical Biology 3:1–18.

Ball, F. 1983. The threshold behaviour of epidemic models. Journal of Applied Probability 20:227–241.

Barenblatt, G. I. 1994. *Scaling Phenomena in Fluid Mechanics*. Cambridge: Cambridge University Press.

Bartlett, M. S. 1955. *Stochastic Processes*. Cambridge: Cambridge University Press.

Bartlett, M. S. 1957. Measles periodicity and community size. Journal of the Royal Statistical Society A 120:48–70.

Bartlett, M. S. 1960. The critical community size for measles in the U.S. Journal of the Royal Statistical Society A 123:37–44.

Barton, N. H. 1992. On the spread of a new gene combination in the third phase of Wright's shifting-balance. Evolution 46:551–557.

Bascompte, J., and R. V. Solé. 1994. Spatially-induced bifurcations in single-species population dynamics. Journal of Animal Ecology 63:256–264.

Bascompte, J., and R. V. Solé. 1995. Appropriate formulations for dispersal in spatially structured models: reply. Journal of Animal Ecology 64:665–666.

Beddington, J. R., C. A. Free, and J. H. Lawton. 1978. Modelling biological control: on the characteristics of successful natural enemies. Nature 273:513–519.

Beddington, J. R., and P. S. Hammond. 1977. On the dynamics of host-parasite-hyperparasite interactions. Journal of Animal Ecology 46:811–821.

Bengtsson, J. 1989. Interspecific competition increases local extinction rate in a metapopulation system. Nature 340:713–715.

Bengtsson, J. 1991. Interspecific competition in metapopulations. In M. Gilpin and I. Hanski, eds., *Metapopulation Dynamics: Empirical and Theoretical Investigations*, 219–237. San Diego: Academic Press.

Bernstein, C., A. Kacelnik, and J. R. Krebs. 1988. Individual decisions and the distribution of predators in a patchy environment. Journal of Animal Ecology 57:1007–1026.

Bernstein, C., A. Kacelnik, and J. R. Krebs. 1991. Individual decisions and the distribution of predators in a patchy environment. II. The influence of travel costs and the structure of the environment. Journal of Animal Ecology 60:205–226.

Biere, A., and J. Antonovics. 1995. Sex-specific costs of resistance to the fungal pathogen *Ustilago violacea* (*Microbotryum violaceum*) in *Silene alba*. Evolution 50:1098–1110.

Black, F. 1966. Measles endemicity in insular populations: critical community size and its evolutionary implication. Journal of Theoretical Biology 11:207–211.

Boerlijst, M. C., M. E. Lamers, and P. Hogeweg. 1993. Evolutionary consequences of spiral waves in a host-parasitoid system. Proceedings of the Royal Society of London B 253:15–18.

Bolker, B. M., and B. T. Grenfell. 1993. Chaos and biological complexity in measles dynamics. Proceedings of the Royal Society of London B 251:75–81.

Bolker, B. M., and B. T. Grenfell. 1995a. Impact of vaccination on the spatial correlation and persistence of Measles dynamics. Proceedings of the National Academy of Sciences 93(22):12648–12653.

Bolker, B. M., and B. T. Grenfell. 1995b. Space, persistence and dynamics of measles epidemics. Proceedings of the Royal Society of London B 348:308–320.

Bolker, B. M., and S. W. Pacala. 1997. Understanding the ecological implications of spatial pattern formation using ensemble models. Theoretical Population Biology. In press.

Bramson, M., R. Durrett, and G. Swindle. 1989. Statistical mechanics of crabgrass. Annals of Probability 17:444–481.

Briggs, C. J. 1993. Competition among parasitoid species on a stage-structured host and its effect on host suppression. American Naturalist 141:372–397.

Brown, J. H., and A. Kodric-Brown. 1977. Turnover rates in insular biogeography: effect of immigration on extinction. Ecology 58:445–449.

Brown, M. W., and E. A. Cameron. 1979. Effects of dispalure and egg mass size on parasitism by the gypsy moth egg parasite, Ooencyrtus kuwani. Environmental Entomology 8:77–80.

Brownlee, J. 1911. The mathematical theory of random migration and epidemic distribution. Proceedings of the Royal Society Edinburgh 31:262–289.

Bucher, E. H. 1992. The causes of extinction of the passenger pigeon. Current Ornithology 9:1–36.

Burdon, J. J. 1987. *Diseases and Plant Population Biology*. Cambridge: Cambridge University Press.

Burnett, T. 1956. Effects of natural temperatures on oviposition of various numbers of an insect parasite (Hymenoptera, Chalcididae, Tenthredinidae). Annals of the Entomological Society of America 49:55–59.

Byers, R. E., R. I. C. Hansell, and N. Madras. 1992. Stability-like properties of population models. Theoretical Population Biology 42:10–34.

Cain, M. L., S. W. Pacala, J. A. Silander Jr., and M. J. Fortin. 1995. Neighborhood models of clonal growth in the white clover, *Trifolium repens*. American Naturalist 145:888–917.

Caley, M. J., and D. Schluter. 1996. The relationship between local and regional diversity. Ecology 78:70–80.

Cantrell, S., and C. Cosner. 1993. Should a park be an island? SIAM Journal of Applied Mathematics 53:219–252.

Carter, R. N., and S. D. Prince. 1981. Epidemic models used to explain biogeographical distribution limits. Nature 293:644–645.

Casas, J. 1988. Analysis of searching movements of a leafminer parasitoid in a structured environment. Physiological Entomology 13:373–380.

Casas, J. 1989. Foraging behaviour of a leafminer parasitoid in the field. Ecological Entomology 14:257–265.

Case, T. J., D. T. Bolger, and A. D. Richman. 1992. Reptilian extinctions: the last ten thousand years. In P. L. Fielder and S. K. Jain, eds., *Conservation Biology*, 91–125. New York: Chapman & Hall.

Caswell, H. 1976. Community structure: a neutral model analysis. Ecological Monographs 46:327–354.

Caswell, H., and J. E. Cohen. 1991. Disturbance and diversity in metapopulations. Biological Journal of the Linnean Society 42:193–218.

Caswell, H., and J. E. Cohen. 1993. Local and regional regulation of species-area relations: a patch occupancy model. In R. E. Ricklefs and D. Schluter, eds., *Species Diversity in Ecological Communities: Historical and Geographical Perspectives*, 99–107. Chicago: University of Chicago Press.

Chatfield, C. 1989. *The Analysis of Time Series: An Introduction*. London: Chapman & Hall.

Chesson, P. L. 1983. Coexistence of competitors in a stochastic environment: the storage effect. In H. H. Freedman and C. Strobeck, eds., *Population Biology: Lecture Notes in Biomathematics 52*, 188–198. New York: Springer.

Chesson, P. 1986. Environmental variation and the coexistence of species. In J. Diamond and T. J. Case, eds., *Community Ecology*, 240–254. New York: Harper and Row.

Chesson, P. L. 1991. A need for niches? Trends in Ecology and Evolution 6:26–28.

Chesson, P. L., and W. W. Murdoch. 1986. Aggregation of risk: relationships among host-parasitoid models. American Naturalist 127:696–715.

Cliff, A. D., P. Haggett, J. D. Ord, and G. R. Versey. 1981. *Spatial Diffusion*. Cambridge: Cambridge University Press.

Cliff, A., P. Haggett, and M. Smallman-Raynor. 1993. *Measles: An Historical Geography of a Major Human Viral Disease*. Oxford: Blackwell.

Cohen, D., and S. A. Levin. 1991. Dispersal in patchy environments: the effects of temporal and spatial structure. Theoretical Population Biology 39(1):63–99.

Cohen, D. S., and J. D. Murray. 1981. A generalized diffusion model for growth and dispersal in a population. Journal of Mathematical Biology 12:237–249.

Comins, H. N., and M. P. Hassell. 1976. Predation in multi-prey communities. Journal of Theoretical Biology 62:93–114.

Comins, H. N., and M. P. Hassell. 1979. The dynamics of optimally foraging predators and parasitoids. Journal of Animal Ecology 48:335–351.

Comins, H. N., and M. P. Hassell. 1987. The dynamics of predation and competition in patchy environments. Theoretical Population Biology 31:393–421.

Comins, H. N., and M. P. Hassell. 1996. Persistence of multispecies host-parasitoid interactions in spatially distributed models with local dispersal. Journal of Theoretical Biology 183:19–28.

Comins, H. N., M. P. Hassell, and R. M. May. 1992. The spatial dynamics of host-parasitoid systems. Journal of Animal Ecology 61:735–748.

Compton, S. G., J. H. Lawton, and V. K. Rashbrook. 1989. Regional diversity, local community structure, and vacant niches: the herbivorous arthropods of bracken in South Africa. Ecological Entomology 14:365–373.

Connell, J. H. 1978. Diversity in tropical rain forests and coral reefs. Science 199:1302–1310.

Connell, J. H. 1983. On the prevalence and relative importance of interspecific competition: evidence from field experiments. American Naturalist 122:661–696.

Cornell, H. V. 1985a. Local and regional richness of cynipine gall wasps on California oaks. Ecology 66:1247–1260.

Cornell, H. V. 1985b. Species assemblages of cynipid gall wasps are not saturated. American Naturalist 126:565–569.

Cornell, H. V. 1993. Unsaturated patterns in specie assemblages: the role of regional processes in setting local species richness. In R. E. Ricklefs and D. Schluter, eds., *Species Diversity in Ecological Communities: Historical and Geographical Perspectives*, 243–252. Chicago: University of Chicago Press.

Cornell, H. V., and R. H. Karlson. 1996. Species richness of reef-building corals determined by regional and local processes. Journal of Animal Ecology 65:233–241.

Cornell, H. V., and J. H. Lawton. 1992. Species interactions, local and regional processes, and limits to the richness of ecological communities: a theoretical perspective. Journal of animal Ecology 61:1–12.

Crawley, M. J. 1990. The population dynamics of plants. Philosophical Transactions of the Royal Society of London B 330:125–140.

Crawley, M. J., and R. M. May. 1987. Population dynamics and plant community structure: competition between annuals and perennials. Journal of Theoretical Biology 125:475–489.

Cresswell, J. E., V. M. Vidal-Martinez, and N. J. Crichton. 1995. The investigation of saturation in species richness of communities: some comments on methodology. Oikos 72:301–304.

Cronin, J. T., and D. R. Strong. 1990. Density-independent parasitism among host patches by Anagrus delicatus (Hymenoptera, Mymaridae) —experimental manipulation of hosts. Journal of Animal Ecology 59:1019–1026.

REFERENCES

Crowell, K. L. 1986. A comparison of relict versus equilibrium models for insular mammals of the Gulf of Maine. In L. R. Heaney and B. D. Patterson, eds., *Island Biogeography of Mammals*, 37–64. London: Academic Press.

Daily, G. C. 1995. Restoring value to the world's degraded lands. Science 269:350–354.

Damman, H. 1993. Patterns of herbivore interaction among herbivore species. In N. E. Stamp and T. M. Casey, eds., *Caterpillars: Ecological and Evolutionary Constraints on Foraging*, 132–169. New York: Chapman & Hall.

Darlington, P. J. 1957. *Zoogeography: The Geographical Distribution of Animals*. New York: John Wiley.

Darwin, C. 1859. *The Origin of Species*. London: Murray.

Dawah, H. A., B. A. Hawkins, and M. F. Claridge. 1995. Structure of the parasitoid communities of grass-feeding chalcid wasps. Journal of Animal Ecology 64:708–720.

DeAngelis, D. L., and L. J. Gross. 1992. *Individual-Based Models and Approaches in Ecology: Populations, Communities and Ecosystems*. New York: Chapman & Hall.

de Jong, G. 1979. The influence of the distribution of juveniles over patches of food on the dynamics of a population. Netherlands Journal of Zoology 29:33–51.

Denno, R. F., M. S. McClure, and J. R. Ott. 1995. Interspecific interactions in phytophagous insects: competition reexamined and resurrected. Annual Review of Entomology 40:297–331.

Diamond, J. M. 1972. Biogeographic kinetics: estimation of relaxation times for avifaunas of Southwest Pacific Islands. Proceedings of the National Academy of Sciences USA 69:3199–3203.

Diamond, J. M. 1984. "Normal" extinctions of isolated populations. In M. H. Nitecki, ed., *Extinctions*, 191–246. Chicago: Unversity of Chicago Press.

Dietz, K. 1976. The incidence of infectious diseases under the influence of seasonal fluctuations. Lecture Notes in Biomathematics 11:1–15.

Djordjevic, Z. V. 1992. Phase transition in the topological properties of the generalized percolation model. Physica A 187:425–435.

Dobson, A. P., and A. M. Lyles. 1989. The population dynamics and conservation of primate populations. Conservation Biology 3:362–380.

Dobzhansky, T., and S. Wright. 1947. Genetics of natural populations. XV: Rate of diffusion of a mutant gene through a population of *Drosophila pseudoobscura*. Genetics 32:303–324.

Doherty, P., and A. Fowler. 1994a. Demographic consequences of variable recruitment to coral-reef fish populations—a congeneric comparison of two damselfishes. Bulletin of Marine Science 54:297–313.

Doherty, P., and A. Fowler. 1994b. An empirical test of recruitment limitation in a coral-reef fish. Science 263:935–939.

Done, T. J., P. K. Dayton, A. E. Dayton, and R. Steger. 1991. Regional and local variability in recovery of shallow coral communities: Moorea, French Polynesia and central Great Barrier Reef. Coral Reefs 9:183–192.

Downes, B. J. 1995. Spatial and temporal variation in recruitment and its effects on regulation of parasite populations. Oecologia 102:501–510.

Drake, J. A., H. A. Mooney, F. di Castri, R. H. Groves, and F. J. Kruger. 1989. *Biological Invasions: A Global Perspective*. Chichester: John Wiley.

Driessen, G., and L. Hemerik. 1991. Aggregative responses of parasitoids and parasitism in populations of drosophila breeding in fungi. Oikos 61:96–107.

Durrett, R. 1988. Crabgrass, measles and gypsy moths: an introduction to modern probability. Bulletin of the AMS 18:117–143.

Durrett, R. 1991. The contact process, 1974–1989. Lectures in Applied Mathematics 27:1–18.

Durrett, R., and S. A. Levin. 1994a. The importance of being discrete (and spatial). Theoretical Population Biology 46:363–394.

Durrett, R., and S. A. Levin. 1994b. Stochastic spatial models: a user's guide to ecological applications. Philosophical Transactions of the Royal Society of London B, 343:329–350.

Durrett, R., and S. A. Levin. 1996. Spatial models for species area curves. Journal of Theoretical Biology 179:119–127.

Durrett, R., and S. A. Levin. 1997. Allelopathy in spatially distributed populations. Journal of Theoretical Biology 184:in press.

Dytham, C. 1995. Competitive coexistence and empty patches in spatially explicit metapopulation models. Journal of Animal Ecology 64:145–146.

Ebenhard, T. 1991. Colonization in metapopulations: a review of theory and observations. Biological Journal of the Linnean Society 42:105–121.

Eber, S., and R. Brandl. 1994. Ecological and genetic spatial patterns of *Urophora cardui* (Diptera: Tephritidae) as evidence for population structure and biogeographical processes. Journal of Animal Ecology 63:187–199.

Eggleston, D. B., and D. A. Armstrong. 1995. Presettlement and postsettlement determinants of estuarine dungeness crab recruitment. Ecological Monographs 65:193–216.

Ehler, L. E. 1987. Distribution of progeny in two ineffective parasites of a gall midge (Diptera: Cecidomyiidae). Environmental Entomology 15:1268–1271.

Ehrlich, P. R., and A. H. Ehrlich. 1981. *Extinction: The Causes and Consequences of the Disappearance of Species*. New York: Random House.

Elton, C. S. 1958. *The Ecology of Invasions by Animals and Plants*. London: Methuen.

Embree, D. G. 1966. The role of introduced parasites in the control of the winter moth in Nova Scotia. Canadian Entomologist 98:1159–1168.

Endler, J. A. 1977. *Geographic Variation, Speciation, and Clines*. Princeton: Princeton University Press.

Engbert, R. 1993. Chance and chaos in seasonally driven epidemics. Ph.D. thesis, Forschungszentrum Julich, GmbH.

Engbert, R., and F. R. Drepper. 1994. Chance and chaos is population biology—models of recurrent epidemics and food chain dynamics. Chaos, Solitons and Fractals 4:1147–1169.

Epperson, B. K. 1993. Recent advances in correlation studies of spatial patterns of genetic variation. Evolutionary Biology 27:95–155.

Ferguson, N. M., R. M. Anderson, and G. P. Garnett. 1996a. Mass vaccination to control chickenpox: the influence of zoster. Proceedings of the National Academy of Sciences of the USA 93:7231–7235.

Ferguson, N. M., D. J. Nokes, and R. M. Anderson. 1996b. Dynamical complexity in age-structured models of the transmission of the measles virus: epidemiological implications at high levels of vaccine uptake. Mathematical Biosciences 138:101–130.

Fine, P. E. M., and J. A. Clarkson. 1982a. Measles in England and Wales. I. An analysis of factors underlying seasonal patterns. International Journal of Epidemiology 11:5–15.

Fine, P. E. M., and J. A. Clarkson. 1982b. Measles in England and Wales. III. Assessing published predictions of the impact of vaccination on incidence. International Journal of Epidemiology 12:332–339.

Fisher, R. A. 1937. The wave of advance of advantageous genes. Annal of Eugenics 7:355–369.

Fitt, B. D. L., P. H. Gregory, A. D. Todd, H. A. McCarney, and O. C. MacDonald. 1987. Spore dispersal and plant disease gradient: a comparison between two empirical models. Journal of Phytopathology 118:227–242.

Foley, P. 1994. Predicting extinction times from environmental stochasticity and carrying capacity. Conservation Biology 8:124–137.

Foley, P. 1997. Extinction models for local populations. In I. Hanski and M. Gilpin, eds., *Metapopulation Biology: Ecology, Genetics and Evolution*, 215–246. San Diego: Academic Press.

Frank, S. A. 1997. Spatial processes in host-parasite genetics. In I. Hanski and M. E. Gilpin, eds., *Metapopulation Biology: Ecology, Genetics and Evolution*, 325–358. San Diego: Academic Press.

Gaines, S. D., and M. D. Bertness. 1992. Dispersal of juveniles and variable recruitment in sessile marine species. Nature 360:579.

Gandhi, A., S. A. Levin, and S. Orszag. 1998. "Critical slowing down" in time-to-extinction: an example of critical phenomena in ecology. Journal of Theoretical Biology. In press.

Garnett, G. P., and B. T. Grenfell. 1992a. The epidemiology of varicella-zoster virus infections: a mathematical model. Epidemiology and Infection 108:495–511.

Garnett, G. P., and B. T. Grenfell. 1992b. The epidemiology of varicella-zoster virus: the influence of varicella on the prevalence of herpes-zoster. Epidemiology and Infection 108:513–528.

Gaston, K. J., and I. D. Gauld. 1993. How many species of pimplines (Hymenoptera: Ichneumonidae) are there in Costa Rica? Journal of Tropical Ecology 9:491–499.

Gause, G. F. 1935. *The Struggle for Existence*. Baltimore: Williams & Wilkins.

Gierer, A., and H. Meinhardt. 1972. A theory of biological pattern formation. Kybernetik 12:30–39.

Gilpin, M. E. 1975. *Group Selection in Predator-Prey Communities*. Princeton: Princeton University Press.

Gilpin, M., and I. Hanski, eds. 1991. *Metapopulation Dynamics: Empirical and Theoretical Investigations*. London: Academic Press.

Gilpin, M. E., and M. E. Soulé. 1986. Minimum viable populations: processes of species extinction. In M. E. Soulé, ed., *Conservation Biology: the Science of Scarcity and Diversity*, 19–34. Sinauer, Mass.: Sunderland Associates.

Godfray, H. C. J. 1994. *Parasitoids: Behavioral and Evolutionary Ecology*. Princeton: Princeton University Press.

Godfray, H. C. J., and M. P. Hassell. 1989. Discrete and continuous insect populations in tropical environments. Journal of Animal Ecology 58:153–174.

Godfray, H. C. J., M. P. Hassell, and R. D. Holt. 1994. The dynamic consequences of the disruption of synchrony between hosts and parasitoids. Journal of Animal Ecology 63:1–10.

Godfray, H. C. J., and S. Pacala. 1992. Aggregation and the population dynamics of parasitoids and predators. American Naturalist 140:30–40.

Godfray, H. C. J., and J. K. Waage. 1991. Predictive modelling in biological control: the mango mealy bug (Rastrococcus invadens) and its parasitoids. Journal of Applied Ecology 28:434–453.

Goldenfeld, N. 1992. *Lectures on Phase Transitions and the Renormalization Group. Frontiers in Physics*. Reading, Mass.: Addison Wesley.

Goss Custard, J. D., R. W. G. Caldow, R. T. Clarke, and A. D. West. 1995. Deriving population parameters from individual variations in foraging behavior. 2. Model tests and population parameters. Journal of Animal Ecology 64:277–289.

Grenfell, B. T., and R. M. Anderson. 1985. The estimation of age-related rates of infection from case notifications and serological data. Journal of Hygiene of Cambridge 95:419–436.

Grenfell, B. T., and A. P. Dobson. 1995. *Ecology of Infectious Diseases in Natural Populations*. Cambridge: Cambridge University Press.

Grenfell, B. T., A. Kleczkowski, C. A. Gilligan, and B. M. Bolker. 1995. Spatial heterogeneity, nonlinear dynamics and chaos in infectious diseases. Statistical Methods in Medical Research 4:160–183.

Griffiths, D. 1997. Local and regional diversity in North American lacustrine fish species. Journal of Animal Ecology. In press.

Grünbaum, D. 1992. Local processes and global patterns: Biomathematical models of bryozoan feeding currents and density dependent aggregations in Antarctic krill. Ph.D. diss., Ithaca, N.Y.: Cornell University.

Grünbaum, D. 1994. Translating stochastic density-dependent individual behavior with sensory constraints to a Eulerian model of animal swarming. Journal of Mathematical Biology 33:139–161.

Gueron, S., and S. A. Levin. 1993. Self-organization of front patterns in large wildebeest herds. Journal of Theoretical Biology 165:541–552.

Gurevitch, J. 1992. A meta-analysis of competition in field experiments. American Naturalist 140:539–572.

Gurevitch, J., L. L. Morrow, A. Wallace, and J. S. Walsh. 1992. A meta-analysis of field competition experiments. American Naturalist 140:539–572.

Gyllenberg, M., and I. Hanski. 1992. Single-species metapopulation dynamics: a structured model. Theoretical Population Biology 42:35–61.

Gyllenberg, M., I. Hanski, and A. Hastings. 1997. Structured metapopulation models. In I. Hanski and M. Gilpin, eds., *Metapopulation Biology: Ecology, Genetics and Evolution*, 93–122. San Diego: Academic Press.

Gyllenberg, M., and D. S. Silverstrov. 1994. Quasi-stationary distributions of a stochastic metapopulation model. Journal of Mathematical Biology 33:35–70.

Hails, R., and M. J. Crawley. 1992. Spatial density dependence in populations of a cynipid gall-former Andricus quercuscalicis. Journal of Animal Ecology 61:567–583.

Hairston, N. G., F. E. Smith, and L. B. Slobodkin. 1960. Community structure, population control, and competition. American Naturalist 94: 421–425.

Haldane, J. B. S. 1948. The theory of cline. Journal of Genetics 48:277–284.

Hamer, W. H. 1906. Epidemic disease in England—the evidence of variability and of persistency of type. The Lancet 1:733–739.

Hanski, I. 1983. Coexistence of competitors in patchy environment. Ecology 64:493–500.

Hanski, I. 1985. Single-species spatial dynamics may contribute to long-term rarity and commonness. Ecology 66:335–343.

Hanski, I. 1990. Dung and carrion insects. In B. Shorrocks and I. R. Swingland, eds., *Living in a Patchy Environment*, 127–145, Oxford: Oxford University Press.

Hanski, I. 1991. Single-species metapopulation dynamics: Concepts, models and observations. Biological Journal of the Linnean Society 42:17–38.

Hanski, I. 1992. Inferences from ecological incidence functions. American Naturalist 139:657–662.

Hanski, I. 1994a. Patch-occupancy dynamics in fragmented landscapes. Trends in Ecology and Evolution 9:131–135.

Hanski, I. 1994b. A practical model of metapopulation dynamics. Journal of Animal Ecology 63:151–162.

Hanski, I. 1994c. Spatial scale, patchiness, and population dynamics on land. Philosophical Transactions of the Royal Society of London B 343:19–25.

Hanski, I. 1996. Metapopulation ecology. In O. Rhodes, R. Chesser, and M. Smith, eds., *Population Dynamics in Ecological Space and Time*, 13–43. Chicago: University of Chicago Press.

Hanski, I. 1997. Metapopulation dynamics: from concepts and observations to predictive models. In I. Hanski and M. Gilpin, eds., *Metapopulation Biology: Ecology, Genetics and Evolution*, 69–72. San Diego: Academic Press.

Hanski, I., and M. E. Gilpin, eds. 1997. *Metapopulation Biology: Ecology, Genetics and Evolution*, San Diego: Academic Press.

Hanski, I., and M. Gyllenberg. 1993. Two general metapopulation models and the core-satellite species hypothesis. American Naturalist 142:17–41.

Hanski, I., and M. Kuussaari. 1995. Butterfly metapopulation dynamics. In P. Price and N. Capuccino, eds., *Population Dynamics: New Approahces and Synthesis*, 149–171. London: Academic Press.

Hanski, I., M. Kuussaari, and M. Nieminen. 1994. Metapopulation structure and migration in the butterfly *Melitaea cinxia*. Ecology 75:747–762.

Hanski, I., A. Moilanen, and M. Gyllenberg. 1996a. Minimum viable metapopulation size. American Naturalist 147:527–541.

Hanski, I., A. Moilanen, T. Pakkala, and M. Kuussaari. 1996b. Metapopulation persistence of an endangered butterfly: a test of the quantitative incidence function model. Conservation Biology 10:578–590.

Hanski, I., T. Pakkala, M. Kuussaari, and G. Lei. 1995a. Metapopulation persistence of an endangered butterfly in fragmented landscapes. Oikas 72:21–28.

Hanski, I., J. Pöyry, T. Pakkala, and M. Kuussaari. 1995b. Metapopulation equilibria in metapopulation dynamics. Nature 377:618–621.

Hanski, I., and E. Ranta. 1983. Coexistence in a patchy environment: three species of Daphnia in rock pools. Journal of Animal Ecology 52:263–279.

Hanski, I., and C. D. Thomas. 1994. Metapopulation dynamics and conservation: a spatially explicit model applied to butterflies. Biological Conservation 68:167–180.

Hanski, I., and D. Y. Zhang. 1993. Migration, metapopulation dynamics and fugitive co-existence. Journal of Theoretical Biology 163:491–504.

Harada, Y., H. Ezoe, Y. Iwasa, H. Matsuda, and K. Sato. 1995. Population persistence and spatially limited social interaction. Theoretical Population Biology 48:65–91.

Harada, Y., and Y. Iwasa. 1994. Lattice population dynamics for plants with dispersing seeds and vegetative propagation. Researches on Population Ecology 36:237–249.

Hardin, D. P., P. Takac, and G. F. Web. 1988a. Asymptotic properties of a continuous-space discrete-time population model in a random environment. Bulletin of Mathematical Biology 26:361–374.

Hardin, D. P., P. Takac, and G. F. Webb. 1988b. A comparison of dispersal strategies for survival of spatially heterogeneous populations. SIAM Journal of Applied Mathematics 48:1396.

Hardin, D. P., P. Takac, and G. F. Webb. 1990. Dispersion population models discrete in time and space. Journal of Mathematical Biology 28:1–20.

Harper, J. L. 1977. *Population Biology of Plants*. London: Academic Press.

Harrison, S. 1989. Long-distance dispersal and colonization in the Bay checkerspot butterfly, *Euphydryas editha bayensis*. Ecology 80:1236–1243.

Harrison, S. 1991. Local extinction in a metapopulation context: An empirical evaluation. Biological Journal of the Linnean Society 42:73–88.

Harrison, S. 1994. Metapopulations and conservation. In P. J. Edwards, R. M. May, and N. R. Webb, eds., *Large-scale Ecology and Conservation Biology*, 111–128. Oxford: Blackwell Scientific Press.

Harrison, S., D. D. Murphy, and P. R. Ehrlich. 1988. Distribution of the Bay Checkerspot butterfly, *Euphydryas editha bayensis*: evidence for a metapopulation model. American Naturalist 132:360–382.

Harrison, S., and A. D. Taylor. 1997. Empirical evidence for metapopulation dynamics. In I. Hanski and M. Gilpin, eds., *Metapopulation Biology: Ecology, Genetics and Evolution*, 27–42. San Diego: Academic Press.

Harrison, S., C. Thomas, and T. Lewinsohn. 1995. Testing a metapopulation model of coexistence in the insect community on ragwort. American Naturalist 145:546–567.

Harvey, P. H., and M. D. Pagel. 1991. *The Comparative Method in Evolutionary Biology*. Oxford: Oxford University Press.

Hassell, M. P. 1978. *The Dynamics of Arthropod Predator-Prey Systems*. Princeton: Princeton University Press.

Hassell, M. P. 1979. Non-random search in predator-prey models. Fortschritte der Zoologie 25:311–333.

Hassell, M. P. 1980. Some consequences of habitat heterogeneity for population dynamics. Oikos 35:150–160.

Hassell, M. P. 1984. Parasitism in patchy environments: inverse density dependence can be stabilizing. IMA Journal of Mathematics Applied in Medicine and Biology 1:123–133.

Hassell, M. P., H. N. Comins, and R. M. May. 1991. Spatial structure and chaos in insect populations. Nature 353:255–258.

Hassell, M. P., H. N. Comins, and R. M. May. 1994. Species coexistence and self-organizing spatial dynamics. Nature 370:290–292.

Hassell, M. P., H. C. J. Godfray, and H. N. Comins. 1993. Effects of global change on the dynamics of insect host-parasitoid interactions. In P. M.

Kareiva, J. G. Kingsolver, and R. B. Huey, eds., *Biotic Interactions and Global Change*, 402–423. Sunderland, Mass.: Sinauer.

Hassell, M. P., J. H. Lawton, and R. M. May. 1976. Patterns of dynamical behavior in single-species populations. Journal of Animal Ecology 45:471–486.

Hassell, M. P., and R. M. May. 1973. Stability in insect host-parasite models. Journal of Animal Ecology 42:693–726.

Hassell, M. P., and R. M. May. 1974. Aggregation of predators and insect parasites and its effect on stability. Journal of Animal Ecology 43:567–594.

Hassell, M. P., and R. M. May. 1988. Spatial heterogeneity and the dynamics of parasitoid-host systems. Annals Zoologici Fennici 25:55–61.

Hassell, M. P., O. Miramontes, P. Rohani, and R. M. May. 1995. Appropriate formulations for dispersal in spatially structured models: comments on Bascompte and Solé. Journal of Animal Ecology 64:662–664.

Hassell, M. P., S. Pacala, R. M. May, and P. L. Chesson. 1991. The persistence of host-parasitoid associations in patchy environments. I. A general criterion. American Naturalist 138:568–583.

Hastings, A. 1980. Disturbance coexistence, history, and competition for space. Theoretical Population Biology 18:363–373.

Hastings, A. 1990. Spatial heterogeneity and ecological models. Ecology 71:426–428.

Hastings, A., and S. Harrison. 1994. Metapopulation dynamics and genetics. Annual Review of Ecology and Systematics 25:167–188.

Hastings, A., and K. Higgins. 1994. Presistence of transients in spatially structured ecological models. Science 263:1133–1136.

Haury, L. R., J. A. McGowan, and P. H. Wiebe. 1978. Patterns and processes in the timespace scales of plankton distributions. In J. H. Steele, ed., *Spatial Pattern in Plankton Communities*, 277–327. New York: Plenum.

Hawkins, B. A., and S. G. Compton. 1992. African fig wasp communities: vacant niches and latitudinal gradients in species richness. Journal of Animal Ecology 61:361–372.

Hediger, H. 1950. *Wild Animals in Captivity*. London: Butterworth.

Hengeveld, R. 1989. *Dynamics of Biological Invasions*. London: Chapman & Hall.

Heppner, F., and U. Grenander. 1990. A stochastic nonlinear model for coordinated bird flocks. In S. Krusna, ed., *The Ubiquity of Chaos*, 233–238. Washington, D.C.: AAAS Publications.

Hethcote, H. W. 1988. Optimal ages for vaccination for measles. Mathematical Biosciences 80:29–52.

Hill, G. E. 1993. House finch (*Carpodacus mexicanus*). In A. Poole and F. Gill, eds., *The Birds of North America*, 1–24. Philadelphia: Academy of Natural Sciences.

Hochberg, M. E., and B. A. Hawkins. 1992. Refuges as a predictor of parasitoid diversity. Science 255:973–976.

Hochberg, M. E., and B. A. Hawkins. 1993. Predicting parasitoid species richness. American Naturalist 142:671–693.

Hogarth, W. L., and P. Diamond. 1984. Interspecific competition in larvae between entomophagous parasitoids. American Naturalist 124:552–560.

Holling, C. S. 1992. Cross-scale morphology, geometry and dynamics of ecosystems. Ecological Monographs 62:447–502.

Holmes, E. E., M. A. Lewis, J. E. Banks, and R. R. Veit. 1994. Partial differential equations in ecology: spatial interactions and population dynamics. Ecology 75:17–29.

Holt, R. D. 1977. Predation apparent competition and the structure of prey communities. Theoretical Population Biology 12:197–229.

Holt, R. D., and J. H. Lawton. 1994. The ecological consequences of shared natural enemies. Annual Review of Ecology and Systematics 25:495–520.

Hooge, P. N. 1990. Maintenance of pair-bonds in the house finch. Condor 92:1066–1067.

Horn, H. S., and R. H. MacArthur. 1972. Competition among fugitive species in a harlequin environment. Ecology 53:749–752.

Hubbell, S. P. 1979. Tree dispersion, abundance, and diversity in a tropical dry forest. Science 203:1299–1309.

Hubbell, S. P. 1995. Towards a theory of biodiversity and biogeography on continuous landscapes. In G. R. Carmichael, G. E. Folk, and J. L. Schnoor, eds., *Preparing for Global Change: A Midwestern Perspective*, 173–201. Amsterdam: SPB Academic Publishing.

Hubbell, S. P. 1997. *A Theory of Bigeography and Relative Species Abundance*. Princeton: Princeton University Press. In press.

Hubbell, S. P., and R. B. Foster. 1986. Biology, chance, and history and the structure of tropical rain forest tree communities. In J. Diamond and T. Case, eds., *Community Ecology*, 314–329. New York: Harper and Row.

Huffaker, C. B. 1958. Experimental studies on predation: dispersion factors and predator-prey oscillations. Hilgardia 27:343–383.

Hugueny, B., and D. Paugy 1995. Unsaturated fish communities in African rivers. American Naturalist 146:162–169.

Huisman, J., and F. J. Weissing. 1994. Light-limited growth and competition for light in well-mixed aquatic environments: an elementary model. Ecology 75:507–520.

Huisman, J., and F. J. Weissing. 1995. Competition for nutrients and light in a mixed water column: a theoretical analysis. American Naturalist 146:536–564.

Hurlburt, S. H. 1990. Spatial distribution of the montane unicorn. Oikos 58:257–271.

Hurtt, G. C., and S. W. Pacala. 1995. The consequences of recruitment limitation: reconciling chance history, and competitive differences between plants. Journal of Theoretical Biology 176:1–12.

Huston, M. A. 1979. A general hypothesis of species diversity. American Naturalist 113:81–101.

Huston, M. A. 1985. Patterns of species diversity on coral reefs. Annual Review of Ecology and Systematics 16:149–177.

Huston, M. A. 1994. *Biological Diversity: The Coexistence of Species on Changing Landscapes*. Cambridge: Cambridge University Press.

Hutchinson, G. E. 1951. Copepodology for the ornithologist. Ecology 32:571–577.

Hutchinson, G. E. 1961. The paradox of the plankton. American Naturalist 95:137–147.

Huth, A., and C. Wissel. 1992. The simulation of the movement of fish schools. Journal of Theoretical Biology 156:365–385.

Ingerson, T. E., and R. L. Buvel. 1984. Structure and asynchronous cellular automata. Physica 10D:59–68.

Ives, A. R. 1988. Aggregation and the coexistence of competitors. Annales Zoologici Fennici 25:75–88.

Ives, A. R. 1991. Aggregration and coexistence in a carrion fly community. Ecological Monographs 61:75–94.

Ives, A. R. 1992. Density-dependent and density-independent aggregation in model host-parasitoid systems. American Naturalist 140:912–937.

Ives, A. R. 1995. Measuring competition in a spatially heterogeneous environment. American Naturalist 146:911–936.

Ives, A. R., and R. M. May. 1985. Competition within and between species in a patchy environment: relations between microscopic and macroscopic models. Journal of Theoretical Biology 115:65–92.

Iwasa, Y., S. A. Levin, and V. Andreasen. 1987. Aggregation in model ecosystems. I. Prefect aggregations. Ecological Modeling 37:287–302.

Iwasa, Y., S. A. Levin, and V. Andreasen. 1989. Aggregation of model ecosystems. II. Approximate aggregations. IMA Journal of Mathematics Applied in Medicine & Biology 6:1–23.

Iwasa, Y., and J. Roughgarden. 1986. Interspecific competition among metapopulations with space-limited subpopulations. Theoretical Population Biology 30:194–214.

Jarosz, A. M., and J. J. Burdon. 1991. Host-pathogen interactions in natural populations of *Linum marginale and Melampsora lini*. II. Local and regional variation in patterns of resistance and racial structure. Evolution 45: 1618–1627.

Johnson, W. C., and T. Webb. 1989. The role of blue jays (*Cyanocitta cristata L.*) in the postglacial dispersal of fagaceious trees in eastern North America. Journal of Biogeography 16:561–571.

Jones, T. H., H. C. J. Godfray, and M. P. Hassell. 1996. Relative movement patterns of a tephritid fly and its parasitoid wasps. Oecologia 106:317–324.

Jones, T. H., M. P. Hassell, and S. Pacala. 1993. Spatial heterogeneity and the population dynamics of a host-parasitoid system. Journal of Animal Ecology 62:251–262.

Kadanoff, L. P. 1966. Scaling laws for Ising model near T_c. Physics 2:263–272.

Kakehashi, N., Y. Suzuki, and Y. Iwasa. 1984. Niche overlap of parasitoids in host-parasitoid systems: its consequence to single versus multiple introduction controversy in biological control. Journal of Applied Ecology 21:115–131.

Kaltz, O., and B. Schmid. 1995. Plant venereal disease: a model for integrating genetics, ecology and epidemiology. Trends in Ecology and Evolution 10:221–222.

Kaneko, K. 1992. Overview of coupled map lattices. Chaos 2:279–282.

Kareiva, P. 1990. Population dynamics in spatially complex environments: theory and data. Philosophical Transactions of the Royal Society of London B 330:175–190.

Karlson, R. H., and L. E. Hurd. 1993. Disturbance, coral reef communities, and changing ecological paradigms. Coral Reefs 12:117–125.

Kauffman, S. 1993. *The Origins of Order*. New York: Oxford University Press.

Kelly, J. K. 1994. The effect of scale dependent processes on kin selection: mating and density dependent regulation. Theoretical Population Biology 46:32–57.

Kendall, D. G. 1965. Mathematical model of the spread of infection. In *Mathematics and Computer Science in Biology and Medicine*, 213–225. London: H. M. Stationery Office.

Kennedy, C. R., and J. F. Guégan. 1994. Regional vs. local helminth parasite richness in British freshwater fish: saturated or unsaturated parasite communities? Parasitology 109:175–185.

Kennedy, C. R., and J. F. Guégan. 1996. The number of niches in intestinal helminth communities of *Anguilla anguilla*: are there enough spaces for parasites? Parasitology 113:293–302.

Kermack, W. O., and A. G. McKendrick. 1927. A contribution to the mathematical theory of epidemics. Proceedings of the Royal Society of London A 115:700–721.

Kermack, W. O., and A. G. McKendrick. 1932. A contribution to the mathematical theory of epidemics. II. The problem of endemicity. Proceedings of the Royal Society of London A 138:55–83.

Kierstead, H., and L. B. Slobodkin. 1953. The size of water masses containing plankton blooms. Journal of Marine Research 12:141–147.

Kindvall, O., and I. Ahlén. 1992. Geometrical factors and metapopulation dynamics of the bush cricket, *Metrioptera bicolor Philippi* (Orthoptera: Tettigoniidae). Conservation Biology 6:520–529.

Kneidel, K. A. 1985. Patchiness, aggregation, and the coexistence of competitors for ephemeral resources. Ecological Entomology 10:441–448.

Kolmogorov, A., I. Petrovsky, and H. Piscunov. 1937. Étude de l'équation de la diffusion avec croissance de la quantitè de matière et son application à un problème biologique. Moscow University Bulletin Series International Section A 1:1–25.

Kot, M. 1989. Diffusion-driven period doubling bifurcations. Biosystems 22:279–287.

Kot, M. 1992. Discrete-time travelling waves: Ecological examples. Journal of Mathematical Biology 30:413–436.

Kot, M., M. A. Lewis, and P. van der Driessche. 1996. Dispersal data and the spread of invading organisms. Ecology 77:2027–2042.

Kot, M., and W. M. Schaffer. 1986. Discrete-time growth-dispersal models. Mathematical Biosciences 80:109–136.

Kouki, J., and I. Hanski. 1995. Population aggregration facilitates coexistence of many competing carrion fly species. Oikos 72:223–227.

Kreitman, M., B. Shorrocks, and C. Dytham. 1992. Genes and ecology: two alternative perspectives using *Drosophila*. In R. J. Berry, T. J. Crawford, and G. M. Hewitt, eds., *Genes in Ecology*, 281–312. Oxford: Blackwell.

Kuulasmaa, K. 1982. The spatial general epidemic and locally dependent random graphs. Journal of Applied Probability 19:745–758.

Kuussaari, M., M. Nieminen, and I. Hanski. 1996. An experimental study of migration in the Glanville fritillary butterfly *Melitaea cinxia*. Journal of Animal Ecology 65:791–801.

Lande, R. 1993. Risks of population extinction from demographic and environmental stochasticity and random catastrophes. American Naturalist 142:911–927.

Laurance, W. F. 1990. Comparative responses of five arboreal marsupials to tropical forest fragmentation. Journal of Mammalogy 71:641–653.

Laurance, W. F. 1991. Ecological correlates of extinction proneness in Australian tropical rain forest mammals. Conservation Biology 5:79–89.

Lawton, J. H. 1990. Local and regional species-richness of bracken-feeding insects. In J. A. Thompson and R. T. Smith, eds., *Bracken Biology and Management*, 197–202. Sydney: Australian Institute of Agricultural Science.

Lawton, J. 1996. The facility at Silwood Park: the value of "big bottle" experiments. Ecology 77:665–669.

Lawton, J. H., S. Nee, A. J. Letcher, and P. H. Harvey. 1994. Animal distributions: patterns and process. In. P. J. Edwards, R. M. May, and N. R. Webb, eds., *Large-Scale Ecology and Conservation Biology*, 41–58. Oxford: Blackwell Scientific Press.

Lawton, J. H., and G. L. Woodroffe. 1991. Habitat and the distribution of water voles: why are there gaps in a species' range? Journal of Animal Ecology 60:79–91.

Lei, G.-C. and I. Hanski. 1997. Metapopulation structure of *Cotesia melitaearum*, a specialist parasitoid of the butterfly *Melitaea cinxia*. Oikos 78:91–100.

Lessells, C. M. 1985. Parasitoid foraging: should parasitism be density dependent. Journal of Animal Ecology 54: 27–41.

Levin, S. A. 1974. Dispersion and population interactions. American Naturalist 108:207–228.

Levin, S. A. 1976. Population dynamic models in heterogeneous environments. Annual Review of Ecology and Systematics 7:287–311.

Levin, S. A. 1978a. Pettern formation in ecological communities. In J. H. Steele, ed., *Spatial Pattern in Plankton Communities*, 433–465. Proceedings of the NATO Conference on Marine Biology, Erice, Sicily, November 1977. New York: Plenum Press.

Levin, S. A. 1978b. Population models and community structure in heterogeneous environments. In S. A. Levin, ed., *Mathematical Association of America Study in Mathematical Biology. II. Populations and Communities. Studies in Mathematics 16*, 439–475. Washington, D.C.: Mathematical Association of America.

Levin, S. A. 1981. Mechanisms for the generation and maintenance of diversity. In R. W. Hiorns and D. Cooke, eds., *The Mathematical Theory of the Dynamics of Biological Populations*, 173–194. London: Academic Press.

Levin, S. A. 1988. Pattern, scale, and variability: an ecological perspective. In A. Hastings, ed., *Community Ecology. Lecture Notes in Biomathematics* 7, 1–12. Heidelberg: Springer-Verlag.

Levin, S. A. 1992. The problem of pattern and scale in ecology. Ecology 73:1943–1967. Reprinted 1995, in J. H. Steele and T. Powell, eds., *Ecological Time Series*, 277–326. New York: Chapman & Hall.

Levin, S. A., and L. Buttel. 1987. Measures of patchiness in ecological systems. Ecosystems Research Center Report No. ERC-130. Ithaca, N.Y.: Cornell University.

Levin, S. A., and R. Durrett. 1996. From individuals to epidemics. Philosophical Transactions of the Royal Society of London B 351:1615–1621.

Levin, D. A., and H. W. Kerster. 1975. The effect of gene dispersal on the dynamics and statics of gene substitution in plants. Heredity 35:317–336.

Levin, S. A., and R. T. Paine. 1974. Disturbance, patch formation, and community structure. Proceedings of the National Academy of Sciences USA 71:2744–2747.

Levin, S. A., and L. A. Segel. 1985. Pattern generation in space and aspect. SIAM Review 27:45–67.

Levins, R. 1968. *Evolution in Changing Environments*. Princeton: Princeton University Press.

Levins, R. 1969. Some demographic and genetic consequences of environmental heterogeneity for biological control. Bulletin of the Entomological Society of America 15:237–240.

Levins, R. 1970. Extinction. Lecture Notes in Mathematics 2:75–107.

Levins, R., and D. Culver. 1971. Regional coexistence of species and competition between rare species. Proceedings of the National Academy of Sciences USA 68:1246–1248.

Lewin, R. 1986. Supply-side ecology. Science 234:25–27.

Lipschitz, M., and M. A. Nowak. 1994. The evolution of virulence in sexually transmitted HIV/AIDS. Journal of Theoretical Biology 174:427–440.

Lloyd, M. 1967. Mean crowding. Journal of Animal Ecology 30:1–30.

Lockhart, A. B., P. H. Thrall, and J. Antonovics. 1996. The distribution and characteristics of sexually-transmitted diseases in animals: ecological and evolutionary implications. Biological Reviews of the Cambridge Philosophical Society 71:415–471.

Loehle, C., and B.-L. Li. 1996. Habitat destruction and the extinction debt revisited. Ecological Applications 6:784–789.

Lomolino, M. V. 1993. Winter filtering, immigrant selection and species composition of insular mammals of Lake Huron. Ecography 16:25–30.

Lotka, A. J. 1925. *Elements of Physical Biology*. Baltimore: Williams and Wilkins.

Lovejoy, T. E., J. M. Rankin, R. O. Bierregaard Jr., K. S. Brown Jr., L. H. Emmons, and M. E. van der Voort. 1984. Ecosystem decay of Amazon forest remnants. In M. H. Nitecki, ed., *Extinctions*, 295–325. Chicago: University of Chicago Press.

Lui, R. 1982a. A nonlinear integral operator arising from a model in population genetics. I. Monotone initial data. SIAM Journal of Mathematical Analysis and Applications 13:913–937.

Lui, R. 1982b. A nonlinear integral operator arising from a model in population genetics. II. Initial data with compact support. SIAM Journal of Mathematical Analysis and Applications 13:938–953.

Lui, R. 1983. Existence and stability of travelling wave solutions of a nonlinear integral operator. Journal of Mathematical Biology 16:199–220.

Ma, S. K. 1976. *Modern Theory of Critical Phenomena*, New York: Benjamin.

MacArthur, R. H. 1960. On the relative abundance of species. American Naturalist 94:25–36.

MacArthur, R. H. 1965. Patterns of species diversity. Biological Reviews 40:510–533.

MacArthur, R. H. 1968. The theory of the niche. In R. C. Lewontin, ed., *Population Biology and Evolution*, Proceedings of the International Symposium, June 7–9, 1967, Syracuse, N.Y., 159–176. Syracuse, N.Y.: Syracuse University Press.

MacArthur, R. H., and R. Levins. 1964. Competition, habitat selection, and character displacement in a patchy environment. Proceedings of the National Academy of Sciences USA 51:1207–1210.

MacArthur, R. H., and E. O. Wilson. 1967. *The Theory of Island Biogeography*. Princeton: Princeton University Press.

MacDonald, G. 1957. *The Epidemiology and Control of Malaria*. Oxford: Oxford University Press.

Mahdi, A., R. Law, and A. J. Willis. 1989. Large niche overlaps among coexisting plant species in a limestone grassland community. Journal of Ecology 76:386–400.

Mandelbrot, B. B. 1977. *Fractals: Form, Chance, and Dimension*. San Francisco: Freeman.

Massot, M., J. Clobert, J. Lecomte, and R. Barbault. 1994. Incubent advantage in common lizards and their colonizing ability. Journal of Animal Ecology 63:431–440.

Matsuda, H., N. Ogita, A. Sasaki, and K. Sato. 1992. Statistical mechanics of population: the lattice Lotka-Volterra model. Progress in Theoretical Physics 88:1035–1049.

May, R. M. 1973a. On the relationship between various types of population models. American Naturalist 107:46–57.

May, R. M. 1973b. *Stability and Complexity in Model Ecosystems*. Princeton: Princeton University Press.

May, R. M. 1975. Patterns of species abundance and diversity. In M. L. Cody and J. M. Diamond, eds., *Ecology and Evolution of Communities*, 81–120. Cambridge, Mass.: Harvard University Press.

May, R. M. 1978. Host-parasitoid systems in patchy environments: a phenomenological model. Journal of Animal Ecology 47:833–843.

May, R. M. 1981. Models for two interacting populations. In R. May, ed., *Theoretical Ecology: Principles and Applications*, 78–104. 2nd ed. Sunderland, Mass.: Sinauer Associates.

May, R. M. 1991. The role of ecological theory in planning reintroduction of endangered species. Symposium of the Zoological Society of London 62:145–163.

May, R. M., and R. M. Anderson. 1983. Epidemiology and genetics in the coevolution of parasites and hosts. Proceedings of the Royal Society of London B 219:281–313.

May, R. M., J. R. Beddington, J. W. Horwood, and J. G. Sheperd. 1978. Exploiting natural populations in an uncertain world. Mathematical Biosciences 42:219–252.

May, R. M., and M. P. Hassell. 1981. The dynamics of multiparasitoid-host interactions. American Naturalist 117:234–261.

May, R. M., J. H. Lawton, and N. E. Stork. 1995. Assessing extinction rates. In J. H. Lawton and R. M. May, eds., *Extinction Rates*, 1–24. Oxford: Oxford University Press.

Maynard Smith, J. 1974. *Models in Ecology*. Cambridge: Cambridge University Press.

McCauley, D. E. 1993. Evolution in metapopulations with frequent local extinction and recolonization. Oxford Surveys in Evolutionary Biology 9:109–134.

McCauley, D. E., J. Raveill, and J. Antonovics. 1995. Local founding events as determinants of genetic structure in a plant metapopulation. Heredity 75:630–636.

McClure, M. S. 1977. Parasitism of the scale insect, Fiorinia externa (Homoptera: Diaspididae) by *Aspidiotiphagus citrinus* (Hymenoptera: Eulophidae), in a hemlock forest: density dependence. Environmental Entomology 6:551–555.

Meagher, T. R., J. Antonovics, and R. Primack. 1978. Experimental ecological genetics in *Plantago*. III. Genetic variation and demography in relation to survival of *Plantago cordata*, a rare species. Biological Conservation 14:243–257.

Memmott, J., H. C. J. Godfray, and I. D. Gauld. 1994. The structure of a tropical host-parasitoid community. Journal of Animal Ecology 63:521–540.

Milby, M. M., and M. E. Wright. 1976. Survival of house sparrows and house finches in Kern County, California. Bird Banding 47:119–122.

Minogue, K. P. 1989. Diffusion and spatial probability models for disease spread. In M. J. Jeger, ed., *Spatial Components of Plant Epidemic Diseases*, 127–143. Englewood Cliffs, N.J.: Prentice-Hall.

Moilanen, A., and I. Hanski. 1995. Habitat destruction and coexistence of competitors in a spatially realistic metapopulation model. Journal of Animal Ecology 64:141–144.

Moilanen, A., and I. Hanski. 1997. A parameterized metapopulation model: the incidence function complemented with environmental factors. Manuscript.

Moilanen, A., A. Smith, and I. Hanski. 1997. Long-term dynamics in a metapopulation of the American pika (*Ochotona princeps*). Manuscript.

Mollison, D. 1977. Spatial contact models of ecological and epidemic spread. Journal of the Royal Statistical Society B 29:283–326.

Mollison, D. 1984. Simplifying simple epidemic models. Nature 310:224–225.

Molofsky, J., R. Durrett, S. Levin, J. Dushoff, and D. Griffeath. 1997. Frequency-dependence in a spatial context. Manuscript.

Moloney, K. A., S. A. Levin, N. R. Chiariello, and L. Buttel. 1992. Pattern and scale in a serpentine grassland. Theoretical Population Biology 39:63–99.

Monge, J. P., P. DuPont, A. Idi, and J. Huignard. 1995. The consequences of interspecific competition between *Dinarmus basalis* (Rond) (Hymenoptera: Pteromalidae) and *Eupelmus vuilleti* (CRW) (Hymenoptera: Eupelmidae) on the development of their host populations. Acta Oecologia 16:19–30.

Mooney, H. A., and J. A. Drake. 1986. *Ecology of Biological Invasions of North America and Hawaii*. Berlin: Springer-Verlag.

Murdoch, W. W., C. J. Briggs, R. M. Nisbet, W. S. C. Gurney, and A. Stewart-Oaten. 1992. Aggregation and stability in metapopulation models. American Naturalist 140:41–58.

Murdoch, W. W., and A. Oaten. 1975. Predation and population stability. Advances in Ecological Research 9:1–131.

Murdoch, W. W., J. D. Reeve, C. B. Huffaker, and C. E. Kennett. 1984. Biological control of olive scale and its relevance to ecological theory. American Naturalist 123:371–392.

Murdoch, W. W., and A. Stewart-Oaten. 1989. Aggregation by parasitoids and predators: effects on equilibrium and stability. American Naturalist 134:288–310.

Murdoch, W., S. Swarbuck, R. Luck, S. Walde, and D. Yu. 1996. Refuge dynamics and metapopulation dynamics: an experimental test. American Naturalist 147:424-444.

Murray, J. D. 1989. *Mathematical Biology*. Heidelberg: Springer-Verlag.

Murray, J. D. 1990. *Mathematical Biology. Biomathematics 19*. Heidelberg: Springer-Verlag.

Nee, S. 1994. How populations persist. Nature 367:123–124.

Nee, S., and R. M. May. 1992. Dynamics of metapopulations: habitat destruction and competitive coexistence. Journal of Animal Ecology 61:37–40.

Nee, S., R. M. May, and M. P. Hassell. 1997. Two-species metapopulation models. In I. A. Hanski and M. E. Gilpin, eds., *Metapopulation Biology: Ecology, Genetics, and Evolution*, 123–148. San Diego: Academic Press.

Neubert, M., M. Kot, and M. A. Lewis. 1995. Dispersal and pattern formation in a discrete-time predator-prey model. Theoretical Population Biology 48:7–43.

Nicholson, A. J. 1933. The balance of animal populations. Journal of Animal Ecology 2:131–178.

Nicholson, A. J., and V. A. Bailey. 1935. The balance of animal populations. I. Proceedings of the Zoological Society of London 3:551–598.

Nowak, M. A., S. Bonhoeffer, and R. M. May. 1994. More spatial games. International Journal of Bifurcation and Chaos 4:33–56.

Oates, M. R., and M. S. Warren. 1990. *A Review of Butterfly Introductions in Britain and Ireland*. Godalming: JCCBI/WWF.

Okubo, A. 1980. *Diffusion and Ecological Problems: Mathematical Models. Lecture Notes in Biomathematics 10*. Heidelberg: Springer-Verlag.

Okubo, A. 1986. Dynamical aspects of animal grouping; swarms, schools, flocks and herds. Advances in Biophysics 22:1–94.

Okubo, A., and S. A. Levin. 1989. A theoretical framework for the analysis of data on the wind dispersal of seeds and pollen. Ecology 70:329–338.

Olivieri, I., Y. Michalakis, and P.-H. Gouyon. 1995. Metapopulation genetics and the evolution of dispersal. American Naturalist 146:202–228.

Olsen, L. F., and W. M. Schaffer. 1990. Chaos versus noisy periodicity: alternative hypotheses for childhood epidemics. Science 249:499–504.

Olsen, L. F., G. L. Truty, and W. M. Schaffer. 1988. Oscillations and chaos in epidemics: a nonlinear dynamic study of six childhood diseases in Copenhagen, Denmark. Journal of Theoretical Population Biology 33:344–370.

O'Neill, R. V., R. H. Gardner, and M. G. Turner. 1992. A hierarchical neutral model for landscape analysis. Landscape Ecology 7:55–61.

Opdam, P., and A. Schotman. 1987. Small woods in rural landscape as habitat islands for woodland birds. Acta Oecologia 8:269–274.

Pacala, S. W. 1986a. Neighborhood models of plant population dynamics. 2. Multispecies models of annuals. Theoretical Population Biology 29:262–292.

Pacala, S. W. 1986b. Neighborhood models of plant population dynamics. 4. Single-species and multispecies models of annuals with dormant seeds. American Naturalist 128:859–878.

Pacala, S. W. 1997. Dynamics of plant communities. In M. C. Crawley, ed., *Plant Ecology*, 532–555. 2d ed. Oxford: Blackwell Scientific.

Pacala, S. W., C. D. Canham, J. Saponara, J. A. Silander, R. K. Kobe, and E. Ribbens. 1996. Forest models defined by field measurements. II. Estimation, error analysis and dynamics. Ecological Monographs 66:1–43.

Pacala, S. W., C. D. Canham, and J. A. Silander. 1993. Forest models defined by field measurements. I. The design of a northeastern forest simulator. Canadian Journal of Forestry 23:1980–1988.

Pacala, S. W., and D. J. Deutschman. 1996. Details that matter: the spatial distribution of individual trees maintains forest ecosystem function. Oikos 74:357–365.

Pacala, S., and M. P. Hassell. 1991. The persistence of host-parasitoid associations in patchy environments. II. Evaluation of field data. American Naturalist 138:584–605.

Pacala, S., M. P. Hassell, and R. M. May. 1990. Host-parasitoid associations in patchy environments. Nature 344:150–153.

Pacala, S. W., and J. A. Silander Jr. 1985. Neighborhood models of plant population dynamics. I. Single-species models of annuals. American Naturalist 125:385–411.

Pacala, S. W., and J. A. Silander Jr. 1990. Field tests of neighborhood population dynamic models of two annual weed species. Ecological Monographs 60:113–134.

Pacala, S. W., and D. Tilman. 1994. Limiting similarity in mechanistic and spatial models of plant competition in heterogeneous environments. American Naturalist 143:222–257.

Paine, R. T. 1980. Food webs: linkage, interaction strength and community infrastructure. The Third Tansley Lecture. Journal of Animal Ecology 49:667–685.

Paine, R. T., and S. A. Levin. 1981. Intertidal landscapes: disturbance and the dynamics of pattern. Ecological Monographs 51:145–178.

Patil, N. G., P. S. Baker, W. Groot, and J. K. Waage. 1994. Competition between *Psyllaephagus yaseeni* and *Tamarixia leucaenae*, two parasitoids of the leucaena psyllid (*Heteropsylla cubana*). International Journal of Pest Management 40:211–215.

Peters, R. H. 1983. *The Ecological Implications of Body Size*. Cambridge: Cambridge University Press.

Pickett, S., and M. L. Cadenasso. 1995. Landscape ecology: spatial heterogeneity in ecological systems. Science 269:331–333.

Pickett, S. T. A., and P. S. White, eds. 1985. *The Ecology of Natural Disturbance and Patch Dynamics*. Orlando, Fla.: Academic Press.

Pimm, S. L., G. J. Russell, J. L. Gittleman, and T. M. Brooks. 1995. The future of biodiversity. Science 269:347–350.

Platt, W., and I. Weis. 1977. Resource partitioning and competition within a guild of fugitive prairie plants. American Naturalist 111:479–513.

Porter, J. W. 1972. Predation by *Acanthaster* and its effect on coral species diversity. Bulletin of the Biological Society of Washington 2:89–116.

Preston, F. 1948. The commonness, and rarity, of species. Ecology 29:254–283.

Pulliam, H. R. 1988. Sources, sinks, and population regulation. American Naturalist 132:652–661.

Rand, D. A., M. Keeling, and H. B. Wilson. 1995. Invasion, stability and evolution to criticality in spatially extended, artificial host-pathogen ecologies. Proceedings of the Royal Society of London B 259:55–63.

Rand, D. A., and H. B. Wilson. 1991. Chaotic stochasticity: a ubiquitous source of unpredictability in epidemics. Proceedings of the Royal Society of London B 246:179–184.

Rand, D. A., and H. B. Wilson. 1995. Using spatio-temporal chaos and intermediate-scale determinism to quantify spatially extended ecosystems. Proceedings of the Royal Society of London B 259:111–117.

Rees, M. 1995. Community structure in sand dune annuals: is seed weight a key quantity? Journal of Ecology 83:857–863.

Rees, M., P. J. Grubb, and D. Kelly. 1996. Quantifying the impact of competition and spatial heterogeneity on the structure and dynamics of a four-species guild of winter annuals. American Naturalist 147:1–32.

Reeve, J. D. 1988. Environmental variability, migration, and persistence in host-parasitoid systems. American Naturalist 132:810–836.

Reeve, J. D., J. T. Cronin, and D. R. Strong. 1994. Parasitoid aggregation and the stabilization of a salt-marsh host-parasitoid system. Ecology 75:288–295.

Renshaw, E. 1991. *Modelling Biological Populations in Space and Time*. Cambridge: Cambridge University Press.

Rhodes, C. J., and R. M. Anderson. 1996. Power laws governing epidemics in isolated populations. Nature 381:600–602.

Richardson, D. M., R. M. Cowling, B. B. Lamont, and H. J. van Hensbergen. 1995. Coexistence of *Banksia* species in southwestern Australia: the role of regional and local processes. Journal of Vegetation Science 6:329–342.

Ricklefs, R. E. 1987. Community diversity: relative roles of local and regional processes. Science 235:167–171.

Ricklefs, R. E., and R. E. Latham. 1993. Global patterns of diversity in mangrove floras. In R. E. Ricklefs and D. Schluter, eds., *Species Diversity in Ecological Communities: Historical and Geographical Perspectives*, 215–229. Chicago: University of Chicago Press.

Robinson, G. R., J. F. Quinn, and M. L. Stanton. 1995. Invasibility of experimental habitat islands in a California winter annual grassland. Ecology 76:786–794.

Roff, D. A. 1994. Habitat persistence and the evolution of wing dimorphism in insects. American Naturalist 144:772–798.

Rohani, P., H. C. J. Godfray, and M. P. Hassell. 1994. Aggregation and the dynamics of host-parasitoid systems—a discrete-generation model with within-generation redistribution. American Naturalist 144:491–509.

Rohani, P., R. M. May, and M. P. Hassell. 1996. Metapopulations and equilibrium stability: the effects of spatial structure. Journal of Theoretical Biology 181:97–110.

Roughgarden, J. 1995. *Anolis Lizards of the Caribbean: Ecology, Evolution, and Plate Tectonics*. Oxford: Oxford University Press.

Roughgarden, J., S. Gaines, and H. Possingham. 1988. Recruitment dynamics in complex life cycles. Science 241:1460–1466.

Roughgarden, J., and Y. Iwasa. 1986. Dynamics of a metapopulation with space-limited sub-populations. Theoretical Population Biology 29:235–261.

Roughgarden, J., T. Pennington, and S. Alexander. 1994. Dynamics of the rocky intertidal zone with remarks on generalization in ecology. Philosophical Transactions of the Royal Society London B 343:79–85.

Roughgarden, J., and F. Smith. 1996. Why fisheries collapse and what to do about it. Proceedings of the National Academy of Sciences USA 93:5078–5083.

Roy, B. A. 1994. The use and abuse of pollinators by fungi. Trends in Ecology and Evolution 9:335–339.

Saarinen, P. 1993. Kalliosinisiiven (*Scolitantides orion*) ekologia ja esiintyminen Lohjalla. M.Sc. thesis, University of Helsinki.

Sakai, S. 1973. A model for group structure and its behavior. Biophysics (Japan) 13:82–93.

Sato, K., H. Matsuda, and A. Sasaki. 1994. Pathogen invasion and host extinction in a lattice structured population. Journal of Mathematical Biology 32:251–268.

Schaffer, W. M., and M. Kot. 1985. Nearly one dimensional dynamics in an epidemic. Journal of Theoretical Biology 112:403–427.

Schaffer, W. M., L. F. Olsen, G. L. Truty, and S. L. Fulmer. 1990. The case for chaos in childhood epidemics. In S. Krasner, ed., *The Ubiquity of Chaos*, 138–166. Washington, D.C.: American Association for the Advancement of Science.

Schenzle, D. 1984. An age-structured model of pre- and post-vaccination measles transmission. IMA Journal of Mathematics Applied in Medicine and Biology 1:169–191.

Schluter, D. 1986. Tests for similarity and convergence in finch communities. Ecology 67:1073–1085.

Schluter, D., and R. E. Ricklefs. 1993. Convergence and the regional component of species diversity. In R. E. Ricklefs and D. Schluter, eds., *Species Diversity in Ecological Communities: Historical and Geographical Perspectives*, 230–240. Chicago: University of Chicago Press.

Schoener, T. E., and D. A. Spiller. 1987. High population persistence in a system with high turnover. Nature 330:474–477.

Schoener, T. W. 1983. Field experiments on interspecific competition. American Naturalist 122:240–285.

Schoener, T. W., and A. Schoener. 1983. The time to extinction of a colonizing propagule of lizards increases with island area. Nature 302:332–334.

Scott, J. M., B. Csuti, and S. Caicco. 1991. Gap analysis: assessing protection needs. In W. E. Hudson, ed., *Landscape Linkages and Biodiversity*, 15–26. Washington, D.C.: Island Press.

Segel, L. A., and J. Jackson. 1972. Dissipative structure: An explanation and an ecological example. Journal of Theoretical Biology 37:545–559.

Serfling, R. E. 1952. Historical review of epidemic theory. Human Biology 24:145–166.

Shaw, D. 1990. An initial exploration of the course of measles epidemics in England and Wales from 1960 to 1963. Undergraduate thesis, Sheffield University.

Shmida, A., and S. Ellner. 1984. Coexistence of plant species with similar niches. Vegetatio 58:29–55.

Shorrocks, B. 1990. Coexistence in a patchy environment. In B. Shorrocks and I. R. Swingland, eds., *Living in a Patchy Environment*, 91–106. Oxford: Oxford University Press.

Shorrocks, B., and J. G. Sevenster. 1995. Explaining local species diversity. Proceedings of the Royal Society of London B-Biological 260:305–309.

Simberloff, D. S. 1984. Mass extinction and the destruction of moist tropical forests. Zhurnal Obshchei Biologii 45:767–778.

Simon, H. A., and A. Ando. 1961. Aggregation of variables in dynamic systems. Econometrica 29:111–138.

Sjögren, P. 1991. Extinction and isolation gradients in metapopulations: the case of the pool frog (*Rana lessonae*). Biological Journal of the Linnean Society 42:135–147.

Sjögren-Gulve, P. 1994. Distribution and extinction patterns within a northern metapopulation of the pool frog, Rana lessonae. Ecology 75:1357–1367.

Sjögren-Gulve, P., and C. Ray. 1996. Using logistic regression to model metapopulation dynamics: large-scale forestry extirpates the pool frog. In D. R. McCullough, ed., *Metapopulations and Wildlife Conservation*, 111–137. Washington, D.C.: Island Press.

Skellam, J. G. 1951. Random dispersal in theoretical populations. Biometrika 38:196–218.

Skorupa, J. P. 1986. Responses of rainforest primates to selective logging in Kibale Forest, Uganda: a summary report. In K. Benirschke, ed., *Primates. The Road to Self-Sustaining Populations*, 57–70. New York: Springer-Verlag.

Slatkin, M. 1973. Gene flow and selection in a cline. Genetics 75:733–756.

Slobodkin, L. B., F. E. Smith, and N. G. Hairston. 1967. Regulation in terrestrial systems, and the implied balance of nature. American Naturalist 101:109–124.

Smale, S. 1976. On the differential equations of species in competition. Journal of Mathematical Biology 3:5–7.

Smith, A. T. 1980. Temporal changes in insular populations of the pika *Ochotona princeps*. Ecology 61:8–13.

Sokal, R. S., and J. F. Rohlf. 1981. *Biometry*. 2d ed. New York: Freeman.

Solé, R. V., J. Bascompte, and J. Valls. 1992. Nonequilibrium dynamics in lattice ecosystems: chaotic stability and dissipative structures. Chaos 2:387–395.

Solé, R. V., S. C. Manrbia, B. Luque, J. Delgado, and J. Bascompte. 1995–96. Phase transitions and complex systems. Complexity 1(4):13–26.

Soper, M. A. 1929. The interpretation of periodicity in disease prevalence. Journal of the Royal Statistical Society A 92:34–61.

Stiling, P. D. 1987. The frequency of density dependence in insect host-parasitoid systems. Ecology 68:844–856.

Stone, L. 1995. Biodiversity and habitat destruction: a comparative study of model forest and coral reef ecosystems. Proceedings of the Royal Society of London B 261:381–388.

Stouffer, P. C., and R. O. Bierregaard Jr. 1995. Use of Amazonian forest fragments by understory insectivorous birds. Ecology 76:2429–2445.

Strong, D. R., J. H. Lawton, and T. R. E. Southwood. 1984. *Insects on Plants: Community Patterns and Mechanisms*. Oxford: Blackwell Scientific.

Stuart, C. T., and M. A. Rex. 1994. The relationship between developmental pattern and species diversity in deep-sea prosobranch snails. In C. M. Young and K. J. Eckelbarger, eds., *Reproduction, Larval Biology and Recruitment in the Deep Sea Benthos*, 118–136. New York: Columbia University Press.

Swinton, J., and R. M. Anderson. 1995. Model frameworks for plant-pathogen interactions. In B. T. Grenfell and A. P. Dobson, eds., *Ecology of Infectious Diseases in Natural Populations*, 280–294. Cambridge: Cambridge University Press.

Taylor, A. D. 1988. Parasitoid competition and the dynamics of host-parasitoid models. American Naturalist 132:417–436.

Taylor. A. D. 1990. Metapopulations, dispersal, and predator-prey dynamics: an overview. Ecology 71:429–433.

Taylor, A. D. 1991. Studying metapopulation effects in predator-prey systems. Biological Journal of the Linnean Society 42:305–323.

Taylor, A. D. 1993. Heterogeneity in host parasitoid interactions—aggregation of risk and the CV^2 greater than 1 rule. Trends in Ecology & Evolution 8:400–405.

Taylor, G. I. 1950. The formation of a blast wave by a very intense explosion. I. Theoretical discussion. Proceedings of the Royal Society Series A 201:159–174.

Taylor, R. A. J. 1978. The relationship between density and distance of dispersing insects. Ecological Entomology 3:63–70.

Terborgh, J. 1974. Preservation of natural diversity: the problem of extinction prone species. BioScience 24:715–722.

Thomas, C. D. 1992. The establishment of rare insects in vacant habitats. Antenna 16:89–93.

Thomas, C. D. 1994a. Difficulties in deducing dynamics from static distributions. Trends in Ecology and Evolution 9:300.

Thomas, C. D. 1994b. Local extinctions, colonizations and distributions: habitat tracking by British butterflies. In S. R. Leather, A. D. Watt, N. J. Mills, and K. F. A. Walters, eds., *Individuals, Populations and Patterns in Ecology*, 319–336. Andover, Mass.: Intercept.

Thomas, C. D., and I. Hanski. 1997. Butterfly metapopulations. In I. Hanski and M. E. Gilpin, eds., *Metapopulation Biology: Ecology, Genetics and Evolution*, 359–386. San Diego: Academic Press.

Thomas, C. D., and S. Harrison. 1992. Spatial dynamics of a patchily-distributed butterfly species. Journal of Animal Ecology 61:437–446.

Thomas, C. D., and T. M. Jones. 1993. Partial recovery of a skipper butterfly (*Hesperia comma*) from population refuges: lessons for conservation in a fragmented landscape. Journal of Animal Ecology 62:472–481.

Thomas, C. D., J. A. Thomas, and M. S. Warren. 1992. Distributions of occupied and vacant butterfly habitats in fragmented landscapes. Oecologia 92:563–567.

Thompson, W. R. 1924. La théorie mathématique de l'action des parasites entomophages et le facteur du hasard. Annales de la Faculté des Sciences de Marseille 2:69–89.

Thrall, P. H., and J. Antonovics. 1995. Theoretical and empirical studies of metapopulations: population and genetic dynamics of the *Silene-Ustilago* system. Canadian Journal of Botany 73 (Suppl.):1249–1258.

Thrall, P. H., J. Antonovics, and D. W. Hall. 1993. Host and pathogen coexistence in vector-borne and venereal diseases characterized by frequency-dependent disease transmission. American Naturalist 142:543–552.

Thrall, P. H., A. Biere, and M. K. Uyenoyama. 1995. Frequency-dependent disease transmission and the dynamics of the *Silene-Ustilago* host-pathogen system. American Naturalist 145:43–62.

Thrall, P. H., and A. M. Jarosz. 1994a. Host-pathogen dynamics in experimental populations of *Silene alba* and *Ustilago violacea*. I. Ecological and genetic determinants of disease spread. Journal of Ecology 82:549–559.

Thrall, P. H., and A. M. Jarosz. 1994b. Host-pathogen dynamics in experimental populations of *Silene alba* and *Ustilago violacea*. II. Experimental tests of theoretical models. Journal of Ecology 82:561–570.

Thrall, P. H., S. W. Pacala, and J. A. Silander. 1989. Oscillatory dynamics in populations of an annual weed species *Abutilon theophrasti*. Journal of Ecology 77:1135–1149.

Tilman, D. 1977. Resource competition between planktonic algae: an experimental and theoretical approach. Ecology 58:338–348.

Tilman, D. 1982. *Resource Competition and Community Structure*. Princeton: Princeton University Press.

Tilman, D. 1988. On the meaning of competition and the mechanisms of competitive superiority. Functional Ecology 1:304–315.

Tilman, D. 1990. Constraints and tradeoffs: toward a predictive theory of competition and succession. Oikos 58:3–15.

Tilman, D. 1994. Competition and biodiversity in spatially structured habitats. Ecology 75:2–16.

Tilman, D. 1997. Community invasibility, recruitment limitation, and grassland biodiversity. Ecology 78:81–92.

Tilman, D., and J. A. Downing. 1994. Biodiversity and stability in grasslands. Nature 367:363–365.

Tilman, D., C. L. Lehman, and C. Yin. 1997. Habitat destruction and deterministic extinction in competitive communities. American Naturalist. In press.

Tilman, D., R. M. May, C. L. Lehman, and M. A. Nowak. 1994. Habitat destruction and the extinction debt. Nature 371:65–66.

Tilman, D., and S. Pacala. 1993. The maintenance of species richness in plant communities. In R. E. Ricklefs and D. Schluter, eds., *Species Diversity in Ecological Communities: Historical and Geographical Perspectives*, 13–25. Chicago: University of Chicago Press.

Tudor, D. W. 1985. An age-dependent epidemic model with application to measles. Mathematical Biosciences 73:131–147.

Turing, A. M. 1952. The chemical basis of morphogenesis. Philosophical Transactions of the Royal Society of London B 237:37–72.

U.S. Congress, Office of Technology Assessment. 1993. *Harmful Non-Indigenous Species in the United States*. Washington, D.C.: Government Printing Office.

Vance, R. 1984. The effect of dispersal on population stability in one-species discrete-space population growth models. American Naturalist 123:230–254.

Vanderlaan, J. D., and P. Hogeweg 1995. Predator-prey coevolution: interactions across different timescales. Proceedings of the Royal Society of London B 259(1354):35–42.

Varley, G. C. 1947. The natural control of population balance in the knapweed gall-fly (*Urophora jaceana*). Journal of Animal Ecology 16:139–187.

Veit, R. R., and M. A. Lewis. 1996. Dispersal, population growth and the Allee effect: dynamics of the House Finch invasion of North America. American Naturalist 148:255–274.

Verboom, J., A. Schotman, P. Opdam, and J. A. J. Metz. 1991. European nuthatch metapopulations in a fragmented agricultural landscape. Oikos 61:149–156.

Veron, J. E. N. 1993. A biogeographic database of hermatypic corals; species of the central Indo-Pacific, genera of the world. Australian Institute of Marine Science Monograph Series 10:1–433.

Volterra, V. 1926. Variazioni e fluttuazioni del numero d'individui in specie animale conviventi. Memorie della Reale Accademia Nazionale dei Lincei (Series 6) 2:31–113.

Volterra, V. 1928. Variations and fluctuations of the number of individuals in animal species living together. Journal du Conseil, Conseil International pour l'Exploration de la Mer 3:3–51.

Waage, J. K. 1983. Aggregation in field parasitoid populations: foraging time allocation by a population of Diadegma (Hymenoptera: Ichneumonidae). Ecological Entomology 8:447–453.

Waage, J. K., and M. P. Hassell. 1982. Parasitoids as biological control agents: a fundamental approach. Parasitology 84:241–268.

Wade, M. J. 1992. Sewall Wright: gene interaction and the shifting balance theory. Oxford Surveys in Evolutionary Biology 8:35–62.

Wahlberg, N., A. Moilanen, and I. Hanski. 1996. Predicting the occurrence of endangered species in fragmented landscapes. Science 273:1536–1538.

Walde, S. 1994. Immigration and the dynamics of a predator-prey interaction in biological control. Journal of Animal Ecology 63:337–346.

Walde, S. J., and W. W. Murdoch. 1988. Spatial density dependence in parasitoids. Annual Review of Entomology 33:441–466.

Weinberger, H. F. 1978. Asymptotic behavior of a model in population genetics. In J. M. Chadam, ed., *Nonlinear Partial Differential Equations and Applications, Lecture Notes in Mathematics 648*, 47–96. Proceedings Indiana 196:7–1977.

Weinberger, H. F. 1982. Long-time behavior of a class of biological models. SIAM Journal of Applied Mathematics 13:353–396.

Wen, B. R., and J. H. Brower. 1995. Competition between *Anisopteromalus calandrae* and *Choetospila elegans* (Hymenoptera: Pteromalidae) at different parasitoid densities on immature rice weevils (Coleoptera: Curculionidae) in wheat. Biological Control 5:151–157.

Wen, B. R., L. Smith, and J. H. Brower. 1994. Competition between *Anisopteromalus calandrae* and *Choetospila elegans* (Hymenoptera: Pteromalidae) in corn. Environmental Entomology 23:367–373.

Wennergren, U., M. Ruckelshaus, and P. Kareiva. 1995. The promise and limitations of spatial models in conservation biology. Oikos 75:349–356.

Werner, E. E. 1986. Species interactions in freshwater fish communities. In J. Diamond and T. J. Case, eds., *Community Ecology*, 344–358. New York: Harper and Row.

Werner, P. A., and W. J. Platt. 1976. Ecological relationship of co-occurring goldenrods (*Solidago: Compositae*). American Naturalist 110:959–971.

Whitlock, M. C. 1995. Variance-induced peak shifts. Evolution 49:252–259.

Whittle, P. 1955. The outcome of a stochastic epidemic—a note on Bailey's paper. Biometrika 42:116–122.

Wiens, J. A. 1997. Metapopulation dynamics and landscape ecology. In I. Hanski and M. Gilpin, eds., *Metapopulation Biology: Ecology, Genetics and Evolution*, 43–68, San Diego: Academic Press.

Williamson, M. 1981. *Island Populations*. Oxford: Oxford University Press.

Williamson, M. 1989. Mathematical models of invasion. In J. A. Drake, ed., *Biological Invasions: A Global Perspective*, 329–350. New York: John Wiley and Sons.

Willson, M. F. 1993. Dispersal mode, seed shadows and colonization patterns. Vegetatio 107/108:261–280.

Wilson, D. S. 1983. The group selection controversy. Annual Review of Ecology and Systematics 14:159–187.

Wilson, E. O. 1992. *The Diversity of Life*. Cambridge, Mass.: Belknap Press of Harvard University Press.

Wilson, E. O., ed. 1988. *Biodiversity*. Washington D.C.: National Academy Press.

Wilson, H. B., M. P. Hassell, and H. C. J. Godfray. 1996. Host-parasitoid food webs: dynamics, persistence and invasion. American Naturalist 148:787–806.

Wilson, K. G. 1983. The renormalization group and critical phenomena. Reviews of Modern Physics 55:583–600.

Wilson, W. G., A. De Roos, and E. McCauley. 1993. Spatial instabilities within the diffusive Lotka-Volterra system: individual-based simulation results. Theoretical Population Biology 43:91–127.

Wolfram, S. 1984. Cellular automata as models of complexity. Nature 311:419–424.

Wootton, J. T., and D. A. Bell. 1992. A metapopulation model of the Peregrine Falcon in California: viability and management strategies. Ecological Applications 2:307–321.

Wright, S. 1931. Evolution in Mendelian populations. Genetics 16:97–159.

Wright, S. 1940. Breeding structure of populations in relation to speciation. American Naturalist 74:232–248.

Wright, S. 1943. Isolation by distance. Genetics 28:114–138.

Index